International Yearbook of Soil Law and Policy

The *International Yearbook of Soil Law and Policy* is a book series that discusses the central questions of law andpolicy with regard to the protection and sustainable management of soil andland. The Yearbook series analyzes developments in international law and newapproaches at the regional level as well as in a wide range of nationaljurisdictions. In addition, it addresses cross-disciplinary issues concerningthe protection and sustainable management of soil, including tenure rights,compliance, food security, human rights, poverty eradication and migration.Each volume contains articles and studies based on specific overarching topicsand combines perspectives from both lawyers and natural scientists to ensure aninterdisciplinary discourse.

The *International Yearbook of Soil Law and Policy* offers a valuable resource for lawyers, legislators, scholars andpolicymakers dealing with soil and land issues from a regulatory perspective.Further, it provides an essential platform for the discussion of new conceptualapproaches at the international, national and regional level.

More information about this series at http://www.springer.com/series/15378

Harald Ginzky • Irene L. Heuser • Tianbao Qin • Oliver C. Ruppel • Patrick Wegerdt
Editors

International Yearbook of Soil Law and Policy 2016

Springer

Editors
Harald Ginzky
German Environment Agency
Dessau, Germany

Irene L. Heuser
IUCN World Commission on Environmental Law
Deputy Chair of Specialist Group on Sustainable Soils and Desertification
Kleinmachnow, Germany

Tianbao Qin
Research Institute of Environmental Law
Wuhan University
Wuhan, China

Oliver C. Ruppel
Faculty of Law
University of Stellenbosch
Stellenbosch, South Africa

Patrick Wegerdt
DG Environment
European Commission
Brussels, Belgium

International Yearbook of Soil Law and Policy
ISBN 978-3-319-42507-8 ISBN 978-3-319-42508-5 (eBook)
DOI 10.1007/978-3-319-42508-5

Library of Congress Control Number: 2017932897

Printed on acid-free paper

This Springer imprint is published by Springer Nature
The registered company is Springer International Publishing AG
The registered company address is: Gewerbestrasse 11, 6330 Cham, Switzerland

Foreword

Soils have been the neglected environmental media for a long time. However, several food crises at the beginning of this century have brought a change in perspective of late. Since then, the indispensable value of soils for humans and for the conservation of the environment as a whole has been increasingly understood and gained much momentum too. Also the language around "soils" has changed: Whereas in previous times, soils were predominantly perceived as "dust and dirt", we have more recently experienced a shift in the public perception where emphatic formulations, like the "earth's fragile skin" or "soils for life", started to reflect the increasingly accepted importance of soils for all life on earth and as keystone element for sustainable development.

It is no wonder that soils have been much more prominently featured in the outcome document of the Rio + 20 Conference "The Future We Want" in 2012 and ultimately in Sustainable Development Goals (SDG) in 2015. In addition, the UN General Assembly declared the year 2015 as the first "International Year of Soils" and has decided that December 5 every year shall be known as "World Soil Day".

An effective conservation of soils particularly by sustainably managing them requires reliable data, scientific knowledge, appropriate management concepts, public engagement, sufficient financial resources, effective legal instruments and good governance.

In this regard, it has been our intention to establish a platform for discussion among legislators, lawyers and policymakers regarding regulatory concepts and approaches on the international, regional and national level. The new *International Yearbook of Soil Law and Policy* aims to bring together the global legal fraternity, academics, judges and local legal practitioners to continue the discourse on soils in the attempt to improve the legal promotion and protection of soils by inter alia enhancing existing governance solutions.

This first volume of the *International Yearbook of Soil Law and Policy* focuses on contents and implementation of the objective of a "land degradation neutral world", which is also prominently addressed in SDG Target 15.3. The Yearbook is divided into four parts. We are particularly pleased that this first volume was prominently blessed with words of welcome by internationally renowned academics and visionary politicians from every continent. This introductory part with specific regional perspectives is followed by one of the recent international (law) developments. This part includes an analysis of the content and structure of the SDGs with regard to soil and land, a discussion of scientific methods relevant for their implementation and reports relevant international (legal) regimes. Latest developments include the decisions made at the 12th Conference of the Parties to the United Nations Convention to Combat Desertification (UNCCD) and the United Nations Framework Convention on Climate Change (UNFCCC) COP21 Paris Agreement. The next part deals with cross-disciplinary issues relevant to the implementation of the SDG like the right to food, land tenure, migration and the "Economics of Land Degradation Initiative". The final part reports on developments in national legislation from various domestic jurisdictions and supranational entities, including Brazil, China, the European Union, Mongolia, Namibia and the United States.

It is our intention that the book series *International Yearbook of Soil Law and Policy* will be perceived as a useful tool for academics, legislators and policymakers working in this field. To this end, also the subsequent volumes shall address emerging questions of law and politics with regard to the protection and sustainable management of soil and land—at the international, national and regional level such as urban sprawl, sustainable agriculture, effectiveness of regional agreements or land as a privately owned good.

We would like to emphasize our sincere gratitude to Springer Publishing House for publishing the new *International Yearbook of Soil Law and Policy* as part of its "legal portfolio", making it thus globally accessible.

Dessau, Germany — Harald Ginzky
Kleinmachnow, Germany — Irene L. Heuser
Wuhan, China — Tianbao Qin
Stellenbosch, South Africa — Oliver C. Ruppel
Brussels, Belgium — Patrick Wegerdt

Contents

Part I
Words of Welcome from All Continents

Greeting to the Launch of the Yearbook from an African Perspective

Sem T. Shikongo

> *Upon this handful of soil our survival depends. Husband it and it will grow our food, our fuel, and our shelter and surround us with beauty. Abuse it and the soil will collapse and die, taking humanity with it*
>
> From Vedas Sanskrit Scripture—1500 BC

1 Introduction

Africa is a continent blessed with a unique blend of culturally rich people with a myriad of traditional knowledge innovations and practices that has contributed significantly to natural resource management, more specifically sustainable land management and biodiversity conservation. The pressures of modern change dictated and driven by the northern hemisphere, have however been too fast for Africa to keep up with due to many urgent priorities that African governments contend with, relative to challenges facing developed countries. At the same time global climate change is making its presence felt ever increasingly all over the planet, including Africa, which may perhaps be the least prepared for the significant challenges climate change will pose. African governments must handle the challenge of adapting and coping with the consequences of climate change on our soils, biodiversity and ability to sustainably manage our land and natural resources. Africa does not have the technical, technological, financial and human resources to face this challenge head on in a coordinated, synergistic manner involving all stakeholders, or do we?

S.T. Shikongo (✉)
Directorate of Tourism and Gambling, Ministry of Environment and Tourism – Namibia, Windhoek, Namibia
e-mail: s_shikongo@hotmail.com

H. Ginzky et al. (eds.), *International Yearbook of Soil Law and Policy 2016*,
International Yearbook of Soil Law and Policy,
DOI 10.1007/978-3-319-42508-5_1

This paper serves as a greeting message for the launch of the International Yearbook of Soil Law and policy which has the overarching context of sustainable development goals focusing in particular on the goal of a "land degradation neutral world". The focus is on the importance of soils, the interaction of soils with *Homo sapiens*, the challenges we face from an African perspective and some of the measures we can undertake to address these challenges.

2 The Planet: A Stifled Cry from Mother Earth

Today the world is in great need of healing. The land and the people are one. We are from the soil and we return to the soil. This the sages have known for a long long time as captured in the sayings from the ancient ones.

The species *Homo sapiens* developed the ability to communicate through language, developed culture and the ability to cultivate and tamed fire. These three elements gave this species an advantage over other species and so started the conquest of ecosystems, species and genes by *Homo sapiens* that saw the emergence of change at a grand scale of planet earth. *Homo sapiens* have pushed the planet to its limits and today we are experiencing the silent cry of the planet and the rest of the living world as it battles to fight back against the relentless onslaught to nature by this species which has forgotten inconveniently its integral linkage to nature.

When we look at soil we see only the dirt but if you look deeper you will see a whole cosmos emerge. There is a rich vibrant society of organisms, elements and atoms working together to create a substance that forms the basic ingredients needed by life. . .a grain of soil can constitute an entire ecosystem.

The 7.1 billion people on this planet, which we call earth and home, depend on the biological products derived from the soil for their food, which are provided by the 2.6 billion farmers whose livelihoods depend on generating food from soil (FAO 2013 cited by Chasek et al. 2014). This biological productivity of economic value (mainly food but also other biological products from the land, e.g. fiber, forage, firewood, medicinal, soil organic carbon and soil biodiversity) and ecosystem services, such as water regulation and soil conservation, provided by the vegetation cover of the soil, as well as by the socio-economic and policy drivers that can change these factors (Chasek et al. 2014). These preceding paragraphs encapsulate the reality that we depend on soils and the interactions around soils for our existence for our well-being both physically, mentally, emotionally and spiritually. Not only us but all biological beings on this planet.

Central to soils and their health and survival is *Homo sapiens*, who holds the key to the future—is he/she really the wise one as we have decided to name him/her? If we look around us we see acts of soil destruction and over exploitation of the land and resources. Meanwhile, the land has to continue taking care of ever increasing demands of man despite the fact that there is a diminishing ability of the land to continue producing. Today the international instrument we have that one can say is closely associated with the soil is the United Nations Convention to Combat Desertification. It has also become known as the one Convention that is primarily important for Africa. Yet according to Chasek et al. (2014) on the ground, implementation of the Convention was and is still fraught with difficulties: the Parties' obligations and the Convention's expectations from parties are not quite clear, the robust financial and political capital necessary for its implementation is still not in place, the major tool for its implementation is still not in place, the National Action Plans (NAPs) are irrelevant to mainstream policy making and development cooperation and in many cases donors address land degradation issues bilaterally rather than under the framework of the Convention (Basett and Talafre 2003 cited in Chasek et al. 2014).

Some critical players in this policy arena of land degradation have been conveniently kept out due to the archaic rules of the United Nations systems. The meaningful involvement of the custodians of soils—the local communities—in defining, identifying, monitoring and responding to desertification/land degradation, namely by participation, continues to be missing for various internal and external reasons (Chasek et al. 2014). This is, nevertheless, one of the UNCCD centerpieces of combating desertification/land degradation. A similar situation occurs in sister conventions. The synergies between sister conventions are not addressed meaningfully and are lost in processes which consume limited financial resources and result in more decisions on paper which many times have nothing to do with on the ground real implementation. The challenge continues and a large percentage of humanity continues to suffer whilst a small percentage runs around the planet in search of pipe dreams to solve the plight of the majority.

3 What Can We Do to Address the Cry of Mother Earth and Her Soils?

3.1 Redefine Desertification to Land Degradation Globally

The time has come for us to completely awaken and move away from the concept of desertification and address the real challenge. The problem we face is land degradation or soil degradation the loss of productivity of the land due to mostly anthropogenic causes notwithstanding the fact that, of course, there are natural causes but even these are remotely linked to activities of *Homo sapiens* in one way or another.

We need a synergistic, pragmatic and adaptive approach if we are to halt, reverse and address the issue of soil degradation. If a grain of soil can constitute a whole ecosystem then surely it follows that there are a myriad of relations that one will have to factor in and consider as one deals with the issues of soils.

Humanity has to re-think the road it is on. One of an inexhaustible drive to control, subdue and subvert all natural things to the will of men in his hunger for power, money and perceived control.

We must however remember the words of those who came before us. "*Upon this handful of soil our survival depends. Husband it and it will grow our food, our fuel, and our shelter and surround us with beauty. Abuse it and the soil will collapse and die, taking humanity with it*".

3.2 Africa Needs to Engage in a Truthful Reality Check

It is important that this launch of this important and much needed Yearbook is being greeted from an African perspective, one of the continents which is perceived as not able to feed itself and the humans that inhabit it. One of Africa's most urgent challenges today is food security, this is amidst all the other urgent developmental challenges that are just as pressing and urgent. To make the matter worse a hungry people are an angry people and this worsen governance and Africa's ability to remain focused and deal with the real issues without following the dreams and topics of others that do not serve our interest. Despite many efforts to assist Africa on her road towards sustainable development with a focus on the international sustainable development agenda, the battle has not been won. The problems and challenges continue on the continent driven by, amongst others, loss of culture, loss of language, loss of traditional knowledge, globalization, conflict and war, unclear land and resource tenure, livelihood and poverty, population growth and use of foreign legal systems that fail to take into account our customary laws and culture. All of these challenges have an impact on Africa's ability to deal with her soils.

Is it perhaps because of the fact that Africa is on a road for which the script she is using is not African—thereby addressing issues from a position of inherent disadvantage given the fact that Africa has given up its competitive and comparative advantage? How can African indigenous knowledge system and true culture come to our aid in this struggle against the onslaught of soil degradation given its deleterious effect on Africans and the rest of the world? Is there a way out? The picture looks bleak but what is the road ahead and what can we conclude?

3.3 Africa Can Lead the Way

Despite African being described as the Dark Continent, the continent of disease, hunger, poverty and war it must be recognized that this is the perception and the story presented by another who is not African. The reality is that Africa is still the least 'messed-up' continent in the world with regards to soils, ecosystems and biodiversity. Africa can be developed to become the "green lung of the World" and serve as a global carbon sink. Africans and those who are friends of Africa will however have to ask themselves some tough questions to establish a true reality check. If we embark upon this road we will need to establish some critical assumptions that must be met in order for such an approach to succeed: These critical conditions amongst others are:

- Africa needs to develop a strategy to capitalize on the competitive and comparative advantages inherent in African culture, traditions indigenous knowledge systems, innovations and practices that are relevant to the conservation and sustainable utilization of soils and biodiversity and combine that with western and other knowledge systems as tools to enable us to make a shift in the right direction
- Africa must invest in building lasting social capital (trust) within Africa and in African institutions and her own people
- Africa must assure that the gains received from initiatives aimed to achieve "neutrality" go directly to the people unlike what has happened to the gains from other African treasures such as minerals, oil and gems.
- The need for the establishment of an Africa wide protocol or convention on soils ensuring buy-in and ownership of all critical players in Africa must be met
- Africans must recognize and encourage the full and effective involvement of major groups essential to the success of soils especially the local communities and the private sector
- Africa must assure that any foreign investment into Africa must add to the African agenda and not priorities determined by other continents.
- The African agenda towards economic and financial independence must make room for cultural independence as well as true recognition of African culture and its contribution to sustainable development.

If these conditions are met and African grown initiatives with assistance from those who have the interest of Africa and her people at heart—are developed and implemented truthfully—then this could become the "neutrality" link for Africa's soil, climate and biodiversity with spin-offs for the global community and the planet.

4 Conclusion: More Tough Questions

In conclusion, perhaps therefore the time has come that Africa starts to take bold steps and takes the future of humanity into its own hands? Perhaps if the international conventions to which Africa is a member but in which the agenda is dictated by others, not in the true interest of Africa, can be forced to change their course by Africa's withdrawal from membership? We need a new direction, we need to learn adaptively and face the reality of the various processes that we have engaged on as Africans. Are we today better off? Are our soils better off? Are we truly enjoying the diversity of the products of our soil or is someone else enjoying at the expense of the African and his/her continent? Which continent today is enjoying the best fruits of the planet in terms of the most diverse organic diet derived from the planet's soils? Why?

We may not have all the answers to the above questions, perhaps through initiatives such as these Yearbook Africa can embark upon a road that can assure that the answers may result in the survival of our soils and all that which is linked to them in the interest of our physical, mental and emotional, and spiritual well-being.

References

Basett C, Talafre J (2003) Implementing the UNCDD: towards a recipe for success. Rev Eur Commun Int Environ Law 12:133–139

Chasek P, Safriel U, Shikongo S, Futran Fuhrman V (2014) Operationalizing zero net land degradation: the next stage in international efforts to combat desertification? J Arid Environ. doi:10.1016/j.jaridenv.2014.05.020

FAO (2013) FAOSTAT: Population: Agricultural population 2013. http://faostat3.fao.org/faostat-gateway/go/to/download/O/OA/E. Accessed 20 Mar 2016

Audit of Soil Governance

Aroldo Cedraz de Oliveira

The world is awakening to the need to take better care of its natural and finite resources, such as soil, which is fundamental to the life of every living being and deserves special treatment.

After all, soil underpins everything we have and is at the center of the main challenges of the planet: food, fiber and energy production, as well as environmental services. It is essential to ecosystem functions as it plays a key role in mitigating climate change effects, maintaining water sources and sustaining biodiversity. Therefore, its misuse can even make life on our planet impossible.

This matter has always been very attractive and challenging for me, which is why I took every opportunity to provoke discussion on the subject during my work on different activities.

Thus, stemming from our experience at the first **Global Soil Week** in late 2012 in Berlin, Germany, a small group of Brazilian scholars and I concluded it would be extremely appropriate to bring this debate to Brazil. After all, in a country with continental characteristics like ours, being a power concurrently agricultural, mineral and environmental, the sustainable use of this important natural resource has special relevance.

At that time, it was understood that the Brazilian Court of Audit (TCU)[1] should lead the debate due to its position in the Brazilian state structure. I, as president of TCU, accepted the challenge of convincing my peers that we should hold an event in Brazil to discuss the issue of land, with emphasis on the issue of governance, in line with our institutional mission of contributing to the improvement of public administration for the benefit of society.

[1]http://portal.tcu.gov.br/cidadao/.

A.C. de Oliveira (✉)
Brazilian Court of Accounts, Brasila, Brazil
e-mail: gabpres@tcu.gov.br

H. Ginzky et al. (eds.), *International Yearbook of Soil Law and Policy 2016*,
International Yearbook of Soil Law and Policy,
DOI 10.1007/978-3-319-42508-5_2

It is worth noting that, from the moment the idea was launched, we received immediate support and engagement from several Brazilian institutions working in areas directly connected to the subject as well as international organizations, such as the Institute for Advanced Studies in Sustainability of Germany and the United Nations Food and Agriculture Organization (FAO). All these institutions worked with us on the design and construction of the "Soil Governance Conference", which took place in Brasilia on 25–27 March 2015.

It was a timely choice to hold the event during the International Year of Soils, decreed by the United Nations (UN), as there was more motivation to deepen the thematic discussion and awareness of society, governors and legislators about the importance of this important natural resource. The forum provided a stage for debates regarding highly complex issues related to degradation, governance and soil management.

At the end of the meeting, the "Letter of Brasilia"[2] was launched. In line with the conference's aim to foster fruitful results for our country, this outcome document aimed to influence the adoption of global measures for the conservation and sustainable use of soil. It includes reflections, considerations and conclusions along the theme of soil, while also sharing with society a number of suggestions for materialized improvement in a number of structural measures to be prioritized by the responsible agents.

The Soil Governance Conference, in conclusion, shares the opportunities for improvement a construction of a permanent forum for discussion about soil, emphasis on soil in the policies that promote development, preparation of new surveys and mapping in more detailed scales, for the planning and use of the territory, ongoing information to society about the threats and costs involved in the current form of soil use and exploitation, revision and consolidation of the several scattered norms and laws about soil and investment in proper education, preparation and qualification of different segments of society that deal directly with this resource.

This important issue was again highlight in other forums I attended, such as the 39th FAO Conference held in Rome on the 10th and 11th of June 2015, which brought together 16 Heads of State and Government, 116 Ministers, and around 1700 participants representing 191 delegations, an unprecedented number in the Organization's history. The relationship between soil, food security and climate change set the tone of the debate, focusing on sustainable management.

On that occasion, the FAO released technical data on the global status of soil resources. According to the Organization, soils are responsible for 95 % of world food production and its sustainable management could increase food production by as much as 58 %.

In addition, it was reported that soils make up 1/4 of the planet's biodiversity and account for the second largest global carbon stock—second only to the oceans. However, approximately 1/3 of this natural resource is degraded worldwide due to

[2]http://www.governancadosolo.gov.br.

erosion, compaction, sealing, salinization, acidification, pollution and depletion of nutrients caused by unsustainable management practices.

Moreover, the extent of areas considered "lost" due to intense degradation increases by about 500,000 km^2 per year—equivalent to the size of Costa Rica. According to the FAO, nature can take up to 400 years to produce 1 cm of soil, so soils' recovery far exceeds the span of multiple generations. Therefore, though often perceived as a given resource, soil is a nonrenewable natural resource and is increasingly threatened.

The Post-2015 Development Agenda of Sustainable Development Goals (SDG) adopted by the UN, which replace the Millennium Development Goals, should guide national policies and international cooperation activities over the next 15 years. Goal 15 recognized the dilemma facing soils worldwide, stating: *"protect, restore and promote the sustainable use of terrestrial ecosystems, sustainably manage the forests, combat desertification, halt and reverse land degradation, halt biodiversity loss."*

Few areas of the world contain a territorial reach and amount of land used for agricultural production as Brazil. However, data from the Ministry of the Environment indicate that 140 million hectares of Brazilian land are degraded, which corresponds to 16.5 % of the national territory. Among these lands, about 30 million hectares of pastures are in some stage of degradation, with very low productivity for animal feed. In recent years, the country has faced the phenomenon of desertification, which is affecting various regions such as the Northeast semi-arid region, the Cerrado of Tocantins, Northern Mato Grosso and Minas Gerais. In addition, the process of increasing arenization has reached the Pampas Gauchos. With the formation of arid areas, the temperature rises and the humidity level of the air decreases, making the soil infertile, thereby threatening decreases in food production and increases in hunger and poverty.

Aware of this, the performance audit by the Brazilian TCU of non-urban soil and land governance actions taken by the Federal Government highlighted matters that were on the agenda in all the forums I have mentioned. The audit also resulted in the diagnosis of the risk to the country's natural heritage and identification of the damage suffered.

Incidentally, it is good to clarify that the TCU's operations in this area are supported by its constitutional competence and its strategic objectives, including to "Improve governance in public organizations and policies."

Additionally, in the institutional framework of the Brazilian state, the TCU is an autonomous body that offers society and the National Congress an independent and technical perspective founded on sound methods and international auditing standards. Like other Supreme Audit Institutions (SAI), it has been offering its expertise to the Brazilian Government regarding the implementation of public policies, including those aimed at the sustainable use of soil.

In this context, the main objectives of the Department of Agriculture and Environment (the TCU unit which coordinated this work) are to perform oversight functions that contribute to the production of food quantity and quality to ensure food security for the people of Brazil and generate exports; to the organization of rural areas and the development of family, economic and social agriculture; and to the maintenance of a healthy environment. All these objectives are strongly linked to soil protection.

In this regard, the audit was extremely timely. In the wake of world events, there is much to learn and put into practice. In general, countries have striven to improve soil protection, be it within their governments or within the scope of action of the SAIs, such as the Brazilian TCU, in pursuit of the SDGs.

The audit touched on five of the SDGs and specifically addressed two of them—goal 2 (Eradicating hunger, achieving food security and improving nutrition and promoting sustainable agriculture) and goal 15 (Protect, restore and promote the sustainable use of terrestrial ecosystems, sustainably manage the forests, combating desertification, arrest and reverse land degradation and halt the loss of biodiversity).

Hence, the importance of doing our part becomes apparent, including seeking to integrate with international organizations, such as International Organization of Supreme Audit Institutions (Intosai), which provided an institutionalized framework for supreme audit institutions to foster development and share knowledge and best practices to improve government auditing worldwide and enhance professional capacities.

On this track, Resolution A/RES/69/228 adopted by the UN General Assembly on 19 December 2014, titled **Promoting and fostering the transparency of public administration by strengthening supreme audit Institutions**, recognizes the role that SAIs can play is conducive to achieving the SDGs and soil-related priorities.

Despite the fact that the majority of international agreements conducted by the UN are not compulsory in nature, it is clear the important role that SAIs can play in monitoring the implementation of SDG. SAIs have different mechanisms to induce the implementation of agreed actions as well as promote transparency and accountability in the sphere of global agreements.

Moreover, such regulation encourages countries and UN institutions to intensify cooperation with Intosai, especially with regards to capacity building of the various actors in order to promote good governance at all levels, ensuring efficiency, accountability, effectiveness and transparency from the strengthening of SAIs.

At the national level, we still have a great responsibility: to contribute to a forward-looking vision, of strategic character, that will be a part of the guidance for government plans, policies and programs. We know that if we are looking 15 years ahead, now we need to think, decide and act in a timely manner because in order for positive effects of state social and economic actions to be transformative, years or even decades of learning, maturing and consolidation are required.

As stated in the "Letter of Brasilia", Brazil does not have a specific soil policy, which was proven in the audit and demonstrates the need for the determinations and recommendations provided in the resulting judgment 1,942/2015-TCU-Plenary.[3]

[3]http://portal.tcu.gov.br/fiscalizacao/agricultura-desenvolvimento-agrario-e-meio-ambiente/destaques/fiscalizacoes-concluidas.htm.

Thus, there are several reasons to continue discussions in order to seek elements that may support state actions in the context of a future National Soils Policy, which is quite necessary for the Brazilian nation.

The question to discuss then is how to tread the virtuous path since the Court's audit provided a view of the effects of the complexity and dispersion of rules. There is clearly an instability of objectives and strategies; unpredictability of resources; lack of definition of competencies; absence of formal coordination instruments; lack of coherence between public programs and actors; lack of adequate systems of monitoring and validation; and difficulty in the implementation of public policies.

These deficiencies are not unique to the Brazilian nation, which is why the United Nations has requested the SAIs to assist in reversing this situation so that countries may better pursue the ideal of eradicating hunger, achieving food security, improving nutrition, and promoting sustainable agriculture. Thus, the aim is to ensure access of all people, particularly the poor and vulnerable persons, e.g., children, to safe, nutritious and sufficient food throughout the year.

In fact, the goals are ambitious, but we must bear in mind that all these ideals are part of a fundamental condition for us to achieve full democracy. A democracy in which economic growth proves lasting, preserving the rights of future generations, and showing itself to be inclusive by improving the quality of life of our citizens in the present in a fair and harmonious way.

Therefore, aware of the gravity of the soil situation, the audit by the Brazilian TCU, is another tool for sharing information with the global society on a monitoring methodology and diagnosis of the soil situation. It will certainly serve as input for the national adoption of practical actions to preserve the natural and finite soil resources of Brazil, which are essential to achieving the SDGs.

Soils Need International Governance: A European Perspective for the First Volume of the International Yearbook of Soil Law and Policy

Jochen Flasbarth

It all started in December 2014 in Berlin at an international workshop organized by the German Environment Agency. A small group of about 25 legal experts from all parts of the world recognized that a platform for discussions on soil governance issues was urgently needed. After further deliberations amongst experts and a lot of preparatory and editorial work we now see the final product: The first volume of the new International Yearbook of Soil Law and Policy, published by Springer, is accessible from a large number of libraries across the world. The German Environment Agency, where I had the honor of being President from 2009 to 2013, has put a great deal of effort into this project. Reason enough to send this message of greeting. Furthermore all the time and personnel required to make this Yearbook happen deserve this to be framed in a wider political context, taking a specific European perspective into account.

2015 was a good year for soils, for several reasons. First of all, 2015 was the first International Year of Soils as declared by the UN General Assembly in December 2013. In addition, the UN General Assembly declared 5 December to be World Soil Day.[1] Throughout 2015 soil events took place in all corners of the world.[2] In Germany the launch and the end of the first International Year of Soils were marked by two international conferences in Berlin. All these events significantly contributed to increased awareness of the necessity of soil protection among policymakers as well as the general public—which is of course good news.

[1] 68th UN General Assembly, December 2013: more information at: http://www.fao.org/fileadmin/user_upload/GSP/docs/iys/World_Soil_Day_and_International_Year_of_Soils__UNGA_Resolution_Dec._2013.pdf.

[2] See http://www.un.org/apps/news/story.asp?NewsID=49520#.VswrLbdMv8M.

J. Flasbarth (✉)
German Ministry for the Environment, Nature Conservation, Building and Nuclear Safety, Berlin, Germany
e-mail: jochen.flasbarth@bmub.bund.de

H. Ginzky et al. (eds.), *International Yearbook of Soil Law and Policy 2016*, International Yearbook of Soil Law and Policy,
DOI 10.1007/978-3-319-42508-5_3

Second, the UN General Assembly approved the universal plan for action "Transforming our world: the 2030 Agenda for Sustainable Development" in September 2015, consisting of 17 Sustainable Development Goals and 169 targets which concretize these goals.[3] This program will be the benchmark for international and national policies up to 2030, setting obligations for all countries—developed and developing countries alike—with regard to sustainable development. For soils, the agenda demands that states will strive for a land degradation-neutral world, meaning there has to be an equal balance between land degradation and restoration/rehabilitation. This is in fact very ambitious. The agenda, including its goals and targets, will have to be implemented at national and international level.

Third, Parties to the Convention to Combat Desertification took several very important decisions at the 12th Conference of the Parties in October 2015 in Ankara, Turkey. They agreed that the UNCCD should de facto include the objective of land degradation neutrality in its mandate and that it should act as the international lead organization in this regard. Furthermore, first requirements for the implementation of this objective were also agreed upon by establishing a universal definition and by defining the necessary steps, measures and indicators. However, these decisions are not legally binding. Ultimately, the commitment of the Parties at national level will determine whether these decisions will have a real impact on the sustainable management of land and soil.

Fourth, the Paris Agreement, which was approved by the Parties to the UNFCCC in December 2015, is an essential step forward with regard to sustainable soil management. The international community agreed in a legally binding way on the 2 °C objective and further ambitious goals (1.5 °C and a neutral greenhouse gas balance for the second half of the century). All states are called on to submit their national contributions taking into account all types of sources, sinks and reservoirs. Details will have to be negotiated and determined. In the long run, however, land use and land use changes will have to be accounted for in these calculations.

Fifth, the Economics of Land Degradation Initiative provided detailed evidence that sustainable management of land and soil is economically a better alternative—as compared to degrade first and repair later.

How was so much progress achieved? What were the driving forces? About 5 years ago the perception of soils started to change. This was in part due to the visionary commitment of the former UNCCD Executive Secretary, Luc Gnacadja, who pushed the ambition of "zero net land degradation". He received support from many institutions, such as the Potsdam Institute for Advanced Sustainability Studies (IASS) and its former director Klaus Töpfer. The IASS established Global Soil Week, which has been conducted three times since then, each involving participation of about 400 experts. In 2011 the German Environment Agency—under my presidency—brought together experts and policy makers in order to promote the idea of sustainable soil management, which ultimately led to the "Call for Action".[4]

[3]See A/RES/70/1, at: http://www.un.org/ga/search/view_doc.asp?symbol=A/RES/70/1&Lang=E.

[4]See https://www.umweltbundesamt.de/sites/default/files/medien/377/dokumente/statement_protecting_soils_for_our_common_future_september_2011_0.pdf.

The World Food and Agriculture Organization (FAO) strongly supported this endeavor, for example by initiating the Global Soil Partnership. This started the ball rolling. . ..

In conclusion, one could state that soil and land management are now generally perceived as a global concern. However, does that mean that everything is alright now? The answer, obviously, is no. There is still enormous pressure on soils worldwide. On the international level, ongoing land degradation is still causing hunger and malnutrition, migration and even war. Refugees from all over the world trying to reach Europe illustrate this human suffering. Also, global issues like climate change and the conservation of biodiversity are adversely affected by the devastation of land and soil. However, land degradation is not the only reason for these phenomena. The causes are multiple and often difficult to understand, like the current catastrophe in Syria. The last four summers prior to the outbreak of civil war were the driest in Syria's history. A lot of the rural population left their homes and their communities, thereby destabilizing Syrian society. This is not to say that land degradation was the cause of the war, but intensified social conflicts paved the way for civil war. To put it another way, we need to understand the interaction between the several causes in order to be able to effectively address them.

Due to the state of its economic development, Europe faces different kinds of major soil-related challenges. In view of two centuries of industrial development and dense population, the major internal concerns in Europe are the ongoing land take, soil contamination and unsustainable agriculture.

Despite the economic, social and political status quo, with the current regulatory system in Europe and in Germany it would probably not be possible to achieve the ambitious objective of a "land degradation-neutral world". The attempt to put in place an overarching regulation at European level—the so-called European Soil Framework Directive of 2006—failed due to a blocking minority of some European member states that considered soils to be a domestic issue. The directive was withdrawn by the European Commission in 2014. In contrast, there is European legislation regarding the protection of water, air and nature. The current 7th Environment Action Programme[5] addresses soil and land issues in several passages, for example stating that binding European regulations for soil quality issues should be considered by the Commission together with the member states.

The situation is similar regarding German soil legislation. The existing Soil Protection Act focuses primarily on the restoration of existing brownfields. Provisions concerning other drivers of land and soil degradation like sealing, urbanization, land take and agriculture are covered in various acts and ordinances. A comprehensive regulatory approach on how to achieve the neutrality objective still needs to be developed. The German Environment Agency has commissioned several studies to support a more sustainable use of soil resources.

[5] See http://ec.europa.eu/environment/action-programme/.

However, it should be stated that both European and German legislation prescribe regulatory approaches that seem promising and that might also serve as models for other states. Two examples:

According to the Industrial Emissions Directive, operators of industrial installations are obliged to ensure that groundwater and soil are not detrimentally affected by the operation of the installation. If this does not hold true after operation at the installation ends, groundwater and soil have to be restored to their original status. Since this is very costly the provision has the effect that operators try to avoid negative impacts on groundwater and soil.

Second, according to the German Nature Conservation Act every negative impact on nature, including soil, has to be offset by the restoration of nature in other places. The act essentially demands the neutrality of human intervention in the environment. Although the implementation of this fundamental regulatory concept is weakened in practice for several reasons (for example, soil is underestimated in relation to living nature), the general approach could serve as a model.

As well as the internal challenges, Europe will have to deal with the transboundary impacts caused by the behavior of its societies with regard to sustainable soil management. It is important to differentiate between two aspects:

Due to its consumption patterns Europe has a significant virtual land and soil import. In other words, we need to use soils from abroad to maintain our standard of living and would not be able to produce the goods consumed internally on our own soils. Europe will therefore have to find answers to how to deal with this transboundary responsibility.

In addition, soil- and land-related activities by European citizens or enterprises abroad, like foreign investments in land, will have to be regulated in order to prevent negative impacts on the environment and local communities.

As we have seen, a lot has been achieved during the last 5 years: increased awareness, the inclusion of the objective of land degradation neutrality in the Sustainable Development Goals, the decisions by the UNCCD Parties and so on. However, whether we will get to see real improvements will depend on implementation. Implementation builds on governance, governance builds on good concepts and ideas. In this regard, the new International Yearbook of Soil Law and Policy could fill a gap by serving as a platform for discussion among regulators, lawyers and policy makers regarding regulatory concepts and approaches on the international, regional and national level. In this regard it is a very valuable tool and a major step forward. Widespread availability will further enhance the valuable impact of the Yearbook.

Different levels of soil governance will have to be addressed. On the international level the implementation of the Sustainable Development Goals as well as of the UNCCD decisions will have to be first on the agenda. Other topics like effective cooperation between existing international regimes (UNFCCC, UNCCD, CBD) are also important. In the longer run it might be necessary to elaborate in more detail on the need for an integrated international regime of sustainable land and soil management. In addition, a better understanding of how the different governance levels—international, regional, national and local—could be coordinated effectively will be needed. Finally, a comparative analysis of the various regulatory approaches at national level could not only deepen the mutual understanding of different challenges in the various regions, but could also serve as a tool to identify innovative and effective approaches with regard to sustainable soil and land management.

The first volume of the International Yearbook of Soil Law and Policy comprehensively fulfils these expectations. It provides information about the latest developments in international law, includes several reports on national legislation which facilitate the above-mentioned comparison, and it is of special importance that crucial cross-cutting topics like land degradation and migration or social and environmental impacts of tenure rights are also tackled. To conclude I would like to offer my wholehearted congratulations on this first volume of the International Yearbook of Soil Law and Policy. I very much look forward to the second one.

North American Soils and World Food

William E. Rees

It has often been said that the 'discovery' and opening up of the New World rescued Europe from early decline. The most obvious basis for this claim is the renewed abundance of resources ranging from gold, silver and other minerals, to timber and guano (natural fertilizer) essential for the growth and maintenance of civilization.

1 Contributions of North American Soils and Agriculture

Perhaps of most importance, however (and increasingly so?), was the prodigious area of newly available agricultural lands and the natural productivity of the soils. North America is particularly noteworthy in this respect. Consider that Canada's relatively small population of 36 million occupies the second largest contiguous national territory on earth, and while much of the nation's land-base is sub-Arctic and boreal forest, the country boasts 67.5×10^6 hectares of agricultural land, including 45.7×10^6 hectares of arable land. The country ranks seventh among countries in terms of total arable land area, and third in terms of arable land per person (1.26 ha). The United States with almost ten times the Canadian population also has six times the area of agricultural land, and 3.8 times as much arable land—indeed, the US ranks first in the world for total area of arable land and fourteenth in arable land per capita (0.54 ha) (Tables 1A and 2). The 360 million people (4.9 % of the world's population) who live in Canada and the US occupy fully 14 % of the world's landmass, an area that includes 10 % of the agricultural land and 15 % of the arable land on Earth (Table 1B).

W.E. Rees (✉)
School of Community and Regional Planning, University of British Columbia, 6333 Memorial Road, Vancouver V6T 1Z2, British Columbia, Canada
e-mail: wrees@mail.ubc.ca

H. Ginzky et al. (eds.), *International Yearbook of Soil Law and Policy 2016*,
International Yearbook of Soil Law and Policy,
DOI 10.1007/978-3-319-42508-5_4

Table 1 North American agricultural land/soil resources in global context

Territory (population)	Total land area	Agricultural land	Arable land area	Percent of land in agriculture	Percent of agricultural land that is arable
A.	Thousands of hectares			Percent	
World (7,400,000,000)	**13,013,475**	**4,967,580**	**1,421,169**	**38**	**29**
Canada (36,300,000)	909,351	67,500	45,660 (rank = 7)	7	68
United States (323,200,000)	916,192	414,778	174,448 (rank = 1)	45	42
B.	Canada + US as percent of world				
	Population	Land area	Agri-land area	Arable land area	
	4.9	14	10	15	

Arable land generally means 'cropland'. Specifically it refers to land under temporary crops (double-cropped areas are counted only once), temporary meadows for mowing or pasture, land under market and kitchen gardens and land temporarily fallow (less than 5 years)
Source(s): FAO (2008), StatsCan (2009), WPC (2016)

Table 2 Per capita 'supply' of agricultural land, world and North America

Territory	2016 population (millions)	Agricultural land per capita (ha)	Arable land/capita (ha)
World	7400.0	0.67	0.19
Canada	36.3	1.86	1.26 (6.6 × global supply/cap; rank = 3)
US	323.2	1.28	0.54 (2.8 × global supply/cap; rank = 14)

Source(s): FAO (2008), WPC (2016)

The global comparison is worth pursuing further. The Earth contains only about 0.19 ha/capita of arable land. Thus, Canada's 1.26 ha/capita and the US's 0.54 ha/capita are 6.6 and 2.8 times greater respectively than global per capita supply (Table 2). Note that Western European agricultural powerhouses, France and Spain, have only 0.27 ha cropland/capita. This average is slightly more that the EU as whole (0.26 ha/capita) and twice the Western European average (0.13 ha/cap) but is still *less than half* the per capita availability of arable land in North America (European data from Nationmaster 2011).[1]

Given the vast areas of land/soils suitable for agriculture of all kinds, North America unsurprisingly makes disproportionate contributions to global food production and international trade in major food commodities. Canada typically ranks

[1]To add perspective, note that it takes about .45 ha (1.1 ac) of US grazing lands and cropland to feel the average American at prevailing US levels of productivity and consumption (2700 cal/day including wastage). Europeans have lower intake, but with much less agricultural land/capita many European countries are net importers of core foodstuffs and livestock fodder.

Table 3 Ranking North America in cereal and meat production

Territory	Cereals production (2013) (thousands of tonnes)	Percent of world output	Rank	Meat production (2004)	Percent of world output	Rank
Canada	66,372	2.4	8	4592	1.8	10
United States	436,554	15.7	2	38,891	15.0	2
Canada + US	441,750	18.1	1	43,483	16.8	2

Sources: World Bank (2013) for cereals, StatsCan (2009) for meat

eighth in total cereals production and the US second, but together the two countries occupy first place. With less than 5 % of the world's population, North America produces almost a fifth of the world's grain and exports a disproportionate share of grain (Table 3).

Indeed, North America (NA) often sets the price at which major grains are traded around the globe and is often the producer that the world depends on in tight supply situations. NA is generally the world's largest exporter of feed grains, particularly maize and soy, of which the US is the world's largest producer.[2] The American price is the benchmark wheat, maize and soybean price (Lyddon 2011; USDA 2016). The US and Canada together rank second behind China in wheat production but are first and second respectively in wheat exports. Canada, with just over half US production but lower domestic consumption, is beginning to rival the US in wheat exports, overtaking the US for the first time in 2014/15 with exports of 24 million tonnes (WP 2016).

Meat production is a similar but fluctuating story. In 2004, Canada ranked 10th and the US second in total meat output, and together these two countries accounted for 17 % of global meat production (Table 3). Canada was the third-ranked pork exporter in 2013. Note that China is first in total meat production but with only 0.083 ha/capita of arable land must source much of its animal feed from offshore—China is the world's largest importer of soy, mostly from Brazil and the US, so it might be said that *much of China's meat is actually produced by soils in the Americas*. The same holds true for meat production in densely populated European and other countries with little agricultural land/capita. The Netherlands, for example, is a major meat producer and second after China in soy imports, mostly for animal feed (see Note 1). While still on animal products, note that the US occupies first rank in chicken output and in the production of cows' milk and cheese, producing a quarter of the world's supply of the latter. (The meat and dairy situation is changing rapidly as developing countries ramp up production.)

[2]Note that 40 % of US and 11 % of Canadian maize output have been diverted to fuel bio-ethanol production in recent years, encouraged by federal subsidies. This is a morally questionable diversion of feed-grain. Moreover, it is doubtful, from a complete systems perspective whether maize ethanol even represents a net energy source.

North America is also a significant contributor to global fruit, vegetable and tree-nut production—the US is fourth-ranked in fruit output and third in vegetable production though its output of vegetables (34 million tonnes in 2013) is dwarfed by number one producer, China (583 million tonnes) (FAO 2015, cited in TFE 2015 and in Geohive 2015).[3] The US is the world's largest producer of several individual fruit crops such as almonds (80 % of world harvest), strawberries, blueberries and the latter's close relative, cranberries; it is second in tomato production and among the top five for several other crops.

Because of a unique coincidence of fertile soils, benign climate (sunny semi-arid in places), historically abundant water and intensive cultivation practices, a single state, California, dominates fruit, vegetable and nut production in the US. The state accounts for about 70 % of total US fruit and tree nut farm value and 55 % of vegetable farm value exclusive of potatoes (USDA 2015). In addition to virtually all US almonds, California is the source of 99 % of US walnuts, 97 % of kiwis, 97 % of plums, 95 % of celery, 95 % of garlic, 90 % of broccoli, 89 % of cauliflower, 71 % of spinach, 69 % of carrots, etc. No other state, or combination of states, matches California's productivity per ha—"lemon yields, for example, are more than 50 % higher than neighboring states. California spinach yield [per ha] is 60 % higher than the national average" (WFP 2013).

Even with California, however, the US is actually a net importer of fruit, tree nuts, and vegetables, a dependence that is growing (and that may deepen rapidly if California production succumbs to climate change—see WFP 2013).

Canada is the number two producer of blueberries and cranberries and the third ranked producer (first-ranked exporter) of pulses but because of its harsher climate and seasonal production has always been a net-importer of fruits and vegetables.

2 Soil Conservation and Management

US land and conservation policies are concerned with patterns of land use ownership and generally favour market-based instruments over strict regulation within an overall environmental governance framework. Much of that framework was erected in response to the severe drought and dust-bowl conditions of the 1930s. For example, Public Law 74-46, the Soil Conservation Act, (April 1935) recognized that "the wastage of soil and moisture resources on farm, grazing, and forest lands . . . is a menace to the national welfare" and directed the Secretary of Agriculture to establish the Soil Conservation Service (SCS) as a permanent agency in the USDA (USDA ND). This legislation was amended and replaced in February 1936 by The Soil Conservation and Domestic Allotment Act (Pub.L. 74–461), with the express purpose of encouraging the use of soil resources in such a manner as to preserve and improve fertility, promote economic use, and diminish the exploitation and

[3]China produces about half the world's vegetables and 20–30 % of its fruit.

unprofitable use of the national soil resources (US Senate 1936). The latter legislation allowed the government to pay farmers to reduce production so as to 'conserve soil' and prevent erosion. (In 1994, Congress changed SCS's name to the Natural Resources Conservation Service [NRCS]). Another important part of the framework is The Food Security Act of 1985 (P.L. 99–198), with its Sodbuster, Swampbuster, and Highly Erodible Lands provisions, making conservation a prerequisite for participation in USDA programs. It also established the Conservation Reserve Program (CRP) to provide rental payments to farmers for removing marginal cropland into grass or trees (USDA ND). This is the world's longest-running formal 'payments for ecosystems' services program (although as food prices have increased faster than CRP payments the amount of land so 'conserved' is in decline).

In Canada, severe wind erosion in the 1930s led to the creation of the Prairie Farm Rehabilitation Administration 1935 to deal with soil erosion and related issues including scarcity of water for agricultural purposes. For the most part, however, Canadian land-use planning is a provincial responsibility and legislation differs widely among provinces.

One particularly innovative development was British Columbia's long-standing Agricultural Land Reserve Program (ALRP) that inhibits development on approximately 4.7 million ha of agricultural land throughout the province. When implemented in 1973, the ALRP was considered the most progressive instrument of its kind in North America. Not until 2005, after millions of hectares of some of Canada's best agricultural soils had succumbed to urbanization, did Ontario create a Greenbelt to similarly protect 0.7 million ha of agricultural and natural lands in the most populated region surrounding Toronto. Programs involving payment for Ecosystem Services are becoming more widespread in Canada but will become fully effective only after being accepted as a necessary component of conservation land use planning frameworks at all levels of government (various sources including FAO 2015).

One practice that has become virtually the norm in both the US and Canada since the 1980s is conservation tillage,[4] including no-tillage production. By 2009 there were about 111 million ha of arable land under no-tillage globally of which 36 % was in North America (USA 26.5×10^6 ha, Canada 13×10^6 ha) (Derpsch et al. 2010). Estimates are that conventional tillage and associated practices led to soil organic carbon losses of up to 30 % in a variety of native Canadian soils on conversion to cropland. US losses are likely similar or higher. No-tillage can reverse the loss of organic matter, aid in water conservation, improve and maintain soil porosity and thus prolong the availability of plant-available soil water in times

[4]Defined here as "a system of planting (seeding) crops into untilled soil by opening a narrow slot, trench or band only of sufficient width and depth to obtain proper seed coverage" The goal should be a state of permanent or continuous zero-tillage in which post-harvest soils remain permanently covered with crop residues from previous cash crops or green manure cover crops which will remain undisturbed on the soil surface after re-seeding (Derpsch et al. 2010).

of drought. The practice can also reduce weed, insect pest and disease incidence while increasing biodiversity.

In Canada, soil loss from the combined effects of wind, water, and tillage decreased in all ecozones between 1981 and 2006. Over that period the proportion of Cropland in the very low risk class (sustainable) increased from 64 to 80 % due to conservation tillage, combined with reduced summer-fallowing and conversion of some erodible landscapes from annual crops to perennial forage and tame pasture (payments for ecosystems services). By 2006, only 10 % of arable remained in the moderate to very high risk classes due to high levels of water erosion in the Mixedwood Plains and Atlantic Maritime ecozones (McConkey et al. 2010).

3 Ecological Impacts of Soil, Live-Stock and Water Mis-Management

Industrial agriculture is arguably the most ecologically damaging of human activities with impacts ranging from large-scale deforestation, and regional landscape degradation through air, soil and water pollution, to climate change and precipitous biodiversity loss (Tilman et al. 2001; FAO 2003; Eshel et al. 2014). Perhaps the most immediately *felt* negative impact of agriculture in North America is widespread nutrient imbalance and resultant pollution of both surface and groundwaters. The major sources are elevated levels of nitrogen and phosphorus in soils from excess fertilizer application and manure produced in large-scale industrial livestock feed-lots. Water quality problems range from eutrophication and algal blooms in Lake Winnipeg in Manitoba, at the northern edge of the agricultural zone, to seasonal hypoxia (resulting in a spreading 'dead zone') in the shallow coastal waters of the Louisiana shelf in the northern Gulf of Mexico, at the southern end of the agricultural zone (FAO 2015).

The case of the corn-belt state of Iowa is particularly interesting—with less than 5 % of the land area in the Mississippi basin, Iowa is the source of 25 % of the nitrate runoff responsible for the gulf of Mexico dead zone and most of this excess is attributed to farming (Manning 2016). The state is a major producer of maize, one of the most intensely fertilized of crops—one might think of the most intensive cultivation as a form of 'field hydroponics'. Adding to nitrate and phosphate loadings, the state also supports 21×10^6 hogs in concentrated factory operations, and 50×10^6 laying chickens similarly 'housed' in concentrated operations of over 100,000 birds each (USDA 2014). The hogs produce manure is equivalent to the bodily waste 45×10^6 million people, or 15 times the state's human population; the chickens also generate more nitrogen-rich manure than Iowa's people. Almost none of this waste is treated before draining into waterways and making it into drinking water supplies. Result? Over 260 cities and towns in Iowa alone have water supply problems and face millions in investment to make tap water drinkable (and even this will do nothing to address the degraded state of waterways or the widening dead zone in the Gulf) (Manning 2016).

Uncertain water supply for agriculture itself is another issue that would fill volumes. To take a typical example, the Great Plains' Ogallala aquifer supports one-fifth of the wheat, corn, cattle and cotton produced in the United States but is currently being depleted faster than it is be recharged by precipitation. Studies suggest 'peak agriculture' as early as 2040 and, while a 20 % reduction in usage might back the peak off to 2070, farmers would have to reduce water consumption by 80 % to balance withdrawals with recharge (Steward et al. 2013; Winter and Foster 2014).

The water supply problem is heightened by concerns about climate change—significant reductions in precipitation would both reduce immediate water availability and worsen concerns about aquifer recharge. For example, hyper-productive California is in the fifth year of record low rainfall. Despite recent rains, more than 70 % of California is still in extreme drought, with nearly half of the state in "exceptional" drought, the most serious category. Reservoirs are at record lows (USDA 2015). In April, 2015 Governor Jerry Brown issued a first-ever executive order mandating statewide conservation reductions in water withdrawals and Californians have actually reduced water use by 26 %.

Clearly, as drought conditions or other extreme climate events become the norm, all afflicted areas will have to make major adjustments to crop-types and farming practices. Otherwise what hope does the world community have to maintain, let alone double food production as is needed to feed the billion people chronically malnourished today; to meet the increasing dietary demands of populations in emerging economies; *and* provide for the additional 2 or 3 billion people expected by late century?

4 Epilogue: The Unsustainable Entanglement of Nations

Which raises a question pertaining to global sustainability. Trade in agricultural products has enriched the diets of billions of people and the pockets food processors, food distributors and even a few farmers all over the world. Nevertheless, we should sound a cautionary note. The massive post-WW II expansion of trade and the specialization that comes with it implicitly assume stable market and climate conditions, uninterrupted growth in both demand and supply, and the indefinite sustainability of deeply interdependent trading relationships. How reliable are such assumptions? Remember, too, that only a handful of nations produce a physical surplus of foodstuffs on a regular basis—the majority of countries with most of the world's people continue to grow while increasing their dependence on off-shore sources of food.

Consider the case of Canada as one of those surplus suppliers. As we have seen, Canada is a major core food exporter (e.g., grain, meat, pulses, oil-seeds) by virtue of a low population and large areas of productive soils. It follows that increasing millions of people around the world are arguably at least partially dependent on Canada's agricultural land base for survival. Indeed, Kissinger and Rees (2009) show that between 1989 and 2007, Canada on average effectively 'exported' 51.4 % of the agricultural land in its prairie provinces (including 65 % of crop-land) and that the area 'exported' expanded significantly from around 20 million ha at the beginning of this period to as high as 34 million ha. Given the increasing volatility of world climate, possible large-scale losses of agricultural lands from desertification and sea-level rise, growing environment-related inter-regional migrations and the heightened geopolitical tensions generated by such trends is it not reasonable to question both the reliability this production and the security of established trading relationships? Kissinger and Rees (2009, p. 2314) observe that:

> prevailing approaches to globalization encourage the increasing entanglement of nations in a web of growth-oriented dependence. This ensures a perfect storm of unsustainability in an ecologically full-world pressing against the biophysical limits of critical life-support systems that are characterized by lag and threshold (tipping-point) effects.

They go on to ask (Kissinger and Rees 2009, p. 2314):

> 1. Is it wise for a nation to commit a significant proportion of its agricultural output and land-base to satisfying off-shore demand (i.e., creating dependent populations) given the increasing risk that domestic production will decline, food imports may similarly be threatened and domestic demand will rise in the foreseeable future?
> 2. Should any region or country allow its prime agricultural lands to be paved over or otherwise degraded on the assumption that it can always import basic foodstuffs from elsewhere?
> 3. Should any country allow itself to become significantly dependent on increasingly uncertain external sources of essential foodstuffs?

And they might well have added:

> 4. Is it not time that the world community devised a long-term strategy to reverse population growth and maintain the human population within the regenerative capacity of ecosystems and life-support services essential for survival?

Perhaps as we commemorate the end of the International Year of Soils (2015) (which also happened to be the warmest year in the instrumental record and a year stained by bloody wars in the Middle East, the greatest refugee crisis in half a century, and rising geopolitical unrest) the world's nations would do well to revisit such existential questions.

References

Derpsch R, Friedrich T, Kassam A, Hongwen L (2010) Current status of adoption of no-till farming in the world and some of its main benefits. Int J Agric Biol Eng 3(1):1–26

Eshel G, Shepon A, Makov T, Milo R (2014) Land, irrigation water, greenhouse gas, and reactive nitrogen burdens of meat, eggs, and dairy production in the United States. Proc Natl Acad Sci 111(33):11996–12001. doi:10.1073/pnas.1402183111

FAO (2003) World agriculture: towards 2015/2030 (An FAO perspective) esp. Chap 12 'Agriculture and the Environment: Changing Pressures, Solutions and Trade-Offs'. Food and Agriculture Organization of the United Nations, Rome. Available at http://www.fao.org/docrep/005/y4252e/y4252e00.htm#TopOfPage

FAO (2008) FAOSTATS. Food and Agriculture Organization of the United Nations. Available at http://faostat.fao.org/site/377/DesktopDefault.aspx?PageID=377#ancor

FAO (2015) Status of the World's soil resources. Food and Agriculture Organization of the United Nations, Rome

Geohive (2015) World fruit production in metric tons (excluding melons). Available at http://www.geohive.com/charts/ag_fruit.aspx

Kissinger M, Rees WE (2009) Footprints on the prairies: degradation and sustainability of Canadian agricultural land in a globalizing world. Ecol Econ 68(2009):2309–2315

Lyddon C (2011) North America. Available at http://www.worldgrain.com/news/news%20home/features/2011/8/north%20america.aspx?cck&cck=1

Manning R (2016) The Trouble with Iowa (Report). Harpers 332, No. 1989: 23–30

McConkey BG, Lobb DA, Li S, Black JMW, Krug PM (2010) Soil erosion on cropland: introduction and trends for Canada, Canadian Biodiversity: Ecosystem Status and Trends 2010. Technical Thematic Report No. 16 (Canadian Councils of Resource Ministers)

Nationmaster (2011) Hectares per Capita: countries compared. Available at http://www.nationmaster.com/country-info/stats/Agriculture/Arable-land/Hectares-per-capita#2011

StatsCan (2009) Food production and the environment: annual statistics section 1: food in Canada, Stats Canada. Available at http://www.statcan.gc.ca/pub/16-201-x/2009000/part-partie1-eng.htm

Steward D, Bruss P, Yang X, Staggenborg S, Welch S, Apley M (2013) Tapping unsustainable groundwater stores for agricultural production in the High Plains Aquifer of Kansas, projections to 2110. In: Proceedings of the National Academy of Sciences 110 (37), doi:10.1073/pnas.1220351110. Available at http://www.pnas.org/content/110/37/E3477.full

TFE (2015) The top five countries that produce the most vegetables. Top five of everything. Available at https://top5ofanything.com/list/0fe4983f/Countries-that-Produce-the-Most-Vegetables

Tilman D, Fargione J, Wolff B, D'Antonio C, Dobson A, Howarth R, Schindler D, Schlesinger WH, Simberloff D, Swackhamer D (2001) Forecasting agriculturally driven global environmental change. Science 292(5515):281–284

USDA (2014) 2014 State agricultural overview—Iowa. Available at http://www.nass.usda.gov/Quick_Stats/Ag_Overview/stateOverview.php?state=IOWA

USDA (2015) California drought: farm and food impacts. US Department of Agriculture. Available at http://www.ers.usda.gov/topics/in-the-news/california-drought-farm-and-food-impacts/california-drought-crop-sectors.aspx

USDA (2016) Cited in global soybean production. Available at http://www.globalsoybeanproduction.com/

USDA. ND. 80 years helping people help the land: a brief history of NRCS. US Department of Agriculture. Available at http://www.nrcs.usda.gov/wps/portal/nrcs/detail/national/about/history/?cid=nrcs143_021392

US Senate (1936) S.rp.1481, Conservation and utilization of the soil resource. Committee on Agriculture and Forestry, Senate, January 16, 1936

WFP (2013) What happens if US loses California food production? Western Farm Press. Available at http://westernfarmpress.com/tree-nuts/what-happens-if-us-loses-california-food-production?page=1

Winter M, Foster C (2014) Ogallala aquifer – lifeblood of the high plains Part I: Withdrawals exceed recharge. CoBank Knowledge Exchange. Available at http://aquadoc.typepad.com/files/ke_ogallalaaquifer_reportpt1-oct2014.pdf

World Bank (2013) Data – cereal production. Available at http://data.worldbank.org/indicator/AG.PRD.CREL.MT/countries/1W?order=wbapi_data_value_2013%20wbapi_data_value%20wbapi_data_value-last&sort=asc&display=default
WP (2016) Western producer. Canada aims to leapfrog U.S. in global wheat trade (Reuters). Available at http://www.producer.com/daily/canada-aims-to-leapfrog-u-s-in-global-wheat-trade/
WPC (2016) World population clock. Available at http://www.worldometers.info/world-population/

Soils Governance, an Australian Perspective

Paul Martin

It is wonderful to be welcoming the first International Yearbook of Soil Law and Policy with this chapter, which looks at soils governance from the perspective of the people who live in Australia. In this chapter I highlight some aspects of the public policy challenge of soils governance in our country, with an emphasis on systemic issues.

The most recent national scientific assessment of Australia's environment paints a picture of the effectiveness of rural soils management that is not heartening. This highlights the need for more effective soils governance, and this brief chapter aims to 'unpack' the challenges that are involved in achieving this.

> Land-management practices have improved during the past few decades, but soil management has to improve significantly to build soil carbon, control acidification and prevent erosion. . . . Few regions have increasing levels of soil carbon, although the potential in the savanna landscapes of northern Australia is significant. Soil acidification affects about half of Australia's agriculturally productive soils. Its severity and extent are increasing, and large areas will become unproductive and degraded unless they are treated. . . . Current rates of soil erosion by water across much of Australia now exceed soil formation rates by a factor of at least several hundred and, in some areas, several thousand. . . . The longer settled agricultural and coastal zones have the highest concentration of impacts on native vegetation. In most of these regions, less than 50 % of native vegetation remains, and vegetation condition generally deteriorates with diminishing remnant extent. Approximately 13 % of native vegetation nationally has been completely converted to other uses. Annual rates of native vegetation clearing averaged around 1 million hectares in the decade to 2010, balanced by the extent of regrowth by the end of the decade. . . . Assessment of land, soil and vegetation condition over the past decade is complicated by the impacts of drought in much of the south and east of the continent. Drought affected these areas from 2000 to 2010 (sometimes known as the millennium drought), although in some areas it

P. Martin (✉)
Australian Centre for Agriculture and Law, University of New England, Armidale, NSW 2351, Australia
e-mail: Paul.Martin@une.edu.au

H. Ginzky et al. (eds.), *International Yearbook of Soil Law and Policy 2016*,
International Yearbook of Soil Law and Policy,
DOI 10.1007/978-3-319-42508-5_5

began as early as 1997. Furthermore, nationally consistent metrics for assessment of vegetation condition are still under development.[1]

Governance is about directing, shaping or controlling human behavior to achieve public objectives (including sustainability and social justice). Soils governance focuses on issues related to soils and land use, and related social issues. Governance analysis requires that we consider people and organisations who serve different roles, with many and often competing interests: citizens and their organisations; business and its institutions, and the public sector with its many roles. Whilst many might see soils governance through the lens of public-sector environmental protection, the most powerful driver for how land is managed, by either the public or the private sector, is the pursuit of production and wealth from the land. It is also easy to think that environmental harms, like those discussed above, caused mainly by the irresponsible behavior of land stewards, particularly farmers. This perception of what causes problems is likely to lead to interventions like regulation and incentives to change the behavior of individuals. However, soil degradation is also a response to systemic forces including those that drive private mis-management. To manage system failure requires reforms that also address the matters that are not within the control of the individual land steward. This suggests that the focus for soils law and policy needs to be upon both improving private stewardship and institutional reform to better align socio-economic systems with the requirements of sustainability.

1 The Effects of Australia's Rural Context

Human behavior is largely shaped by context. This determines where rewards can be found, what signals and information trigger and inform decisions, and how different things are valued when individuals or organisations make decisions. The context includes natural conditions such as the state of soils and climate, the economic system that determines the rewards and the signals to which economic man will respond, the social system that shapes attitudes and social supports, the political system, and the industrial system that creates the technological and managerial options that land managers have available to them to carry out their activities. Three aspects of socio-economic context are particularly important when considering soils stewardship in rural Australia: the limited labour and economic density of many rural areas; the impacts of the economic dynamics of commodity markets on the fin capacity of rural land stewards; and the limited capacity of governments to make up for these incapacities.

Australia covers 7.688 million square kilometers, with a population of only 23.5 million people (ABS 2014). The country's biodiversity, landscape and climate are diverse, with vast tracts of desert, rich farmland, a temperate climate in the South

[1]State of the Environment Committee, State of the Environment Report 2011 (Independent report to the Australian Government Minister for Sustainability, Environment, Water, Population and Communities, DSEWPaC, 2011), p. 270.

and a tropical climate in the north. Its' population density is roughly equivalent to Mongolia or Iceland (around one hundredth of the density of the UK, one tenth of the USA, or one fiftieth of the density of China) (The World Bank 2015). More than 80 % of this small population lives within 50 km of the coast and 66 % live in capital cities (ABS 2014). The result is that population density declines away from the coastal fringe. Australia has 456 million hectares of agricultural land, constituting 59 % of the landmass, and much agriculture occurs in the sparsely populated inland. A lack of available labour determines what agricultural activities are feasible and in part explains why Australian farming is concentrated in broadacre cropping and grazing rather than intensive farming. Australian farms tend to be capital rather than labour intensive. The harmful impacts of broadacre farming can include large scale land clearing, over-grazing, and cropping practices that deplete the soil. Agriculture is increasingly diverse, and agricultural land use decreased by 18 million hectares (4 %) between 1992–1993 and 2005–2006, one factor being an increase in land dedicated to conservation.[2]

Agricultural harms to the environment are not generally reflected in the prices of the commodities that are produced from using (or destroying) natural resources. Australian governments do not actively address this missing markets problem, unlike Europe and the USA (using subsidies and regulation to 'incentivise' land stewardship), though some farmers participate in environmental stewardship and consumer branding programs. Because the price or cost signal of the harms that are being done is weak the landholders' economic motivation to reduce these is not as strong than it needs to be, though in the medium to long term aspects of soils deterioration can reduce the capital value of land. The dynamic of commodity markets inexorably drives down landholder margins from primary production, which reduces the capacity of land stewards to afford landscape protection and restoration work. The final system consideration is the frequency of drought, which exacerbates land management failings and creates significant economic pressure on farmers (reducing their economic capacity for good land stewardship). Related forms of disadvantage, including poor health status and limited formal education, also help explain the reluctance of farmers to take up innovations in farming practice and technology that could reduce the adverse impacts of farming practices on soils. Whilst not excusing poor stewardship, economic pressure arising from system factors does help to explain it.[3]

[2]ABARES, 'Towards national reporting on agricultural land use change in Australia' (2013).

[3]This is illustrated by the fact that despite working long hours 'the average weekly disposable income of farmers in 2009–2010 ($568) was considerably lower than that of people working in other occupations ($921)' (ABS 2012a). Low income does not mean a lack of wealth. Farming is capital intensive, though the income returns on invested capital are low. The average wealth of farming households in 2009–2010 was $1.3 million, roughly three times the average of all households. Poverty as a characteristic of farming is not unique to Australia. Rural residents are around 55 % of the total population in developing countries. More than 70 % were living in extreme poverty in 2011, compared to 55 % of the total population.

Australia's soils conservation and restoration work has traditionally relied on a combination of landholder and government action. The government role has included scientific extension and advice to landholders, soils research, scrutiny of things like illegal land clearing or other harmful actions, and direct investment in on-ground works. Once powerful and well-resourced soil conservation services carried out on-ground works and provided leadership and support to farmers and other land users, but recent years have seen a seemingly inexorable reduction in government 'front line' soils and (other rural sustainability) investment. There are many reasons why this is so but an important consideration is declines in the capacity for public funding. Australia's ageing population and a bipartisan commitment to capping public debt and the public sector share of GDP (which policy has contributed to Australia's sound economic position) have meant that Australian governments are under ever-increasing pressure to reduce spending. All indications are that this pressure will increase. The most recent Intergenerational Report indicates that Australia's underlying cash deficit under current legislated arrangements will increase from 2014–2015 to around 6 % of GDP by 2054–2055 with net debt increasing to almost 6 % of GDP. This has significant implications for public soils governance, suggesting that any policy that depends on substantial government investment must be considered to have a limited chance of being successfully adopted. A return to a golden age of public investment in soils governance, much as it might be desired, seems unlikely for economic reasons alone.

2 Mixed Problem Archetypes

How a policy maker perceives a problem will largely determine what types of policy prescriptions are proposed. To attribute the failures of soils governance in Australia to deficient farmer stewardship appears to be a 'simple, neat, plausible, and wrong[4]' diagnosis, that is likely to lead to over-simplistic 'solutions'. System causes and thus systemic reforms need to be considered. The need to address the broader institutional changes needed to achieve effective soils governance is reinforced by consideration of the different types of problems ("archetypes") that are involved.

1. Individual stewardship. The conceptually simplest problem archetype is the failure of land managers as stewards of the resource they control. Unsustainable farming, pollution or ill-advised urban or industrial development are often attributed to stewardship failures, and the causes diagnosed as greed or ignorance. This simplistic characterisation is consistent with factionalised political rhetoric. Regulation, education or extension, private incentives, or financial or other supports are the policy mechanisms that are advocated in response to this

[4]"*The Divine Afflatus*," A Mencken Chrestomathy, chapter 25, p. 443 (1949).

characterisation. However, as we have considered this simple archetype has underlying system causes, and a failure to address underlying causes makes it likely that similar problems will continue to be caused by logical responses to socially counterproductive system signals, and by limits to the capacity of land stewards to do what is required to manage soils sustainably. Addressing the system issues brings into consideration the powers and limits of government, not least of all fiscal limits. Many conventional prescriptions ignore these increasingly pressing constraints, and are unlikely to be feasible without an innovative fiscal strategy.

2. Collective action problems. Some issues require collective rather than individual action. This may be because the harm arises from the accumulation of impacts that may individually not be significant, or because remedies cannot be achieved without collective action. For example, maintaining natural windbreaks or ground-cover over a sufficiently large area to reduce wind erosion requires cooperation by many landholders. The cooperation may require changes to farming practices and capital investment. Individual action, whilst necessary for such strategies to succeed unless a sufficient number of land stewards become involved. In Australia as in many other countries, civil and property rights, the limited feasibility of public oversight and the nature of human behaviour, mean that the power of governments to force cooperation is very limited. Public policy can only facilitate cooperation by landscape-scale planning, supported by regulation, incentives and public investments to the extent that this is feasible. Collective action problems involve complex behaviour and the management of social systems.
3. Coordinated action problems. More difficult still are issues where collective action is necessary but insufficient unless it is well-coordinated. An example is controlling or remediating salinity by managing water use and deep-rooted vegetation on a number of properties, which may be some distance apart. Another example is creating a market mechanism such as a carbon credit instrument to reward 'minimum till' agriculture. The requirement for coordination adds a significant institutional dimension to the policy task. In Australia the increasing diversity of land uses in rural areas can make coordination more difficult. Private incentives to cooperate (or not) will often vary with the land use. For example, whilst salinity might be an urgent problem for a cropping enterprise it may be inconsequential for a grazing enterprise if salt-tolerant vegetation can be grown, or many mean nothing to a mining enterprise. The institutional arrangements for coordinated action can call on specialist skills and systems that will not emerge automatically, and may require high levels of bureaucratic as well as community management.

Soils governance is a complex multi-dimensional system issue (of the type sometimes referred to as 'vicious' problems). It is inevitable that soils sustainability will continue to be a major problem for Australia. Given changes to context many conventional solutions will not be viable, for economic and political reasons. Significant innovations in public policy will be necessary to overcome the challenges, and for this reason a year-book that can promote constructive debate is welcome.

There are opportunities to address the challenges. Notwithstanding the socio-economic difficulties facing agriculture and investment the environment, Australia remains a very rich country, on a per-capita basis equivalent to the USA and Great Britain. A philosophical commitment to sustainable land management permeates government, civil society and agricultural institutions, though converting this to practical action remains problematic. The role of soils in reducing carbon emissions or sequestering carbon is an obvious area for policy that can provide some economic incentive for constructive action. Other market-based opportunities also exist, such as biodiversity incentives and the strategic use of water markets to achieve better land management. Voluntary programs, including philanthropic investment and voluntary stewardship initiatives, continue to increase.

References

ABS (2014) Feature article: capital cities: past present and future. http://www.abs.gov.au/AUSSTATS/abs@.nsf/Previousproducts/3218.0Feature%20Article22012-13?opendocument&tabname=Summary&prodno=3218.0&issue=2012-13&num=&view=. Accessed 31 Mar 2015

World Bank (2015) Databank Australia, Population; http://data.worldbank.org/indicator/SP.POP.TOTL

Striving for Land-Soil Sustainability: Some Legal Reflections

Bharat H. Desai and Balraj K. Sidhu

1 Introduction

Humankind is facing one of the biggest tests for striking a proper balance between developmental needs and environmental imperatives. However, ironically, much of the development in the world today is not sustainable. Many of our economic, monetary and trade policies in sectors such as energy, agriculture, forestry and human settlements tend to induce and reinforce non-sustainable development patterns and practices. Apart from irreversible squandering of our biological capital, we are confronting a growing problem of environmental refugees-people who are forced to abandon their traditional habitat as the land cannot sustain them. In fact, the human quest to conquer nature through science and technology has brought us on to the present brink.[1]

The threats to our eco-system essentially emanate from human activities in almost every sector. Among major threats that merit our attention, state of soils especially desertification and drought has emerged as one of the "urgent concern of

[1]For detailed analysis on ramification of developmental spree on environment, see Desai (1986, pp. 27–29; 1990, pp. 1065–1072; 1992, pp. 589–596).

B.H. Desai (✉)
Centre for International Legal Studies, School of International Studies, Jawaharlal Nehru University, New Delhi, India

Centre for Advanced Study on Courts & Tribunals, Amritsar, India
e-mail: desai@jnu.ac.in

B.K. Sidhu
Centre for Advanced Study on Courts & Tribunals, Amritsar, India
e-mail: casct@hotmail.com

H. Ginzky et al. (eds.), *International Yearbook of Soil Law and Policy 2016*,
International Yearbook of Soil Law and Policy,
DOI 10.1007/978-3-319-42508-5_6

international community".[2] In fact process of desertification "affects about one-sixth of the world's population, 70 percent of all drylands, amounting to 3.6 billion hectares, and one quarter of the total land area of the world".[3] The cost of land degradation is estimated to be about US dollars 490 billion per year and some 40 % of this occurs in "areas in with highest incidence of poverty".[4]

Soils are a key to sustain life on the Earth as they "constitute the foundation for agricultural development, essential ecosystem functions and food security".[5] Soil is not only crucial for food production but also a significant regulating factor in mitigation of climate change through carbon sequestration. Productive land and soil are critical natural capital assets underscoring the vital nexus of food, renewable energy and water security.[6] The continued and unabated land and soil degradation has increased vulnerability of communities that rely heavily on land as their main source of livelihood, particularly the rural poor in developing countries. The major form of land degradation in most of Asia is the soil degradation especially due to the "heavy pressure on natural resources for livelihoods".[7] The growing importance of protecting the soil integrity to implement and attain goals of poverty eradication and sustainable development led to revision of the 1981 World Soil Charter in 2015. The *Intergovernmental Technical Panel on Soils* underscored the need for an update:

> The 13 principles listed in the charter are still valid, but need to be updated and revised in light of new scientific knowledge gained over the past 30 years, especially with respect to new issues that emerged or were exacerbated during the last decades, like soil pollution and its consequences for the environment, climate change adaptation and mitigation and urban sprawl impacts on soil availability and functions. New priorities for action as well as follow-up activities should be identified, taking stock of past experiences and learning from the failures and mistakes that have resulted in a still persistent global problem of soil degradation and unsustainable use of available soil resources.[8]

The challenge lies in retaining the capacity of soil to function with natural or managed ecosystem, sustain plant and animal productivity, maintain air and water quality and support human health and habitation.

---

[2]Preamble to the United Nations Convention to Combat Desertification in Countries Experiencing Serious Drought and/or Desertification, Particularly in Africa. http://www.unccd.int/en/about–the-convention/Pages/Text-Prologue.aspx.

[3]Para 12.2 in chapter 12 (Managing fragile ecosystems: Combating desertification and drought) of Agenda 21; see United Nations, Agenda 21: The United Nations Programme of Action from Rio. New York, 1993.

[4]UNCCD, *Land Degradation Neutrality: Resilience at Local, National and Regional Levels* (Bonn, UNCCD, 2014), p. 4.

[5]The United Nations General Assembly resolution 68/232 adopted on 20 December 2013; see preamble to the UN Doc. A/RES/68/232 of 7 February 2014.

[6]See UNCCD Secretariat: *A Stronger UNCCD for a Land-Degradation Neutral World* (Bonn: Germany), 2013. For further discussion, see Stringer (2008a, b), pp. 138–144.

[7]See Article 2 of Annex II on Regional Implementation Annex for Asia, n. 2, p. 56.

[8]FAO, *Global Soil Partnership-World Soil Charter,* 39th Session, 6–13 June, 2015, Rome. http://www.fao.org/3/a-mn442e.pdf.

2 Deforestation and Soil Degradation

At a time when we are facing impending threat of global warming and climate change, the global green cover is fast disappearing. The green cover is a vital part of sour eco-system and helps preventing the earth to be too hot for us. The forests are diminishing at a rate of 15 million hectares per year. Most of this loss is occurring in humid parts of Africa, Asia and Latin America.[9] It is now estimated that in 2030, the demands for energy will rise 50 %, food 45 %, and water 30 %. As a result, continuing land degradation due to growing population pressure could drive 700 million people out of their homes.[10] The major losses in the tropical forests take place for their lumber as well as for plantations, pastures, crops and other development purposes. The increasing pressure on green cover is not only because of growing population but also the need to earn foreign exchange by most of the Asian countries. Some of the heavily debt-ridden Asian countries are forced to fell trees, which feeds the appetite of industrialized countries. The process of felling and torching of forests is so harmful that it takes away with it the genetic resources and leads to increase in flooding, landslides, soil degradation and a host of other problems.

Though some belated efforts have been made at the international level, however, there is hardly any significant impact in halting the fast erosion of green cover. As such we need to ponder over some crucial questions: Is it possible for the South and South-East Asian countries to stop the practice of shifting cultivation? Can we give incentives in the form of swapping of debt for preservation of forests? Can we not think in terms of a global moratorium on timber trade and depend upon alternatives to our timber needs? When the green cover disappears, the land is exposed. It paves the way for gradual soil degradation and deforestation.

3 Creeping Desertification

Desertification and land degradation occur in numerous ways, such as soil erosion, the deterioration of soil biology and the loss of natural vegetation. In fact some of the unsustainable land use practices such as, over cultivation and overgrazing only accentuate this process. Desertification of which human being is an agent as well as a victim makes the habitat too harsh to survive—pushing the people *en masse* elsewhere just for survival or better living conditions. The enormity of the problem can be seen from the fact that two-fifths of Africa's non-deserted land resources is being turned into desert, as does one-third of Asia's and one-fifth of Latin America's. The process of desertification is most prominent in arid and semi-arid

[9]UNEP, *Global Environment Outlook 2000* (Nairobi: UNEP, 1999), p. 851. http://www.grida.no/publications/other/geo2000/?src=/geo2000/.

[10]See n. 4, p. 2.

lands.[11] Out of a total land area of 4.3 billion hectares, Asia contains some 1.7 billion hectares of arid, semi-arid and dry sub-humid land. It stretches from the Mediterranean coast to the shores of the Pacific.[12] It seems that in the slow process of growing desertification, political, economic and social factors are more important than low rainfall. The subsistence farmers have no option but to resort to over cultivation and the nomads indulge in extensive grazing only for bare survival.

The effects of wide-spread deforestation and desertification spill into neighbouring countries too. The problem of growing desertification, land degradation and drought has accentuated poverty and inequality thereby unleashing resource-wars in the affected countries.[13] Asia remains the most severely affected continent in terms of number of people. The threat being posed by growing desertification, poses a serious challenge. We have witnessed the pathetic sight of human swarms crossing frontiers in countries like Myanmar, Laos, Nepal and others. What can the international community do on a long-term basis to halt the desertification and land use practices leading to soil degradation in these helpless countries?

The process of legal regulation of state behaviour is now manifesting in various spheres of international life. The multilateral law-making process has emerged as a predominant method of regulating state behaviour on a global *problematique*. The multilateral environmental agreements (MEAs) have emerged as a unique technique that encompasses flexibility, pragmatism, in-built law making mechanism as well as step-by-step consensual approach to norm-setting.[14] With the new scientific understanding that desertification and associated problems of drought and land degradation could take an alarming proportion globally, the United Nations Convention to Combat Desertification[15] was adopted in 1994 as an outcome of the Rio Earth Summit (1992). As it is clear from the title of the Convention that it concerns essentially with the desertification and does not cover land/soil degradation in general. It defines the term 'desertification' as "land degradation in arid, semi-arid and dry sub-humid areas resulting from various factors including climatic variations and human activities".[16] The Convention seeks to prevent and/or reduce land degradation, rehabilitate partly degraded land, reclaim desertified land and mitigate the effects of drought.[17] The UNCCD's Regional Implementation Annex

---

[11]For general discussion, see Kassas (1995), pp. 115–128.

[12]Addressing desertification, land degradation and drought in Asia (UNCCD: Bonn). http://www.unccd.int/en/regional-access/Asia/Pages/alltext.aspx.

[13]For general discussion, see De Jong (2013), pp. 155–177.

[14]For detailed discussion on multilateral environmental agreements (MEAs), see Desai (2010); Also see Desai (2006), pp. 1–3.

[15]See the United Nations Convention to Combat Desertification in Countries Experiencing Serious Drought and/or Desertification, Particularly in Africa; see http://www.unccd.int/en/about-the-convention/Pages/Text-Prologue.aspx. For further discussion on UNCCD, see Kjellen (2003), p. 127; Bryant (1996), p. 445; Stringer (2008a, b), pp. 2065–2074.

[16]Article 1(a) UNCCD.

[17]Article 1(b) & (c) UNCCD.

for Asia[18] recognizes the unique challenges from the Asian region. It calls for appropriate activities at the national, sub-regional and regional level in the form of coordinated and integrated action programs.

It encourages the countries to submit "National Action Programmes" (NAPs) outlining the national policy and institutional initiatives planned to combat desertification as well as indication on funding requirement for implementation.[19] It requires all parties to engage through political, practical and financial investment in the maintenance of productive land by improving affected ecosystems and living conditions. It seems the "UNCCD was designed in the hope that countries could rely on voluntary programs that employed a "bottom-up" strategy that would be driven by international assistance from "non-affected" developed countries to "affected" countries that lacked the resources to implement a clear National Action Program".[20] However, the uncertainties concerning funding the initiatives under UNCCD has weakened the impact of the Convention on the ground. In comparison to other Rio Conventions, namely, Convention on Biological Diversity and United Nations Framework Convention on Climate Change, UNCCD has received scant international visibility as well as meagre funding/aid.[21]

4 SDGs and Land Degradation Neutrality

The economic and social significance of a good land management including soil and its contribution to economic growth and social progress is recognized in paragraph 205 of the "Future We Want".[22] In paragraph 206, it underlined "the need for urgent action to reverse land degradation. In view of this we will strive to achieve a land-degradation neutral world ...".[23] The UN Sustainable Development Summit (UNSDS) held in New York became an epoch-making event to finally adopt a blueprint of Sustainable Development Goals (SDGs) to be addressed in a 15 year cycle.[24] The UN Secretary-General Ban Ki-moon, in a timely chastisement ahead of the final gavelling of the SDGs, underscored that "We have a big, bold agenda before us—now we must work to make it real in people's lives everywhere".[25]

[18]See Article 2 of Annex II on Regional Implementation Annex for Asia, n. 2, p. 56.

[19]Article 9 & 10 UNCCD.

[20]Tal (2015), p. 25.

[21]For more discussion, see Conliffe (2011), pp. 44–63.

[22]The United Nations General Assembly resolution 66/288 adopted 27 July 2012. https://documents-dds-ny.un.org/doc/UNDOC/GEN/N11/476/10/PDF/N1147610.pdf?OpenElement.

[23]*Id.*

[24]See *Transforming our World: the 2030 Agenda for Sustainable Development,* The United Nations General Assembly resolution 70/1 adopted 25 September 2015. http://www.un.org/ga/search/view_doc.asp?symbol=A/RES/70/1&Lang=E.

[25]The Secretary-General, Remarks at UNSD Summit for the adoption of the post-2015 Development Agenda, New York, 25 September 2015. https://sustainabledevelopment.un.org/content/documents/20229sg.pdf.

These goals are essentially a sequel to the Millennium Development Goals (MDGs) adopted at the UN Millennium Summit held in 2000. They are set to be realized in an incremental, persuasive and leisurely way. The Goals provide ominous warning that—without decisive course correction—the galloping human population by 2050 will require natural resources equivalent of three planets to sustain current lifestyles.

Ironically, stocktaking on realization of MDGs has not yielded much to cheer about. It underscores grim reality of a world that has so much pain, hunger, ill health, illiteracy and suffering notwithstanding abundant resources as well as human and technological prowess. The factsheet released by the UN shows crude reality on some of the benchmarks (such as gender equality in primary school enrolment of girls, especially in South Asia), where no change has been seen during 1990–2012 periods!

It is in this backdrop that advent of SDGs, as a plan of action for people, planet, prosperity, peace and partnership represents a renewed quest to fine tune the global goalposts. It represents tragedy of human quest for material progress and remarkable transition from bland millennium "development" goals to "sustainable development" goals. The usage of prefix 'sustainable' adds value and seeks to moderate our greed for unimpeded growth. Are there any limits to it? In a way, unabashed developmental spree, rise in human misery and graphic chasm of inequality during MDGs 15 year journey (2000–2015) would have a sobering effect for the challenge of realizing SDGs in the next 15 years (2015–2030).

Though listed under the rubric of SDGs, the goals include challenges as diverse as inequality in living standards, infrastructure, water & sanitation, women empowerment, access to modern energy, quality education, healthy lives, poverty and hunger. SDG 15 aims to "protect, restore and promote sustainable use of terrestrial ecosystems, sustainably manage forests, combat desertification, and halt and reverse land degradation and halt biodiversity loss".[26] The goal is further divided into nine targets. In fact SDG 15.3 specifically mandates that "By 2030, combat desertification, restore degraded land and soil, including land affected by desertification, drought and floods, and strive to achieve a land degradation-neutral world".[27] It remains to be seen as to how far this goal will be realized.

In fact, the target of *land degradation neutral world* (LDNW) has attracted considerable attention in the policy making circles as well as civil society. It is gradually emerging as a concept and a global derivative of "zero net land degradation".[28] A UNCCD brief highlighted that the goal would mean that:

---

[26]See *Transforming our World: the 2030 Agenda for Sustainable Development,* The United Nations General Assembly resolution 70/1 adopted 25 September 2015. http://www.un.org/ga/search/view_doc.asp?symbol=A/RES/70/1&Lang=E. For more information on SDG 15. https://sustainabledevelopment.un.org/sdg15.

[27]*Ibid.*

[28]There has been churning on having a protocol under UNCCD on Zero Net Land Degradation, see Tal (2015), pp. 25–32.

> By way of reducing degradation processes and scaling up restoration activities at community and/or landscape levels, the objective is to maintain and improve the quantity and quality of productive land to enhance the flow of ecosystem services for current and future generations...The transition to a land-degradation neutral world would stop this vicious cycle by coming to terms with nature through a paradigm shift from *'degrade-abandon-migrate'* to *'protect-sustain-restore'*[29] (emphasis added).

At the recent 12th session of Conference of Parties to UNCCD held in Turkey (2015), the land degradation neutrality (LDN) emerged as an organizing principle that Parties to the Convention would be required to implement according to their specific national circumstances and development priorities.[30] The intergovernmental working group (IWG) formed to follow up on the Rio + 20 outcomes related to LDN in its report formulated a science-based definition of LDN as "Land Degradation Neutrality [in affected areas][in arid, semi-arid and dry sub-humid areas] is a state whereby the amount and quality of land resources, necessary to support ecosystem functions and services and enhance food security, remains stable or increases within specified temporal and spatial scales and ecosystems".[31] The text is bracketed due to failure of members to arrive at consensus with regard to universal application of the definition. The IWG considered the adoption of LDN target within the ambit of SDGs both a challenge and opportunity for the Convention. As wittingly remarked by the present Executive Secretary Ms. Monique Barbut "land will fill a place as the Queen of both international development and climate change actions. Unlike other pieces on the board, the Queen provides cover and supports our ambitions on many fronts; food, energy, water, climate, biodiversity, jobs, migration and security".[32] The adoption of LDN target has put UNCCD on par with Biodiversity and Climate Change conventions as they have their own targets.

The goal of LDNW seems to be quite futuristic. It could give a new lease of life to UNCCD.[33] In fact the former Executive Secretary of UNCCD, Luc Gnacadja, has suggested to move in the direction of "zero-net rate of land degradation". It could serve as panacea for global society grappling with challenges of land

---

[29]See UNCCD Secretariat: *A Stronger UNCCD for a Land-Degradation Neutral World* (Bonn: Germany), 2013, p. 7.

[30]See Decision ICCD/COP(12)/L.4; available at: http://www.unccdcop12.gov.tr/en/Home/Icerik/101.

[31]*Id.*

[32]UNCCD COP12: From the UN General Assembly to the Paris Climate Change Conference; see http://www.unccd.int/en/about-the-convention/the-bodies/the-cop/COP_12/Pages/default.aspx.

[33]Although the Convention came into force in 1996, its implementation has been tardy and there has been no scientifically verifiable assessment how much amount of land degradation/desertification/drought has been reversed in the drylands during these 20 years of Convention. For further discussion, see Bassett and Talafre (2003), pp. 133–139. More recently, Canada pulled out of UNCCD citing "membership in this convention was costly for Canadians and showed few results, if any for the environment". http://www.ctvnews.ca/politics/canada-only-un-member-to-pull-out-of-droughts-and-deserts-convention-1.1214065.

degradation as well as giving a new mandate to UNCCD.[34] It has proposed an innovative technique to achieve the goal that:

> If the continuing loss of fertile land can be offset by restoring already degraded land, and the annual rate of reclamation equals the annual degradation rate, then a zero net rate of land degradation is attained and the area of global fertile land remains stable.[35]

In view of daunting challenge that lies ahead, the institutionalized international cooperation under the umbrella of UNCCD needs more flesh and concrete actions in terms of scientific study, concerted international legal mechanism and policy responses at domestic level to make it work on the ground. In the quest for land degradation neutrality, the inter-governmental efforts need not be squandered in debates on ambiguities of jargon.

5 Concluding Remarks

The current patterns of institutionalized global cooperation, especially to address desertification and drought, require closer scrutiny in the backdrop of rapidly deteriorating global environment, expectations of the sovereign states, growing fatigue effect in funding contributions and questions regarding effectiveness/adequacy of the tools & techniques adopted to address the *problematique*. In the specific context of renewed quest for land-soil sustainability, the LDNW remains amorphous as it must meet the needs and aspirations of teeming millions in the world. This assumes great significance in view of extreme poverty and misery in the Asian region.

In is this context that advent of the *International Yearbook of Soil Law and Policy* will bring together timely and concerted scholarly discourse to address the challenge of land degradation in the coming decades. The Yearbook will provide a relevant platform to put forward cutting-edge ideas, proposals and analysis that would help the sovereign states, the UN system, the UNCCD, various global forms of cooperation in the field of desertification/drought, donor agencies as well as practitioners to fine-tune appropriate policy, legal and institutional responses in cross-sectoral efforts to address the challenge of land/soil degradation. We commend efforts of the editors-in-chief and extend our best wishes for success of the Yearbook.

---

[34] See Statement of Luc Gnacadja, Executive Secretary UN Convention to Combat Desertification, "Moving to Zero-Net Rate of Land Degradation", UN Headquarters, New York, 26 March 2012. http://www.unccd.int/Lists/SiteDocumentLibrary/secretariat/2012/UNCCD%20ES%20Statement%20at%20PR%20in%20NY%20on%2026%20March%202012.pdf.

[35] *Id.*

References

Bassett C, Talafre J (2003) Implementing the UNCCD: towards a recipe for success. RECIEL 12 (2):133–139

Bryant T (1996) A new approach to an old problem: the Convention to Combat Desertification. Environ Plan Law J 13:445

Conliffe A (2011) Combating ineffectiveness: climate change bandwagoning and the UN Convention to Combat Desertification. Glob Environ Polit 11(3):44–63

De Jong DD (2013) From engines for conflict into engines for sustainable development: the potential of international law to address predatory exploitation of natural resources in situations of internal armed conflict. Nordic J Int Law 82:155–177

Desai BH (1986) Destroying the global environment. Int Perspect 5:27–29

Desai BH (1990) Managing ecological upheavals: a third world perspective. Soc Sci Med (10):1065–1072

Desai BH (1992) Threats to the world eco-system: a role for the social scientists. Soc Sci Med 35 (4):589–96

Desai BH (2006) Understanding the complex web of multilateral environmental agreements: a note on long term view for the Indian Ministry of Environment & Forests, Bonn, 8 June 2006, pp 1–3 (on file with the authors)

Desai BH (2010) Multilateral environmental agreements: legal status of the secretariats. Cambridge University Press, New York

Kassas M (1995) Desertification: a general overview. J Arid Environ 30:115–128

Kjellen B (2003) The saga of the Convention to Combat Desertification: the Rio/Johannesburg process and the global responsibility for the drylands. RECIEL 12:127

Stringer LC (2008a) Reviewing the International Year of Deserts and Desertification 2006: what contribution towards combating global desertification and implementing the UN Convention to Combat Desertification? J Arid Environ 72:2065–2074

Stringer LC (2008b) Can the UN Convention to Combat Desertification guide sustainable use of the world's soil? Front Ecol Environ 6(3):138–144

Tal A (2015) The implications of environmental trading mechanisms on a future Zero Net Land Degradation Protocol. J Arid Environ 112:25–32

Part II
Recent Developments of Soil Regulation at International Level

International Soil Protection Law: History, Concepts and Latest Developments

Ben W. Boer, Harald Ginzky, and Irene L. Heuser

1 Introduction

Soil is the fundamental component of all of the world's land.[1] In the short term, as a natural resource, soil is non-renewable.[2] However, it plays a vital role in the Earth's ecosystem, being the principal basis for terrestrial biodiversity as well as supplying

---

The three authors are members of the Sustainable Use of Soil and Desertification Specialist Group of the 'IUCN World Commission on Environmental Law'. Dr Heuser is Deputy Chair of the Group.

The authors are grateful to Assoc. Prof. Dr Ian Hannam for providing detailed suggestions for this chapter. Any remaining errors are the responsibility of the authors.

[1] 'Land' in this chapter is defined as in Article 1, 'Use of Terms', United Nations Convention to Combat Desertification (hereafter "UNCCD"): '...the terrestrial bio-productive system that comprises soil, vegetation, other biota, and the ecological and hydrological processes that operate within the system'.

[2] Food and Agriculture Organization, 'Soil is a finite resource, meaning its loss and degradation is not recoverable within a human lifespan.' http://www.fao.org/resources/infographics/infographics-details/en/c/278954/.

B.W. Boer (✉)
Research Institute of Environmental Law, Wuhan University, Wuhan, China

University of Sydney, Sydney, Australia
e-mail: ben.boer@sydney.edu.au

H. Ginzky
German Environment Agency, Dessau, Germany
e-mail: harald.ginzky@uba.de

I.L. Heuser
IUCN World Commission on Environmental Law, Deputy Chair of Specialist Group on Sustainable Soils and Desertification, Kleinmachnow, Germany
e-mail: Irene.Heuser@googlemail.com

H. Ginzky et al. (eds.), *International Yearbook of Soil Law and Policy 2016*,
International Yearbook of Soil Law and Policy,
DOI 10.1007/978-3-319-42508-5_7

the vast majority of the world's food. In addition, it is now being more broadly recognised as having an essential role in regulating the global climate. Thus its ecological, economic and cultural functions are central to both human well-being as well as the health of the environment in general.[3]

For centuries, all soil functions were maintained without much difficulty. However, problems began to arise around the beginning of the twentieth century when the pressure of development activity started to more seriously disrupt the ecological functions of the soil. Now, around one third of the world's arable land is affected by degradation.[4] It is also known that degraded land has the effect of increasing the number of people in poverty in developing countries.[5] This chapter argues that one of the roles of environmental law is to fight the causes of land degradation and especially to influence people's behaviour in order to change unsustainable land management practices.

It seems to be a common perception from the point of view of international and national law that soils are the poor cousin to other environmental media, in particular water and the atmosphere.[6] However, this perception is changing, with the UN General Assembly declaring the year 2015 as the first 'International Year of Soils'. In addition, the UN General Assembly has decided that the 5th of December of every year will be known as 'World Soil Day'.[7] This decision underlines the fact that there is an increasing political awareness that soil degradation needs to be treated on the same level as the loss of biodiversity and the adverse effects of climate change, in order to promote a higher level of soil conservation through sustainable land management.[8]

This chapter provides an overview of the history, concepts, milestones and latest developments with regard to the international law and policy aspects of soil conservation, including progress made through major environmental conferences in the past few decades. It then explores the most recent developments, including the land and soil-related aspects found in 'The Future We Want', the Outcome Document of the UN Conference on Environment and Development in 2012, and in the Sustainable Development Goals approved by the UN General Assembly in September 2015. Furthermore it analyses the existing international law provisions relevant in this context. In addition, the chapter argues that soil, as the substantive constitutive element of land, should be regarded as a common concern of humankind, and puts forth reasons why international regulation is required. The chapter concludes with an outlook to the future.

---

[3] Boer and Hannam (2003), p. 149; Ginzky (2016), pp. 2–4.

[4] Milman (2015), p. 1.

[5] Barbier and Hochard (2016).

[6] Boer and Hannam (2015), pp. 1–13; Boer and Hannam (2003), pp. 149–163.

[7] UN General Assembly, December 2013: more information under: http://www.fao.org/fileadmin/user_upload/GSP/docs/iys/World_Soil_Day_and_International_Year_of_Soils__UNGA_Resolution_Dec._2013.pdf.

[8] See Boer (2014), p. 293.

2 Milestones of Soil Protection at the International Level

Soil protection law has developed in several phases. These phases are marked by the relevant international conferences over the past four decades, interspersed with various soft and hard law instruments, as outlined in the following sections.

2.1 *Stockholm Conference 1972*

The United Nations Conference on the Human Environment in Stockholm in 1972 put global environmental issues on the international agenda for the first time. The Stockholm Declaration on the Human Environment, arising from the conference, proclaimed *inter alia* that 'the natural resources of the earth, including (...) land (...), must be safeguarded for the benefit of present and future generations through careful planning or management, as appropriate' and that 'the non-renewable resources of the earth must be employed in such a way as to guard against the danger of their future exhaustion (...)'.[9] The Action Plan for the Human Environment and the establishment of the United Nations Environment Programme, as part of the conference outcomes, have significantly influenced the direction of international environmental policy, and is also seen as marking the beginning of modern international environmental law. The conference can thus also be regarded as marking the first phase of international soil protection law.

The Action Plan was followed by several other non-binding documents ('pré-droit' or 'soft law') which, due to their legal status, did not include responsibilities to protect soils or other binding provisions (whereas binding instruments as agreements between States (treaties or conventions) have a mandatory character).[10] At the regional level, the Council of Europe was the first international organization specifically to tackle the problem of soil protection by adopting its 'European Soil Charter' in 1972.[11] It included the objective for Member States to promote the protection of soils against damage from natural or human causes and their rehabilitation, and it served as an initial role model for European and international soil protection law, thereby raising awareness of the need for soil protection policies.[12]

A few years later, two non-binding instruments were adopted at the international level, calling for action to protect soils: the 'World Soil Charter' from the Food and Agriculture Organisation (FAO) in 1981[13] and the 'World Soils Policy' from the

[9] United Nations (1972), Principles 2 and 5.

[10] 'Soft law' is a recognition of the fact that international law has an evolutionary character, in that it may become 'hard law' as part of a binding legal instrument.

[11] Council of Europe (1972), 1 passim.

[12] Heuser (2005), p. 348.

[13] FAO (1982), see http://www.fao.org/3/az897e.

United Nations Environment Programme (UNEP) in 1982,[14] serving as conjunctive instruments to encourage international cooperation in the rational use of soil resources. They were reflected and put into concrete terms by the UNEP 'Environmental Guidelines for the formulation of National Soil Policies'.[15] The '1981 World Soil Charter' was to encourage international cooperation concerning the sustainable use and protection of soils. It acknowledged the fragility of the world's soil resources and pointed out that soil protection was essential in protecting the food-producing capacity of land resources. The 'World Soils Policy' included more technically relevant aspects and general measures preventing excessive land consumption as well as promoting international co-operation for rational use of soil resources.

The UN 'World Charter for Nature', adopted by the General Assembly of the United Nations on 28 October 1982, indirectly established soil protection objectives through inclusion of land in its general principles of conservation.[16]

The ASEAN Agreement on the Conservation of Nature and Natural Resources,[17] was also introduced in the 1980s, but never came into force.[18] It is an example of a regional agreement that dealt with the development and coordination of national conservation strategies, including aspects of soil biodiversity and the protection of natural ecosystems. The Agreement included soil conservation measures to rehabilitate eroded or degraded soils, establish soil policies, control soil erosion and improve the fertility of soils.[19]

2.2 *Rio Conference 1992*

An essential marker in the second phase of international soil protection law was the United Nations Conference on Environment and Development (UNCED) in Rio de Janeiro in 1992, which led to three legally binding agreements (known as the Rio Conventions) which are relevant to the protection and sustainable use of soils: the United Nations Framework Convention on Climate Change (UNFCCC), the Convention on Biological Diversity (CBD), both signed in Rio in 1992, and the UN Convention to Combat Desertification (UNCCD), which was adopted in Paris in

---

[14]United Nations Environment Programme, 8/10—World Soils Policy, 12th meeting 29 April 1980.

[15]UNEP (1983), p. 1 ff.

[16]United Nations, General Assembly, World Charter for Nature, A/RES/37/7, 48th plenary meeting, 28 October 1982; see especially Article I: '4. Ecosystems and organisms, as well as the land, marine and atmospheric resources that are utilized by man, shall be managed to achieve and maintain optimum sustainable productivity, but not in such a way as to endanger the integrity of those other ecosystems or species with which they co-exist.'

[17]See http://environment.asean.org/agreement-on-the-conservation-of-nature-and-natural-resources/.

[18]See Boer (2015), pp. 251–281.

[19]See Article 7.

1994. The UNCCD is the only legally binding global agreement directly dealing with the promotion of bio-productive land, albeit focusing primarily on desertification processes and mitigating the effects of drought.

Two other important outcomes of the Rio Conference 1992 are Agenda 21, an extensive blueprint for action to achieve sustainable development worldwide,[20] and the Rio Declaration on Environment and Development,[21] which proclaimed in Principle 2 that States have the 'sovereign right to exploit their own resources pursuant to their own environmental and developmental policies as well as the responsibility to ensure that activities within their jurisdiction do not cause damage to the environment of other States'. Furthermore, the right to development was required to be fulfilled so as to 'equitably meet developmental and environmental needs of present and future generations' (Principle 3). The 1992 Summit has influenced all subsequent UN conferences and instruments on environmental aspects of sustainable development, including the document 2015 Transforming our World: A New Agenda for Global Action, incorporating the Sustainable Development Goals and their associated Targets. The latter document is dealt with further below.[22]

During this era, at the regional level, the Alpine Convention of 1991[23] (which entered into force in 1995) included the general obligation (Article 2) for Parties to pursue a comprehensive policy to protect the European Alps. Its Soil Conservation Protocol of 1998[24] records that soil is a characteristic element of nature and landscape and an integral part of the ecological balance; it contains the principles of prevention, polluter pays, and the prudent and sustained use of resources.[25]

2.3 *World Summit on Sustainable Development 2002*

The World Summit on Sustainable Development in Johannesburg in 2002 produced the Johannesburg Declaration on Sustainable Development. The Declaration recognises that 'desertification claims more and more fertile land' (para. 13). Further, the Summit's Plan of Implementation[26] includes a number of references to land degradation. Paragraph 41 in particular is important in the context of international

---

[20]Especially Chapter 10: 'Integrated Approach to the Planning and Management of Land Resources', https://sustainabledevelopment.un.org/content/documents/Agenda21.pdf.

[21]See http://www.unep.org/documents.multilingual/default.asp?documentid=78&articleid=1163.

[22]See below, Fn 37.

[23]See http://www.alpconv.org/en/convention/framework/default.html.

[24]Version of 2005: Official Journal of the European Union, L 337 of 22.12.2005, p. 29.

[25]See the article of Markus (2016) in this volume.

[26]United Nations (2002) Plan of Implementation of the World Summit on Sustainable Development 1992, United Nations (2002) available at http://www.un.org/esa/sustdev/documents/WSSD_POI_PD/English/WSSD_PlanImpl.pdf.

law, calling for strengthening the implementation of the UNCCD 'to address causes of desertification and land degradation in order to maintain and restore land and to address poverty resulting from land degradation.'

The achievements of the Council of Europe, which initiated international attention to the problems of soil protection, as mentioned above, went further than any other policy document in the field at that time. Moreover, the 'Revised European Charter for the Protection and Sustainable Management of Soil',[27] which was a revised version of the European Soil Charter of 1972, was adopted on 28 May 2003 by the Committee of Ministers of the Council of Europe. It provided a significant impetus for the transformation of Europe's environmental policy, pointed to the inclusion of new legal and environmental concepts and standards and in particular included a list of instruments for the sustainable management of soil.

2.4 *Rio+20 Conference 2012*

While the Johannesburg Summit's Plan of Implementation contained important provisions in respect to land degradation, the 2012 United Nations Conference on Sustainable Development (Rio+20 Conference) can be said to mark the beginning of the third phase of international soil protection law.

The outcome document of Rio+20, 'The Future We Want' states:

> We recognize the economic and social significance of good land management, including soil, particularly its contribution to economic growth, biodiversity, sustainable agriculture and food security, eradicating poverty, women's empowerment, addressing climate change and improving water availability. We stress that desertification, land degradation and drought are challenges of a global dimension and continue to pose serious challenges to the sustainable development of all countries, in particular developing countries. (...)
>
> We recognize the need for urgent action to reverse land degradation. In view of this, we will strive to achieve a land-degradation-neutral world in the context of sustainable development.[28]

A significant step forward in these provisions is the common recognition of the 'economic and social significance of good land management' and that land degradation in all forms is a challenge 'of a global dimension'. Most importantly is the joint commitment 'to strive for a land-degradation-neutral world', which means that the balance between on-going land degradation and land restoration has to be zero. Setting this quantitative, instead of mere qualitative requirement is challenging, but the progress made and the actions taken can then be measured. In order to comply with this objective, three categories of measures are needed: first, measures to reduce land degradation, second, measures with regard to land restoration and

[27]Council of Europe (2003), p. 1 ff. Concerning further details see: Heuser (2005), p. 356 ff.

[28]United Nations (2012), The Future We Want, http://www.uncsd2012.org/content/documents/727The%20Future%20We%20Want%2019%20June%201230pm.pdf

rehabilitation and third, measures to plan and manage the achievement of the objective.[29]

2.5 *The Sustainable Development Goals 2015*

On 25 September 2015, the UN General Assembly adopted 'the 2030 Agenda for Sustainable Development'.[30] The Agenda includes 17 SDGs and 169 accompanying Targets. Importantly, the Agenda stipulates commitments for both developing and developed countries. The SDGs replace the Millennium Development Goals of the early 2000s.[31] The SDGs are intended to stimulate the international and national political agenda to the year 2030.

The adoption of a specific Target on land degradation neutrality is a manifestation of the increasing political awareness that soil needs to be protected more effectively.[32] SDG 15 and Target 15.3 are the relevant provisions.[33] SDG 15 reads as follows:

> Protect, restore and promote sustainable use of terrestrial ecosystems, sustainably manage forests, combat desertification, and halt and reverse land degradation and halt biodiversity loss.

Target 15.3 specifies:

> . . . by 2030, combat desertification, and restore degraded land and soil, including land affected by desertification, drought and floods, and strive to achieve a land-degradation neutral world.[34]

Although the Outcome Document and the 2030 Agenda, including the SDGs, are not legally binding, they are politically important as they were endorsed by all States. In addition, they are all potentially measurable, as indicated above. In addition, the increased public awareness for soil issues is underlined by the decision, mentioned above, of the UN General Assembly to declare 2015 the first International Year of Soils and 5 December of every year the UN Day of Soils.

---

[29] Dooley et al. (2015), pp. 209–210.

[30] United Nations (2015), p. 1.

[31] Millennium Development Goals, available at http://www.un.org/milleniumgoals/.

[32] Transforming our World: the 2030 Agenda for Sustainable Development, United Nations (2015), p. 1; this document includes the 17 Sustainable Development Goals.

[33] A thorough analysis of all SDG with regard to soil management is provided by Ehlers in this volume.

[34] Target 15.3 was inspired in part by a policy document produced by the UNCCD Secretariat *Zero Net Land Degradation:* A Sustainable Development Goal for Rio+20; UNCCD (2012).

The chapter of Minelli et al. in this volume provides further information how to operationalize the neutrality objective.

Another significant development in the last few years was the Global Soil Partnership,[35] launched in 2011 as an initiative of the FAO and the European Commission. Its stated aim was to support and facilitate joint efforts towards sustainable management of soil resources for food security and to update the 1982 World Soil Charter. It brings together governments, regional organizations and other stakeholders at various levels and intends to implement a voluntary system of global governance. It is supported by the Intergovernmental Technical Panel on Soils (ITPS), which provides scientific and technical advice on global soil issues. In 2015, coinciding with the International Year of Soils, member states of FAO unanimously endorsed the Revised World Soil Charter[36] as a vehicle to promote and institutionalize sustainable soil management at all levels. These initiatives also included the introduction of a Global Soil Week, which has become a major policy platform for soils.[37]

2.6 *Ideas and Models for a Stand-Alone International Legal Soil-Related Instrument*

The discussion in this part of the chapter focuses mainly on the implementation of SDG 15 and Target 15.3. Conceptual approaches contained in various instruments with regard to protection and sustainable management of soils, such as the early Tutzing proposal for a 'Convention on Sustainable Use of Soil' in 1998,[38] and the 'Revised European Charter for the Protection and Sustainable Management of Soil'[39] should be considered as background and a context for these elements of the SDGs. However, it must also be remembered that a specific study about soil threats by the Scientific Advisory Board of the German Federal Government entitled *Globale Umweltveränderungen*[40] had recommended a Soil Declaration or a global Soil Convention as long ago as 1994. Furthermore, the Specialist Group on Sustainable Use of Soils and Desertification of the World Commission on Environmental Law of the International Union for Conservation of Nature (IUCN) has proposed several versions of a protocol on the protection and sustainable use of soils. The most recent version focuses on the concept of Zero Net Land Degradation (ZNLD), linked either to the CBD or the UNCCD, with the objective of translating ZNLD into legally enforceable provisions at international and national levels.

[35] Homepage: http://www.fao.org/globalsoilpartnership/en.

[36] FAO (2015), p. 1 ff.

[37] See Global Soil Week, http://globalsoilweek.org.

[38] Held et al. (1998).

[39] Council of Europe (2003), p. 1 ff.

[40] Wissenschaftlicher Beirat der Bundesregierung Globale Umweltveränderungen, Welt im Wandel – Die Gefährdung der Böden, Bonn 1994.

Boer and Hannam have also proposed the amendment of Article 30 of the UNCCD to provide for insertion of protocol or the use of Article 31 as the basis for establishing a set of soil law guidelines. Most recently, they have suggested the drafting of a global instrument for the conservation and sustainable use of soil.[41] A global convention or protocol to an existing convention along these lines would fill the existing gap in the still very fragmented international treaty law.[42] Further in-depth discussions of the advantages of such a soil-specific international instrument and of the precise contents are required from legal, scientific and political perspectives.

3 Conventions/Hard Law

Provisions related in some way to soil protection can be found in the three Rio Conventions: the United Nations Framework Convention on Climate Change 1992[43] (UNFCCC) and its Kyoto Protocol, the Convention on Biological Diversity 1992[44] (CBD) and the Aichi Targets under the CBD, and especially the United Nations Convention to Combat Desertification in Countries experiencing Serious Drought and/or Desertification, particularly in Africa 1994 (UNCCD).[45]

This section will analyse whether and how the three Conventions could accommodate an effective regulatory regime for the protection and sustainable use of soils—either on their own terms or collectively. For the purposes of this analysis, the functions of soils should be taken into account and whether they are regulated and managed. These functions include *inter alia* the provision of a healthy environment for human settlement and ecosystems for animals and plants, maintenance of biodiversity, purification functions for surface and groundwater, carbon sinks, and the guarantee of food security. The following provides a brief overview.[46]

---

[41]Boer and Hannam (2015). See also Ginzky (2016), pp. 23–26. A legal expert group which was established during the first Global Soil Week in 2013 has considered these options. A discussion paper is available under: http://www.iass-potsdam.de/sites/default/files/files/gsw_leg_discussion_paper_options_unccd_endg_0.pdf.

[42]See the following section.

[43]Text of UNFCCC available at www.unfccc.int.

[44]Text of CBD available at www.cbd.int.

[45]Text of UNCCD available at www.unccd.int.; the Ramsar Convention on Wetlands of International Importance especially as Waterfowl Habitats also deals with soil related issues. However, as its scope of application is limited to wetlands, the chapter will not deal with it in detail. Text of the Ramsar Convention available at http://www.ramsar.org/.

[46]More details, particularly with regard to the latest developments are provided in the chapters by Streck and Gay on the UNFCCC, of Wolff and Kaphengst on the CBD and of Sara Minelli et al. on the UNCCD in this volume.

3.1 UNFCCC, Kyoto Protocol and Paris Agreement

The objective of the UNFCCC under Article 2 is the 'stabilization of greenhouse gas concentrations in the atmosphere at a level that would prevent dangerous anthropogenic interference with the climate system'. The UNFCCC has been almost universally adopted, with 197 Contracting Parties.[47]

A soil-related obligation is found in Article 3.3, which requests the Contracting Parties to 'cover all relevant sources, sinks and reservoirs of greenhouse gases' in their national climate change policies. Soils are the world's second largest natural reservoir of carbon after the oceans. They are therefore reservoirs within the meaning of the UNFCCC. They could also be regarded as sinks if, for example through the rehabilitation of wetlands, additional carbon is sequestered. The commitments of the Contracting Parties are stipulated in Article 4 of the UNFCCC, inter alia to establish national inventories and report these to the Conference of the Parties and to formulate and implement programs with the required measures. Developed countries are obliged to take appropriate measures to mitigate climate change according to Article 4 paragraph 2 (a) of the UNFCCC.[48]

The Kyoto Protocol[49] adopted in 1997 (coming into force in 2005) determined mandatory emission reduction targets for developed countries for the first time.[50] According to Article 2, developed countries have an obligation to protect and enhance sinks and reservoirs. In 2011, Contracting Parties decided that both the drainage and the restoration of wetlands by water logging should be included as land-use change in the calculation of emission reduction by developed countries according to Article 3.4 of the Kyoto Protocol.[51]

Article 12 of the Kyoto Protocol introduced the Clean Development Mechanism (CDM). The CDM allows developed countries to implement emission reduction measures in developing countries and to consider the emission reductions achieved with regard to their obligation under the Kyoto Protocol. Whereas afforestation projects have been accepted as a CDM, this was not the case for projects directed at fostering carbon sequestration in soils.[52]

---

[47] 'About UNFCCC', available at http://newsroom.unfccc.int/about/.

[48] According to Article 3.1 UNFCCC states have 'common but differentiated responsibilities and respective capabilities'.

[49] Text is available at http://unfccc.int/resource/docs/convkp/kpeng.pdf.

[50] The first commitment period ended in 2011. New reduction obligations were agreed in 2012. See the text of the Doha amendments available at http://unfccc.int/files/kyoto_protocol/application/pdf/kp_doha_amendment_english.pdf.

[51] 2/CMP.7, Nr. 11 available at: http://unfccc.int/meetings/durban_nov_2011/meeting/6245/php/view/decisions.php.

[52] Id. 5/CMP.1, confirmed by 2/CMP.7.

The Paris Agreement which was adopted in December 2015 is a major step forward with regard to climate change policy.[53] The Paris Agreement sets ambitious objectives, *inter alia* (in summary):

- Holding the increase in the global average temperature to 'well below' 2 °C above pre-industrial levels, and to pursue efforts to limit the temperature increase to 1.5 °C above pre-industrial levels (Article 2.1 (a)
- To achieve a balance between anthropogenic emissions by sources and removals by sinks of greenhouse gases in the second half of this century (Article 4.1).

In order to implement these objectives, Contracting Parties must set and communicate 'nationally determined contributions' (NDCs), which will be reviewed every 5 years (Article 4.9) in order to ensure that the objectives of the Paris Agreement will be achieved.

The Paris Agreement does not contain specific soil or land-related obligations. However, a guideline for the counting of the NDCs (Article 4.13) must be dealt with by an ad hoc Working Group. It aims to ensure that all reservoirs and sinks are appropriately taken into account and to avoid double counting.[54] It will be seen whether and how land use and land-use change will have to be methodologically included in the accounting according to the guideline which will be developed by the ad hoc Working Group.

In addition to these official requirements, 'voluntary contributions for timely implementation' of the Paris Agreement are encouraged.[55] Such contributions include initiatives for land-based climate change mitigation measures such as that promoted by the French Ministry of Agriculture to increase soil carbon stocks.[56]

It is thus submitted that the UNFCCC, the Kyoto Protocol and the Paris Agreement clearly mandate initiatives for taking into account carbon sequestration in soil through various agricultural, pastural and forestry practices. All other soil functions, such soil biodiversity or purification, do not directly fall within the mandate of these instruments. The Kyoto Protocol foresees specific soil-related requirements in such areas as afforestation projects and in the drainage and flooding of wetlands as CDMs. While the UNFCCC, the Kyoto Protocol and the Paris Agreement have not established a comprehensive regime with regard to land-related climate change measures, there is certainly room to do so within the language of these instruments.[57]

[53]The text of the Paris Agreement is available at http://unfccc.int/resource/docs/2015/cop21/eng/l09r01.pdf.

[54]See Paras. 31 and 93(f) of Draft Decision, Adoption of the Paris Agreement', FCCC/CP/2015/L.9/Rev.1 available at https://unfccc.int/resource/docs/2015/cop21/eng/l09r01.pdf.

[55]Id. Para. 140.

[56]See: 'Join the 4% Initiative Soils for food security and climate', further information available at http://agriculture.gouv.fr/join-40-initiative-soils-food-security-and-climate-0.

[57]For further details concerning UNFCCC and its related protocols see the chapter by Streck and Gay (2016) in this volume, see also Hannam and Boer (2002), p. 64.

3.2 Convention on Biological Diversity

The objectives of the 1992 Convention on Biological Diversity[58] are 'the conservation of biological diversity, the sustainable use of its components and the fair and equitable sharing of the benefits arising out of the utilization of genetic resources'. The crucial term 'biological diversity' is defined in Article 2 as 'the variability among living organisms from all sources including, *inter alia*, terrestrial, marine and other aquatic ecosystems and the ecological complexes of which they are part: this includes diversity within species, between species and of ecosystems.' Since the term includes also 'terrestrial ecosystems', it is submitted that all soil functions fall within the objectives of the CBD. Moreover, soil's biological functions such as habitat for animals and plants, as reservoirs and sinks for carbon, or its purification function, fall under the objective 'conservation of biological diversity'. The 'use' functions of soil, such as habitat for humans, the production of food and plants that are the basis for renewable energy can be regarded as 'use of its components'. In that sense, the scope of application of the CBD is broad enough for the comprehensive regulation of all relevant aspects of the protection and the sustainable use of soils.[59]

However, the obligations of the Contracting Parties according to the CBD are not very specific. In addition to the obligation for international cooperation according to Article 5, the Contracting Parties must establish national strategies, plans or programmes for the conservation and sustainable use of biological diversity and these must encompass all necessary measures (Article 6). Moreover, Contracting Parties 'shall, as far as possible and as appropriate' put into effect 'in-situ' measures, including a system of protected areas, and the rehabilitation of degraded ecosystems (Article 8 (a) and (f)). According to Article 14 (a), an environmental impact assessment shall be conducted if projects are likely to have significant adverse effects on biological diversity, with a view to avoiding or minimizing such effects; these effects could of course include effects on soil health and productivity. However, it is within the discretion of the Contracting Parties when to apply these measures because the Convention does not clearly state obligatory implementation conditions.

Thus, the CBD text does not contain specific provisions with regard to the protection and sustainable use of soils.[60] However, soil aspects are reflected in several programs of the CBD, among others, agricultural biodiversity, dry and sub-humid lands biodiversity and forest biodiversity.[61] In addition, the Strategic

---

[58] Available at http://www.cbd.int/convention/parties/list/. The CBD has 196 Contracting Parties; see List of Parties, https://www.cbd.int/information/parties.shtml. The US is not yet Party to CBD.

[59] Hannam and Boer (2002), p. 64.

[60] Boer and Hannam (2003), p. 154.

[61] Further information under: http://www.cbd.int/agro/ as well as http://www.cbd.int/drylands/ and http://www.cbd.int/forest/.

Plan for Biodiversity, adopted in 2010 contains 20 'non-binding' Aichi-targets.[62] Several of the Aichi targets clearly involve soil issues, for example Goal B, which requires parties to 'Reduce the direct pressures on biodiversity and promote sustainable use', while Target 7 under this Goal states: 'By 2020 areas under agriculture, aquaculture and forestry are managed sustainably, ensuring conservation of biodiversity'. However, although the Aichi Targets refer to aspects of soil, they do not place legally binding obligations on States Parties.[63]

In summary, the objectives of the CBD would enable the comprehensive regulation of all aspects of soil, even though the Convention does not include soil-specific obligations.

3.3 *UNCCD*

The UNCCD is in force almost universally due to its 195 Contracting Parties. The objective of the UNCCD under Article 2 is 'to combat desertification and mitigate the effects of drought in countries experiencing serious drought and/or desertification, particularly in Africa, through effective action at all levels (…)'.

The geographical scope of application of the UNCCD is, however, limited to drylands and thus to about 40 % of the terrestrial surface of the earth.[64] Since the objective of the Convention is to combat desertification and to mitigate the effects of droughts (Article 1 (d)), the definition of 'desertification' is crucial for its the geographical scope. According to Article 1 (a) 'desertification' is defined as 'land degradation in arid, semi-arid and dry sub-humid areas …'. Article 1(g) defines these as areas in which the ratio of annual precipitation to potential evapotranspiration falls within the range from 0.05 to 0.65.

Nevertheless, Regional Annexes III and V of the UNCCD refer to areas which are affected by 'desertification' but do not comply with the scientific criteria stipulated in Article 1 (g) of the Convention.[65] The question that could be asked is whether this fact could be regarded as a subsequent agreement within the meaning of Article 31.3 of the Vienna Convention on the Law of Treaties,[66] thus functioning to formally extend the geographical scope of application to all

62 Decision 2/X, 19.-20. October 2010, available at http://www.cbd.int/sp/targets/.

63 More specific information is found in the chapter by Wolff and Kaphengst (2016) in this volume.

64 Boer and Hannam (2003), p. 153; Ginzky (2016), p. 17.

65 Boer and Hannam (2003), p. 154; Ginzky (2016), p. 18; also the 'Ten year strategic plan and framework to enhance the implementation of the Convention' of 2007 (UNCCD, Decision 3/COP. 8, available at http://www.unccd.int/Lists/SiteDocumentLibrary/10YearStrategy/Decision%203COP8%20adoption%20of%20The%20Strategy.pdf seems to be an attempt to expand the scope of the Convention's application as it stipulates desertification, drought and land degradation as separate issues.

66 The text of the convention is found at: https://treaties.un.org/doc/Publication/UNTS/Volume%201155/volume-1155-I-18232-English.pdf.

terrestrial areas.[67] However, since many Contracting Parties strictly oppose the formal extension,[68] a unanimous view of the Contracting Parties was not achieved, which is a precondition for formalizing a subsequent agreement.

An additional weakness of the UNCCD is that there are virtually no obligations relating to specific measures. Affected countries are only obliged to establish 'National Action Plans', according to Articles 9 and 10.

The UNCCD Secretariat was one of the major drivers to include the objective of a 'land degradation neutral world' (hereafter 'LDN') in the Rio+20 Outcome Document, 'The Future We Want'.[69] Since then the UNCCD has dealt with the question of what role UNCCD should have and how the objective of LDN could be implemented.

In October 2015 the Contracting Parties took four decisions in that regard:

- 3/COP.12: Integration of the Sustainable Development Goals and targets into the implementation of the United Nations Convention to Combat Desertification and the Intergovernmental Working Group report on land degradation neutrality
- 8/COP.12: Addressing particular regional and national conditions
- 9/COP.12: Leveraging of synergies among the Rio Conventions and promoting partnerships with other international agencies and bodies
- 7/COP.12: Intergovernmental Working Group on the future strategic framework of the Convention.

Decision 3/COP.12 established a first and fundamental guideline on how LDN is to be implemented by the Contracting Parties. The Decision foresees a general scientific definition for the term 'land degradation neutrality', clarifies the necessary measures to be undertaken and the indicators for measuring LDN. Decision 8/COP.12 arguably extends *de facto* the geographical scope of the Convention by 'noting that a significant proportion of land degradation occurs beyond arid, semi-arid and dry sub-humid areas' and by recognizing 'that Parties may use the UNCCD to guide their policies relating to DLDD and voluntary targets when striving to achieve LDN at national and subnational levels'. Therefore it enables all Contracting Parties, independently of whether their territories include drylands, to apply all LDN-related mechanisms.

Decision 9/COP.12 calls for intensified cooperation with other international bodies. Decision 7/Cop.12 establishes an Intergovernmental Working Group to work on the strategic perspectives of the UNCCD. The mandate of the Working Group requested the Group to consider the SDGs, especially SDG 15 and its Target 15.3 on LDN for the forthcoming strategic plan of UNCCD.

---

[67] See the request of the Annex V-countries to clarify this issue, see ICCD/COP(12)/16; available at http://www.unccd.int/Lists/OfficialDocuments/cop12/16eng.pdf.

[68] As an indication of this view, the Terms of Reference for the Intergovernmental Working Group on Land Degradation Neutrality expressly limited the scope to 'arid, semi- arid and dry sub-humid areas', see ICCD/COP(12)/4.

[69] See Fn 31, above.

In summary, the above four decisions are of crucial importance for the UNCCD although they are not legally binding. In fact the Convention can be seen to incorporate LDN in its substantive mandate and has declared itself as the lead body at the international level for the implementation of LDN.[70]

However, as noted, strictly speaking, the Convention is legally applicable only to drylands, even taking into account the most recent decisions taken during COP 12. The provisions require obligations which continue to appear quite vague. In addition, measures by affected countries often depend on financial support by developed countries or international donor institutions. The latest steps are obviously of great importance. Since they are not legally binding obligations, their success will depend on the willingness of the Contracting Parties of UNCCD to implement them voluntarily.[71]

4 Regional Agreements

At the regional level, there are a number of binding legal instruments with relevance for the protection and sustainable use of soil. Many of these instruments address soil protection only in a limited way. A small number of regional instruments include specific articles for soil as well as other articles that can support the objective of sustainable use of soil, as explained below.[72]

The African Convention for the Conservation of Nature and Natural Resources[73] entered into force in 1972 and was revised in 2003. The revised Convention includes fundamental rules for the conservation and sustainable management of land resources in the regions of the African Union, but not all of the 54 African States have ratified the convention yet. According to Article II, it pursues the following objectives: 'to enhance environmental protection, to foster the conservation and sustainable use of natural resources and to harmonize and coordinate policies in these fields with a view to achieving ecologically rational, economically sound and socially acceptable development policies and programmes'. It includes soil as a natural resource along with water, flora and fauna and non-renewable

[70]For further details of the recent developments under UNCCD see the chapter by Minelli et al. in this volume.

[71]The 2030 Development Agenda and its consequences was also discussed at the latest Global Soil Week 2015, http://globalsoilweek.org/thematic-areas/sustainable-development-goals.

[72]Other legal instruments on regional level, not canvassed here, with some relevance to soil are the Convention for the Protection of the Natural Resources and Environment of the South Pacific Region 1986, Benelux Convention 1983, Protocol Concerning Specially Protected Areas and Biological Diversity in the Mediterranean 1999.

[73]See http://au.int/en/sites/default/files/treaties/7763-file-african_convention_conservation_nature_and_natural_resources.pdf.

resources. Article IV[74] stresses the necessity of adopting and implementing preventive measures and to apply the precautionary principle. The most soil-relevant provision is Article VI, which obliges States to take effective measures to prevent land degradation, and in particular to adopt measures for the conservation and improvement of the soil, especially to combat its erosion and misuse as well as the deterioration of its physical, chemical and biological or economic properties. It includes obligations to establish specific land use plans (Article VI 3 (a)) as well as nature conservation areas (Article XII). When implementing agricultural practices and agrarian reforms, States are obliged to improve soil conservation and introduce sustainable farming and forestry practices, to control erosion caused by land misuse and mismanagement and to control pollution caused by agricultural activities (Article VI 3(b)). While it has great relevance for nature protection, the revised Convention only establishes general requirements for the protection and sustainable use of soils. Nevertheless, it has the *potential* to contribute considerably to the prevention of soil biodiversity loss and organic matter decline in the African Union.

The ASEAN Agreement on the Conservation of Nature and Natural Resources,[75] was also introduced in the 1980s, but never came into force.[76] It is an example of a regional agreement that dealt with the development and coordination of national conservation strategies, including aspects of soil biodiversity and the protection of natural ecosystems. The Agreement includes soil conservation measures to rehabilitate eroded or degraded soils, establish soil policies, control soil erosion and improve the fertility of soils.[77]

The only legally binding regional instrument that focuses specifically on soil protection and sustainable use of soil is the Convention concerning the Protection of the Alps of 1991 (Alpine Convention)[78] and its Protocol on the Implementation of the Alpine Convention in the area of Soil Protection of 1998 (Alpine Convention, Soil Protection Protocol, ACSPP[79]). The Alpine Convention is an international agreement made between eight European states and the European Union. It seeks the protection and sustainable development of a globally significant mountain region, and therefore has a decidedly environmental orientation. One of the main goals of the ACSPP is the safeguarding of the multi-functional role of soil, based on the concept of sustainable development. In Article 2, the ACSPP establishes the general obligation for the States to pursue a comprehensive policy to protect the European Alps by applying the principles of prevention, 'polluter pays' and the

[74]'The Parties shall adopt and implement all measures necessary to achieve the objectives of this Convention, in particular through preventive measures and the application of the precautionary principle, and with due regard to ethical and traditional values as well as scientific knowledge in the interest of present and future generations.'

[75]See http://environment.asean.org/agreement-on-the-conservation-of-nature-and-natural-resources/.

[76]See Boer (2015), pp. 251–281.

[77]See Article 7.

[78]http://www.alpconv.org/en/convention/framework/default.html.

[79]http://www.alpconv.org/en/convention/protocols/Documents/SoilProtocolEN.pdf.

prudent and sustained use of resources. The ACSPP, which is based on an ecosystem perspective, recognises the Alpine region for its ecological diversity and highly sensitive ecological systems and that its functional capacity must be preserved. According to Article 2 (d), parties have a specific obligation concerning soil conservation, especially to reduce soil damage by using agricultural and forestry methods that do not harm soil, cause minimal interference to soil and control soil erosion. The ACSPP also restricts soil-sealing practices and promotes minimal interference with soil rehabilitation. It includes a number of specific obligations for the Parties to observe with regard to legal and administrative measures to protect soil. It also promotes coordination and cooperation between institutions and territorial authorities to develop synergies for soil protection (Article 4). It aims to support international cooperation among institutions on soil research (Article 5). It could also be of assistance in the implementation of appropriate measures at the national and regional level. For that reason, examination of the legal mechanisms which reflect the Protocol's obligations in the current versions of the Swiss Environmental Protection Act, the Soil Protection Act of Liechtenstein and the Soil Protection Acts in certain Austrian states would be instructive in terms of effectiveness of implementation. As the only specific legally binding instrument for soil, the ACSPP could be regarded as a stimulus for the overall improvement of the global and national law for soil.[80]

In summary, similar to the situation at the international level, few regional instruments have so far focused on the protection and sustainable use of soils in an effective way.[81] There are some exceptions, such as the Alpine Convention and its comprehensive ACSPP. Some regional agreements have the same approach as the CBD and address sustainable use and protection of soil e.g. as part of the nature protection regime (for example, the African Convention).[82]

5 Land Degradation As a Common Concern of Humankind

This section examines why land degradation should be considered as a common concern of humankind. First, the concept of a 'common concern of humankind' must be explained. Several multilateral environmental treaties include formulations which incorporate this concept. Although the formulations differ in the various agreements, the contents are seen as more or less equivalent. Formulations used are 'the interest of the nations of the world',[83] 'in the interest of mankind'[84] or 'an

[80] For further details about the Soil Protocol see the chapter of Markus (2016).

[81] See further, Hannam and Boer (2002).

[82] Another example is the ASEAN Agreement on the Conservation of Nature and Natural Resources mentioned in Sect. 2.1 above. For an overview of the most relevant regional agreements see Hannam and Boer (2002), 66 passim.

[83] International Convention for the Regulation of Whaling 1946.

[84] Antarctic Treaty 1959.

irreplaceable part of the natural systems of the earth which must be protected for this and the generations to come.'[85]

Scholars have in general come to a more consistent understanding of the main content of the 'common concern of humanity'. For example, Birnie, Boyle and Redgwell state:

> If 'common concern' is neither common property nor common heritage, and if it entails a reaffirmation of the existing sovereignty of states over their own resources, what legal content, if any, does this concept have? Its main impact appears to be that it gives the international community of states both a legitimate interest in resources of global significance and a common responsibility to assist in their sustainable development. Moreover, insofar as states continue to enjoy sovereignty over natural resources and the freedom to determine how they will be used, this sovereignty is not unlimited or absolute, but must now be exercised within the confines of the global responsibilities set out principally in the Climate Change and Biological Diversity Conventions, and also in the Rio Declaration and other relevant instruments.[86]

In other words the concept of 'common interest of humankind' refers to resources under the jurisdiction of states. They have a right to exercise their sovereignty over these resources. However, this right is not absolute, as these resources are of global significance and places an obligation on nations to undertake actions, including the implementation of conservation and protection strategies, to achieve sustainable development. The responsibility to act to this end limits the sovereign right of states over their domestic resources.[87] In that sense, the concept of 'common concern of humankind' is based on the discussion on 'global goods' or 'global commons', which primarily focuses on resources beyond national sovereignty, such as the atmosphere or the high seas.

From a scientific point of view, there is consensus that land degradation is a global concern due to the need to maintain all the essential functions of land and soils, such as food production, as carbon sinks, and as the maintenance of terrestrial biodiversity.[88] From a legal point of view, this is somewhat more complicated, especially as the UNCCD does not use the 'common concern of humankind' formulation.[89] However, recent developments must also be taken into account. The Outcome Document of Rio+20 states that 'desertification, land degradation and drought are challenges of a global dimension and continue to pose serious challenges to the sustainable development of all countries'.[90] This is underlined by including the objective of a 'land degradation neutral world' in SDG 15, as noted above, with the respective targets emphasizing the global significance of land degradation. Finally the decisions taken in 2012 within UNCCD and UNFCCC

---

[85] 1973 Convention on International Trade in Endangered Species of Wild Fauna and Flora; see further Boer (2014), pp. 295–296.

[86] Birnie et al. (2009), p. 130.

[87] See Boer (2014), p. 299.

[88] See the arguments under Sect. 5.

[89] See further, Boer (2014), p. 301.

[90] The Future We Want (2012), Para 205.

Conferences of the Parties underline the increasing perception of the world community that land degradation is a global issue. *De facto* then, from a political point of view the fundamental idea or concept of 'common concern of humankind' that state sovereignty is limited by a state's international responsibility is now starting to be acknowledged in relation to land degradation by the community of states.

6 The Need of Soil Protection Law at the International Level

It is submitted that the need for legal provisions with regard to the protection and sustainable use of soil is indisputable. At the national level this is demonstrated by the fact that many jurisdictions have enacted soil conservation legislation, beginning in the early 1900s. However, international provisions demand a specific justification. In the following, it is argued that international provisions as well as an international regime for soil protection are reasonable and highly justified.

We put forward three main rationales for international regulation of soils.

- If soil degradation directly or indirectly results in transboundary effects, international action is warranted.[91]
- If soil degradation and achieving land degradation neutrality can be more effectively addressed through international cooperation (technical, scientific, operational, economic or political),[92] thus adding strength to national approaches, international regulation makes sense.
- If soil degradation can potentially cause political or even military conflicts between states,[93] the case for international regulation is even stronger.

With regard to the first point, although soils are generally immobile,[94] the conservation and sustainable use of soil, as well as the degradation of soil, can nevertheless have transboundary effects of obvious relevance for other global goods like climate or biodiversity.[95] However, this is also true for the other soil functions. Fertile soils are the basis for adequate food production and thereby the prevention of hunger and malnutrition. Although these are localised or national

[91] Birnie et al. (2009), pp. 8–12.

[92] Birnie et al. (2009), p. 43.

[93] Birnie et al. (2009), p. 80, with regard to the general justification of international cooperation in case of potential conflicts.

[94] Except when moved by human intervention causing the acceleration of natural processes, such as landslides (by removing vegetation or altering natural drainage) and wind erosion (caused by overgrazing), or erosion of loose soil and run-off into water courses after rain, whereby soil can be transported across borders by international rivers.

[95] Fromherz (2012), p. 57; Boer (2014), 300 and *passim*; Wyatt (2008), p. 169.

phenomena, they are essentially of international importance from the point of view of political stability as well as human rights, especially the human right to food. In addition, soil degradation is often influenced by the global economy, for example through foreign land investments or the conditions of international trade law.[96] It is for these kinds of reasons that the Outcome Document of Rio+20 'The Future We Want' declared desertification, land degradation, and drought as '. . .challenges of a global dimension and continue to pose serious challenges to the sustainable development of all countries. . .'.[97]

The second reason noted above is the necessity of well-organised international cooperation. The protection and sustainable use of soils is factually very complicated and demanding, because there are various threads and causes which are partly interlinked. Land degradation is caused by numerous factors, including inappropriate agricultural land use, soil sealing by building infrastructure and urbanization. For effective solutions, a great deal of scientific knowledge and technological know-how is required. International cooperation allows for the exchange of scientific and technological assistance and for the setting of minimum requirements with regard to the protection of soils.[98] Purely national approaches will rarely be as effective, particularly because developing countries often lack the required expertise and financial capacity to address the problems.

The third and most pivotal reason for international regulation is the potential risk of international conflicts. The use and abuse of soil resources can cause political conflict across borders, especially when two or more countries with a common boundary are faced with drought and other forms of degradation, and the respective nations compete for access to fertile land. The legitimacy of foreign land investments or of unilateral trade restrictions due to alleged unsustainable land management might also be disputed amongst the states involved. Moreover, the management of human migration caused by the effects of land degradation and the financial or political responsibility of countries could be a subject matter for conflict between states, which is clearly illustrated by the ongoing political crisis in Europe due the high numbers of refugees.[99]

International regulation can range from soft-law decisions to legally binding provisions potentially combined with an institutional regime. Several justifications can be put for mandatory regulation and the establishment of an international organization for the protection and sustainable use of soils.

- First: The importance and the urgency of the challenges could justify mandatory regulation that would allow for clear and indisputable obligations and rights of the states involved. In order for this to be effective, some kind of compliance mechanism, as found under the climate change regime, might be considered.

[96]See Fromherz (2012), p. 105.

[97]The Future We Want (2012), Para 205.

[98]Wyatt (2008), p. 192.

[99]See Mastrojeni (2016), in this volume.

- Second: Since soil aspects concern, partly at least, the reconciliation of conflicting interests, mandatory regulation seems much more promising. In general, voluntary arrangements may not effectively solve such conflicts.
- An institutional regime would be beneficial as it would enable the joint development and implementation of solutions. Through the regular meeting of Contracting Parties and other subsidiary bodies (such as a scientific advisory board or the Global Soil Partnership mentioned above) it would be possible, for example, to jointly establish technical guidelines under a new international instrument on soils. In addition such institutions would allow for enhanced cooperation between developed and developing countries.

Two arguments might be made against an international regime for soil. First, that the international regulation of soils is too complex, because land is often under private ownership (in its various forms) and international regulation might be perceived to interfere with property rights by some nations. This argument is not convincing for two main reasons. It is generally accepted that most environmental regulation, by its very nature, can justifiably limit development activities of individuals or corporate entities. Therefore soil regulation cannot be singled out in this respect. Further, although land is often privately owned, its soil functions address public needs such as the assurance of food supply. Therefore soils could be regarded as 'hybrid' common goods like forests, privately owned on the one hand and embodying public values on the other.[100]

Second, it could be argued that a framework for international regulation is not feasible because there are very different types of soil resources and the economic, cultural and political circumstances in the various countries differ significantly. This argument is also not convincing because the variety of natural conditions is not soil-specific, in the same way that the conservation of biodiversity differs according to region and locale. Of course, differences in national circumstances can be taken into account when drafting the international provisions, essentially by providing for sufficient flexibility with regard to national implementation.

As a further aspect, the appropriate level of regulation has to be considered. Governing the soil requires international and national collaboration and cooperation between governments, organisations, local authorities, industries and citizens to ensure implementation of coherent policies that encourage practices and methodologies that regulate the use of the soil resources, in order to avoid conflicts between users. Multi-level governance is an approach which describes the integration of various political levels: International law could set a framework for sustainable use of soil (see Sect. 5 above). But of course international obligations must be implemented by the states. More detailed regulation of conservation responsibilities needs to be established at the national level, taking into account specific national conditions and priorities. For this reason, municipalities and/or competent

[100]With regard to international forest regulation see Eikermann (2014), 4 seq.

authorities at the local level are often in charge of implementing national or sub-national provisions and of promoting the involvement of all relevant stakeholders.

7 Conclusion

The safeguarding of soils has until recently been politically and legally neglected, although a range of conceptual proposals has been generated over the past couple of decades. With the Rio+20 Conference in 2012 as well as the establishment of the Post-2015 Development Agenda with the adoption of Sustainable Development Goal 15 and its Land Degradation Neutral Target, some small but very significant steps have been taken. The decisions taken by the Conferences of the Parties of UNFCCC and UNCCD, detailed above, have further promoted this political progression.

This chapter observes that the world community is now beginning to perceive land and soil degradation as a 'common concern of humankind', with the implication that the right of states to use their resources is limited by the obligation to maintain the ecological functions of soils in the interests of all nations.

In addition, the chapter has shown that international provisions on the sustainable use of soil are needed, and are justified primarily because of the transboundary importance of the ecological services of soils, the need for better cooperation and management and in order to avoid international conflicts over access to adequate soil resources.

While there may still be little demonstrable appetite for additional international provisions or even a specific international body on soil management, closer cooperation between analysts in the relevant disciplines of law, political science, economics and natural sciences would no doubt be an important basis for the generation of further political momentum to address the fundamental concerns canvassed here.

References

Barbier EB, Hochard JP (2016) Does land degradation increase poverty in developing countries? PLoS One. Available at http://dx.doi.org/10.1371/journal.pone.0152973

Birnie P, Redgwell C, Boyle A (2009) International law and the environment, 3rd edn

Boer B (2014) Land degradation as a common concern of humankind. In: Lenzerini F, Vrdoljak F (eds) International law for common goods, S. 289, 300 ff

Boer B (2015) Introduction to ASEAN regional environmental law. In: Scholtz W, Verschuuren J (eds) Regional environmental law: transregional comparative lessons in pursuit of sustainable development, pp 251–281

Boer B, Hannam I (2003) Legal aspects of the sustainable soils. Rev Eur Comp Int Environ Law 12 (2):149–163

Boer B, Hannam I (2015) Developing a global soil regime. Special edition, Int J Rural Law Policy 1–13

Council of Europe (1972) Committee of Ministers, Resolution (72) 19, European Soil Charter, adopted by the Committee of Ministers on 30 May 1972 at the 211th meeting of the Ministers' Deputies, p 1 ff

Council of Europe (2003) Committee for the activities of the Council of Europe in the field of biological and landscape diversity, COP-DBP (2003) 10, Revised European Charter for the Protection and Sustainable Management of Soil, adopted by the Committee of Ministers of the Council of Europe at its 840th meeting on 28 May 2003

Dooley E, Robert E, Wunder S (2015) Rechtliche Instrumente zur Operationalisierung des Ziels der Land Degradation Neutral World, Zeitschrift für Umweltrecht, p 209

Eikermann A (2014) Der Wald im internationalen Recht: Defizite, Regelungsoptionen und Mindestanforderungen, Gutachten im Auftrag des Bundesamts für Naturschutz

Food and Agriculture Organization of the United Nations (1982) World Soil Charter, November 1982, Text: http://www.fao.org/3/az897e

Food and Agriculture Organization of the United Nations (2015) 39th Conference, Rome 6–13 June 2015, Global Soil Partnership - World Soil Charter, Doc. C 2015/31 of April 2015, 1

Fromherz NA (2012) The case for a global treaty on soil conservation, sustainable farming, and the preservation of agrarian culture. Ecol Law Quart 39:57–122

Ginzky H (2016) Bodenschutz weltweit – Konzeptionelle Überlegungen für ein internationales Regime. Handbuch Boden, pp 1–32

Hannam I, Boer B (2002) Legal and institutional frameworks for sustainable use of soils. IUCN Environmental Policy and Law Paper No. 45

Held M, Kümmerer K, Odendahl K (1998) Tutzinger Projekt 'Ökologie der Zeit', Böden als Lebensgrundlage erhalten!, Vorschlag für ein 'Übereinkommen zum nachhaltigen Umgang mit Böden (Bodenkonvention)', Tutzing

Heuser I (2005) Europäisches Bodenschutzrecht. Berlin

Markus T (2016) The Alpine convention's soil conservation protocol: a model regime? Yearbook of Soil Law and Policy

Mastrojeni G (2016) Soil degradation and migrations in the age of the global environmental crisis: a policy-making perspective. Yearbook of Soil Law and Policy

Milman O (2 December 2015) Earth has lost a third of arable land in past 40 years, scientists say. The Guardian. http://www.theguardian.com/environment/2015/dec/02/arable-land-soil-food-security-shortage

Minelli S, Erlewein A, Castillo V (2016) Land degradation neutrality and the UNCCD: from political vision to measurable targets. International Yearbook of Soil Law and Policy

Streck C, Gay A (2016) The role of soils in international climate change policy. International Yearbook of Soil Law and Policy

UNCCD Secretariat Policy Brief (2012) Zero Net Land Degradation: A Sustainable Development Goal for Rio+20; see http://www.unccd.int/Lists/SiteDocumentLibrary/Rio+20/UNCCD_PolicyBrief_ZeroNetLandDegradation.pdf

United Nations (1972) 21st Plenary Meeting on 16 June 1972, Declaration of the United Nations Conference on the Human Environment, see http://www.unep.org/documents.multilingual/default.asp?documentid=97&articleid=1503

United Nations (1982) General Assembly, World Charter for Nature, A/RES/37/7, 48th plenary meeting, 28 October 1982

United Nations (2002) Plan of Implementation of the World Summit on Sustainable Development

United Nations (2012) General Assembly, Resolution adopted by the General Assembly on 11 September 2012, 66/288, The Future We Want

United Nations (2015) General Assembly, Seventieth Session, No. 11688, Agenda items 15 and 116, Resolution adopted by the General Assembly on 25 September 2015, 'Transforming our world: the 2030 Agenda for sustainable development', A/RES/70/1, p 1

United Nations Environment Programme (1983) Environmental guidelines for the formulation of national soil policies, Nairobi 1983
Wolff F, Kaphengst T (2016) The UN Convention on biological diversity and soils: status and future options. International Yearbook of Soil Law and Policy
Wyatt A (2008) The dirt on international environmental law regarding soils: is the existing regime adequate? Duke Environ Law Policy Forum 19:165

Chances and Challenges in Using the Sustainable Development Goals as a New Instrument for Global Action Against Soil Degradation

Knut Ehlers

1 Introduction to the SDGs

In September 2015 after more than 3 years of international negotiations the United Nations (UN) General Assembly adopted the Sustainable Development Goals (SDGs) as part of the "2030 Agenda for sustainable development".

The idea to develop a global set of goals for the time after the 2015 was launched in the run up to the United Nations Conference on Sustainable Development in Rio de Janeiro, Brasil in 2012 (Rio + 20 Conference) by the Columbian and Guatemalan government. The proposal was further negotiated and finally agreed upon at the Rio + 20 Conference. At the time, no concrete goals were set or themes defined, but a roadmap for a process to agree SDGs was set out and two very important basic requirements for the SDGs were already formulated: The SDGs were to "address and incorporate in a balanced way all three dimensions of sustainable development and their interlinkages." and they were to be "global in nature and universally applicable to all countries".[1]

So even though the SDGs build on the Millennium Development Goals (MDGs) adopted in 2000, they are meant to be more than just a follow up. Unlike the MDGs the SDGs are meant as a truly universal set of goals applying to developed, developing and emerging countries alike. Also the fact that the SDGs strive to capture the complexity of an economically, ecologically and socially sustainable development makes them a more comprehensive set of goals than the MDGs. Probably due to this broader scope there are now 17 SDGs and 169 associated targets compared to 8 MDGs with 18 associated targets. Even more than these sheer

[1] United Nations (2012), paras 246 and 247.

K. Ehlers (✉)
German Federal Environment Agency, Dessau, Germany
e-mail: knut.ehlers@uba.de

H. Ginzky et al. (eds.), *International Yearbook of Soil Law and Policy 2016*,
International Yearbook of Soil Law and Policy,
DOI 10.1007/978-3-319-42508-5_8

numbers the wording of the SDGs and their underlying targets provide an indication of the struggle to fully grasp the different aspects of sustainable development and the effort to harmonize the three core elements of sustainable development: economic growth, social inclusion and environmental protection. As a result the SDGs and especially their targets often address several cognate subjects at once (examples will be given below), whereas the MDG targets were focused on single issues. This makes the understanding, interpretation and communication of the SDGs more difficult that one probably would wish for a set of goals that was meant to be “concise”, “easy to communicate”, “limited in number” and “aspirational”.[2]

Nevertheless the SDGs, fill out an important gap when it comes to cooperation and coordination for sustainable development on global level for the next 15 years. To what extent the SDGs will actually translate into action remains open. As an instrument of soft law immediate and globally uniform action seems rather unlikely. Instead the current set of SDGs represent a common ground in defining what is to be prioritized in the context of sustainable development. Building on this common understanding discussions will now mature dealing with actions which can be derived from the SDGs on global, regional and national levels. This article is meant as a contribution to these discussions and aims at outlining the importance soils have for the SDGs and vice versa. It furthermore addresses challenges, interlinkages and open questions that will need to be resolved in order to make the SDG implementation a fruitful process for soil and land protection.

2 The Rio + 20 Conference: Paving the Way for Global Action Against Soil Degradation

Land and soil degradation issues were already addressed at the Rio + 20 conference and the Rio + 20 outcome document “The future we want” refers to it in relation to two important aspects. The first one is that they are recognized as “challenges of a global dimension and continue to pose serious challenges to the sustainable development of all countries”.[3] The recognition of the global dimension when it comes to soil and land degradation issues has been a major step ahead as soil and land are commonly understood as typical domains of local and national sovereignty.[4] Unlike air, water, or climate issues, soil and land don’t cross borders. Instead they are often the terrain that we use to define our borders. Moreover soils are often private property, usually owned and always managed locally. Thus, they are a typical domestic resource. This made and still makes it very hard to reach international or even regional agreements on soil protection.[5] Now what made it possible to

[2]United Nations (2012), para 247.

[3]United Nations (2012), para 205.

[4]See the chapter by Boer et al. (2016).

[5]Montanarella (2015).

address soil degradation in a global context at the Rio + 20 Conference? First, there was the 2007/2008 world food price crisis with dramatically increasing food prices causing instability and unrest in several countries. This was accompanied with a drastic increase of large scale investments in fertile soil of globally acting companies, governments and individuals mostly in developing countries. This phenomena, often referred to as "land grabbing" opened the eyes of the global community that while soils are a domestic resource, money and power are not. Next to those global trends triggered by the markets, concerns on the transnational access to water and the global governance of climate enabled a new view on soils: It was finally recognized that while soils are in fact a domestic resource, vitally important soil functions such as carbon storage, water storage and purification and their role in fostering biodiversity and producing food, feed, fiber and biofuel are of global importance.

The second important aspect of soil and land degradation that was discussed and agreed upon was the introduction of a new theme or even leitmotif on land degradation: It was agreed "to strive to achieve a land-degradation-neutral world".[6] This was introduced by UNCCD in the run up to the Rio + 20 conference as the "Zero net land degradation target".[7] In principle, the concept of a "land-degradation-neutral world" is quite simple: It means to prevent any further degradation to the functional capacities of land and soils. This can be achieved by reducing land and soil degradation to a minimum and compensating for any unavoidable degradation by implementing restoration and rehabilitation measures (see Fig. 1).[8,9] For more detailed information on the current discussions on the interpretation of land degradation neutrality in the frame of UNCCD see chapter by Minelli et al. (2016).

3 References to Soils in the SDGs

Soil protection is addressed several times in the now agreed goals and associated targets, with Goal 15 (Protect, restore and promote sustainable use of terrestrial ecosystems, sustainably manage forests, combat desertification, and halt and reverse land degradation and halt biodiversity loss) being particularly pertinent as it already references the need to "halt and reverse land degradation" in its title. This is given more concrete expression in Target 15.3:

> By 2030, combat desertification, restore degraded land and soil, including land affected by desertification, drought and floods, and strive to achieve a land-degradation-neutral world. (United Nations 2015)

---

[6]United Nations (2012), para 206.

[7]UNCCD (2012).

[8]Lal et al. (2012).

[9]Ehlers (2013).

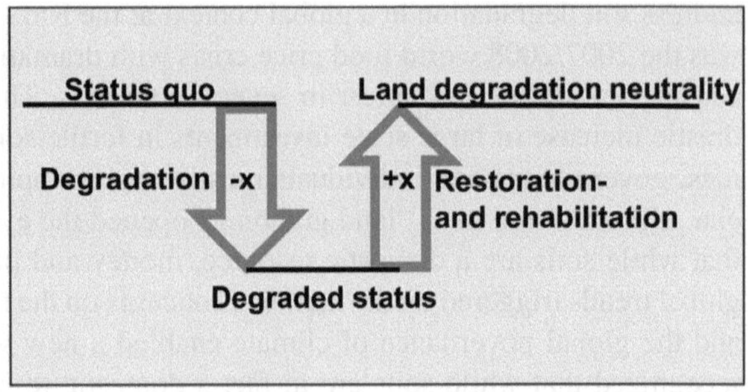

Fig. 1 The concept of Land degradation neutrality (adapted from Ehlers 2013)

Target 15.3 thus mainstreams in the SDGs the concept of a "land degradation neutral world" as developed at Rio + 20 as a leitmotif for soil protection.

Furthermore, associated with Goal 2, which relates to food security and sustainable agriculture, is the following Target 2.4.:

> By 2030, ensure sustainable food production systems and implement resilient agricultural practices that increase productivity and production, that help maintain ecosystems, that strengthen capacity for adaptation to climate change, extreme weather, drought, flooding and other disasters and that progressively improve land and soil quality.(United Nations 2015)

This target is a typical example for an SDG-target that addresses not one single issue but several cognate issues at once. So far it remains open how much impacts these "multi-purpose targets" will actually develop when it comes to making progress at single issue level. Nevertheless, the targets highlight the importance of soils for food security while at the same time making reference to achieving food security within the carrying capacity of supporting ecosystems.

A third reference to the issue of soil can be found in Goal 3 on human well-being under Target 3.9.:

> By 2030, substantially reduce the number of deaths and illnesses from hazardous chemicals and air, water and soil pollution and contamination. (United Nations 2015)

Therefore, three essential goals explicitly reference the importance and significance of soils for sustainable development (see Fig. 2). While Goal 15 directly highlights natural resource protection, Goals 2 and 3 are more strongly targeted at human needs and the functions of soils in this regard.

However, these explicit mentions of soils in sustainable development goals and targets do not cover all the implications soils have for the SDGs. There are in fact a number of additional goals that will either significantly impact on the sustainable utilization of soils, or the achievement of which will depend on whether or not soils and the ecosystem services they provide will be sufficiently protected and supported (see Fig. 3).

Goals 7, 11 and 12 will therefore impact on whether or how soils will be sustainably utilized in future, given that bioenergy generation from renewable agricultural resources as well as our general patterns of consumption of food and biomass crops very strongly impact on the level of agricultural land use pressure.

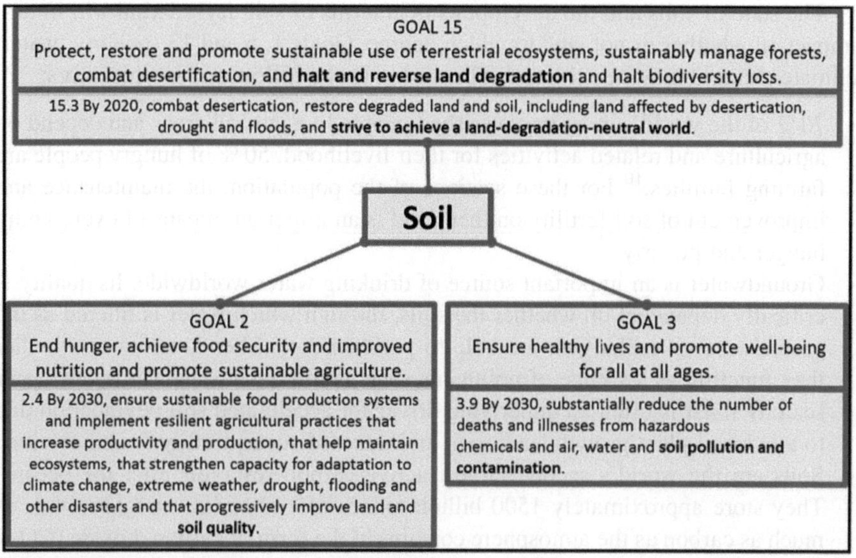

Fig. 2 Explicit mentions of aspects of soil in SDGs in Goals 2, 3 and 15 (text source: United Nations 2015)

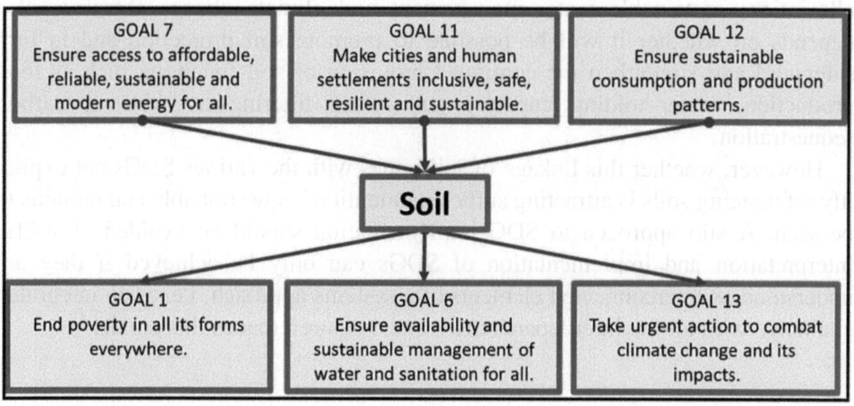

Fig. 3 Implicit linkages between soil protection and other sustainable development goals (text source: United Nations 2015)

The sealing of soil surfaces as a result of building construction is another important factor as this form of land use results in the complete loss of ecological soil function. Therefore, soil protection policies cannot be effective while being counteracted by (at least the outcomes of) bioenergy, consumption and settlement policies. In the interest of the requirement for policy coherence, these policy areas should be taken into account in the development of an overall strategy on soil protection.

The state of soils and the developments in terms of soil degradation will in turn impact on whether or not, and to which degree, Goals 1, 6 and 13 (poverty, water, climate change) may be achieved. This is due to the following circumstances:

- 70 % of the world's most impoverished people live in rural areas and depend on agriculture and related activities for their livelihood. 50 % of hungry people are farming families.[10] For these sections of the population, the maintenance and improvement of soil fertility on their land is an important means of overcoming hunger and poverty.
- Groundwater is an important source of drinking water worldwide. Its quality is critically dependent on whether the soils, through which water is filtered as the groundwater is recharged, are able to perform their filter function or whether they function as a source of pollutants such as nitrates instead. In return water scarcity itself is often an underlying driver for accelerated soil degradation due to a reduced plant growth leading to an increased susceptibility to soil erosion.
- Soils are the world's second largest active carbon reservoir after the oceans. They store approximately 1500 billion tonnes of carbon, i.e. roughly twice as much as carbon as the atmosphere contains in the form of carbon dioxide (CO_2). Whether soils are acting as carbon sinks or carbon sources largely depends on how the land is used.[11]

The degree to which it will be possible to achieve the SDGs on poverty alleviation, sustainable water management and climate change therefore also depends on whether it will be possible to promote soil protection and in turn safeguard and strengthen the continued provision of soil functions such as food production, water-holding capacity and water filtering as well as carbon sequestration.

However, whether this linkage of soil issues with the various SDGs not explicitly referencing soils is attracting sufficient attention is questionable and remains to be seen. A silo approach to SDG implementation should be avoided. A useful interpretation and implementation of SDGs can only be achieved if they are understood as interconnected elements of a systems approach, i.e. as an integrated challenge with all the interdependencies and cross-references this entails.[12]

[10] The Hunger Project (2009).

[11] European Commission (2011).

[12] Mueller et al. (2015).

4 Challenges in Implementing the SDGs: The Search for Indicators

Similar to the approach taken with the MDGs, a set of indicators will be used to monitor progress on achieving the SDGs. Therefore, the indicators' primary role is to make the achievement of set objectives quantifiable and auditable as part of a follow-up and review process. Therefore, monitoring of indicators is a crucial component in implementing the SDGs. But prior to that, the indicators are important in defining and giving concrete expression to the SDG targets. Therefore, the definition of indicators to monitor the SDGs is not only a technical process in order to find out what can be monitored. In fact, the definition of indicators is just as much a political issue as it is a technical issue. This is because some of the targets are non-specific and broadly worded as to be pursuing several goals rather than one specific goal. Target 2.4 mentioned above is a good example of the apparent attempt, as part of the negotiations, to find a compromise between different positions which ultimately resulted in an inconsistent wording which requires more focus and interpretation to be able to be implemented and monitored. The next three sub-chapters will deal with specific challenges in monitoring (a), the current negotiations on land and soil indicators (b) and the issue of leakage effects in respect to the SDG implementation (c).

4.1 Hard to Monitor: The Multifunctionality of Soils

Even just focusing on monitoring aspects on soil degradation issues is not trivial. Defining indicators to monitor soil and land degradation is especially hard, as soil and land are multifunctional. Their most apparent function is the fact that soils are the basis for all terrestrial biomass production including the production of agricultural commodities such as food, feed, fiber and biofuel. Next to this biomass production function, soil do serve as sources for raw materials such as sand and clay and as a platform for us to build houses and infrastructure on. Furthermore, they form the basis for above and below ground habitats and biodiversity. Especially below ground biodiversity largely remains terra incognita with billions of microbes in a gram of topsoil and most likely myriads of unknown species. Soils serve as archives of geological and archaeological heritage and they play a vital role for all life on earth by storing, filtering, transforming a buffering substances such as nutrients, water. In that respect soils do play an important role when it comes to mitigating or accelerating climate change as they can act as a source or sink for atmospheric carbon.

All of these functions are of importance, but they serve different interests as they provide private, common and public goods (see Fig. 4). These three types of goods are commonly differentiated by two criteria: rivalrousness ("whether the

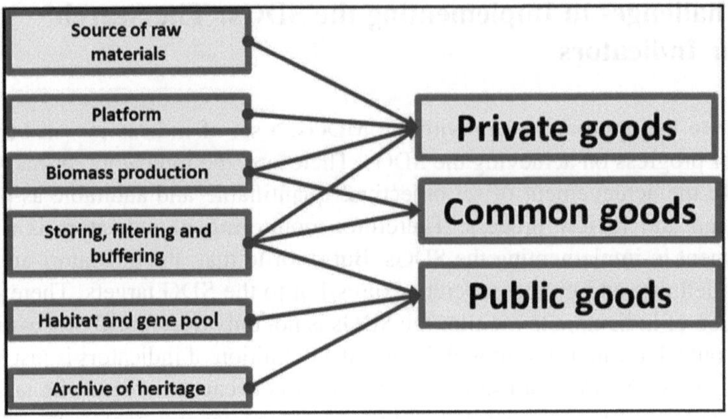

Fig. 4 The role of soil functions in providing private (excludable and rival), common (non-excludable but rival) and public (non-excludable and non-rival) goods

consumption of a good by one person precludes its consumption by another person"[13]) and excludability ("whether it is possible to prevent people (consumers) who have not paid for it from having access to it"[13]). The production of private goods, which are excludable and rival, is mainly in the interest of private landowners and landusers such as farmers. The production of common goods (non-excludable but rival) such as safe drinking water and public goods (non-excludable and non-rival) such as the role of soil carbon sequestration to mitigate climate change are within the interests of the broader public. Weighing up those interests will eventually be a part in the definition of indicators and the implementation of the soil related SDGs.

This process will be difficult as different functions of soils may be in conflict with one another. Efforts to enhance soil functions to one particular end may impede their functions for another. Peatland drainage is an example for such a trade-off relationship between soil functions: It enhances the soils' production function but reduces soil carbon sequestration. Intensification of agricultural production using fertilizers and pesticides is another one as it increases the biomass production but reduces soil's role in fostering biodiversity.

4.2 *The Current State of Negotiations on Indicators*

The difficulties in coming up with clear and directly applicable indicators on soil degradation become evident when looking at the current state of the international negotiations on SDG indicators. The UN Inter-Agency and Expert Group on SDG Indicators (IAEG-SDGs) is currently dealing with this issue. The IAEG-SDGs has

[13] Wikipedia (2016).

been tasked with the development of a list of global indicators for global SDG monitoring, to be presented in draft form in March 2016.[14] The indicators for Targets 2.4 and 15.3 are particularly relevant to the issue of soil. One of the indicators for Target 2.4. is:

Percentage of agricultural area under sustainable agricultural practices.[14]

This indicator must be more clearly defined, given that the views as to what constitutes "sustainable agricultural practices" are widely divergent.[15] The proposed indicator for Target 15.3 is:

Percentage of land that is degraded over total land area.[14]

Again, this indicator is not sufficiently precise, as there is currently no globally accepted definition of the term "land degradation" but a multitude of views coloured by local perspectives and subjective interests due to the above mentioned reasons. Due to this need for further clarification and interpretation of the indicators, the IAEG Secretariat is currently proposing to attribute both indicators as Tier 3 Indicators meaning that these are indicators "for which there are no established methodology and standards or methodology/standards are being developed/tested."[16]

It is doubtful whether it will be possible to successfully establish priorities, interpret and to fully technically refine the SDG targets at the global level. Instead the required refinement of indicators might not take place at the international level but at the national level where it may be easier to develop a common understanding in the process of balancing site-specific and needs-based interests including aspects of monitoring capacity. This would be consistent with the reference contained in the 2030 Agenda for Sustainable Development which states that different national realities, national policies and national priorities play an important role in the implementation of the SDG Agenda and that therefore national policy space will be respected. Additionally, regional implementation processes e.g. on European level might play a role. The 2030 Agenda for Sustainable Development at least does explicitly recognize the value of such regional intergovernmental processes for SDG implementation.

Independent of the level at which the further interpretation and refinement for SDG implementation will ultimately take place, what is certain is that only clear indicators will render SDGs effective and fit for implementation. In this context it is important to ensure that potential national definitions of indicators do not only take specific local interests into account but that due consideration is also given to important global aspects of soil degradation. An example would be the release of CO_2 resulting from the decomposition of soil organic carbon. There is therefore a need for a certain amount of international coordination going beyond the

[14]IAEG (2016).

[15]Caspari et al. (2015).

[16]IAEG Secretariat (2016).

compromises found so far. Especially when it comes to the issue of soil, it will be important to propose indicators that are most likely to do justice to the multifunctional character of soils. Such indicators would be cross-sectoral in nature and would not only help to further refine the explicitly soil-related SDG targets but they would also serve to highlight the multitude of implicit linkages between aspects of soil and SDGs respectively, thus helping to broaden acceptance.

4.3 Leakage Effects: National Responsibilities Beyond National Borders

As mentioned above soils are local resource with partly localized and partly non-localized soil functions. Therefore, national implementation steps for the protection of national soils need to take both, national and global interests into account. But even beyond the SDG implementation for the national soils, successful implementation of the SDGs will also need a certain amount of awareness on how nation's actions in regard to soil related issues will also influence soil related issues in other countries (leakage effects). Facing this responsibility will be especially important for countries that already today are highly dependent on extraterritorial soil use in order to fulfil their needs. In the EU, for instance, the per capita soil demand for arable land is 0.31 ha[17] whereas the available arable land capita on global level is 0.2 ha and for Europe 0.21 ha respectively.[18] Both with a declining trend (see Fig. 5).

These data stress the responsibility countries with a high land demand have for land and soil use not only within their national boundaries but also for land and soils they indirectly lay claim on. The best way to cope with this responsibility would be to adjust a nation's demand for fertile soil to those levels that can be sustained globally (e.g. around 0.2 ha/capita) as proposed by UNEP[17]. This will have strong implications on consumption pattern and thus can only be reached in the medium or even long run. In the meantime nations with a high demand for fertile soils at least can support other nations in achieving the soil related SDGs by providing technical and financial support for capacity building.

5 Outlook

The global negotiations on sustainable development goals and targets have been concluded. The results of global negotiations on indicators appear to be close to conclusion. In all probability the ball will very soon be in the court of national or regional policy development. For each country, therefore, it will come down to its

[17] UNEP (2014).

[18] Worldbank (2016).

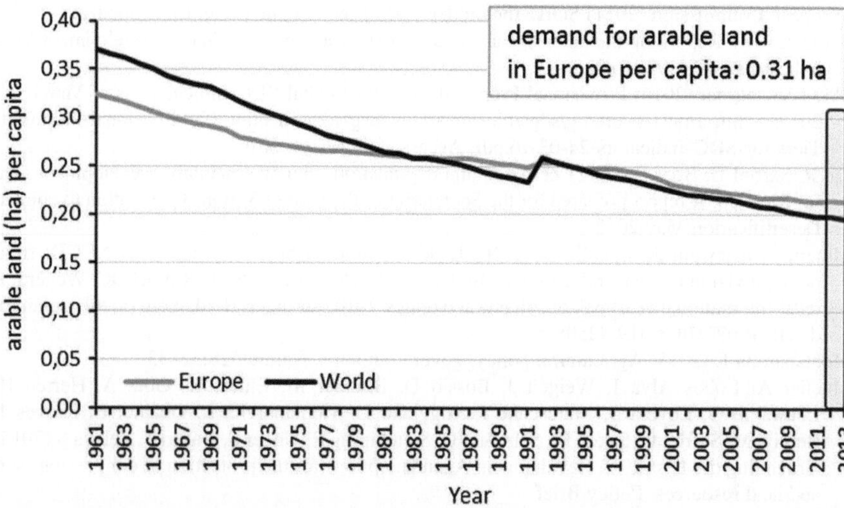

Fig. 5 Arable land per capita (globally and within Europe) versus the demand for arable land in Europe (data source: Worldbank 2016; UNEP 2014)

own interpretation of the soil-related SDGs and the measures it sets out on the basis of this process.

In this context, it will be important for nations to ensure a holistic and interconnected approach in implementing the SDGs which are directly and indirectly linked to soil issues. Paying respects to the various private, public, local, national and global stakeholders and weighing their interests according to local natural conditions will be necessary.

The main aim of the SDG will not prevail unless bolt action is taken within the next 15 years. Successful implementation of the soil related SDGs thus demands high standards and a lot is at stake. The SDGs were adopted by all UN Member States and therefore hopefully will benefit from a high level of political commitment through the next years.

References

Boer B, Ginzky H, Heuser IL (2016) International soil protection law – history, concepts and latest developments. In: Ginzky H, Heuser IL, Qin T, Ruppel OC, Wegerdt P (eds) International yearbook of soil law and policy 2016. Springer, Heidelberg (in this volume). doi:10.1007/978-3-319-42508-5

Caspari T, van Lynden G, Bai Z (2015) Land degradation neutrality: an evaluation of methods. UBA-Texte 62/2015

Ehlers K (2013) Land degradation neutrality: implications for a soil related sustainable development goal. Local land & soil news 44/45, I13, pp. 8–10

European Commission (2011) SOIL: the hidden part of the climate cycle. Luxembourg

IAEG (2016) Report of the inter-agency and expert group on sustainable development goal indicators E/CN.3/2016/2

IAEG Secretariat (2016) Provisional Proposed Tiers for Global SDG Indicators as of March 24, 2016. http://unstats.un.org/sdgs/files/meetings/iaeg-sdgs-meeting-03/Provisional-Proposed-Tiers-for-SDG-Indicators-24-03-16.pdf. Accessed 29 March 2016

Lal R, Safriel U, Boer B (2012) Zero net land degradation - a new sustainable development goal for Rio + 20. A report prepared for the Secretariat of the United Nations Convention to combat Desertification May 2012

Minelli S, Erlewein A, Castillo V (2016) Land degradation neutrality and the UNCCD: from political vision to measurable targets. In: Ginzky H, Heuser IL, Qin T, Ruppel OC, Wegerdt P (eds) International yearbook of soil law and policy 2016. Springer, Heidelberg (in this volume). doi:10.1007/978-3-319-42508-5

Montanarella L (2015) Agricultural policy: govern our soils. Nature 528:32–33

Mueller A, Lobos Alva I, Weigelt J, Bossio D, Brander M, Ehlers K, Götz A, Herren H, Louwagie G, Lovera S, Matuschke I, Miggiano L, Montanarella L, Obaikol E, Reeves J, Bergöö M, Safriel U, Sakoh M, Schm\idt O, Sietchiping R, Unger S, Vaturi T, Zelaya S (2015) Grounding the Post-2015 Development Agenda: options for the protection of our precious soil and land resources. Policy Brief

The Hunger Project (2009) Know your world: facts about hunger and poverty. http://www.thp.org/knowledge-center/know-your-world-facts-about-hunger-poverty/. Accessed 18 November 2015

UNCCD (2012) Zero net land degradation - a sustainable development goal for Rio+20; UNCCD Secretariat policy brief, May 2012

UNEP (2014) Assessing global land use: balancing consumption with sustainable supply. A report of the working group on land and soils of the international resource panel. In: Bringezu S, Schütz H, Pengue W, O'Brien M, Garcia F, Sims R, Howarth R, Kauppi L, Swilling M, and Herrick J

United Nations (2012) The future we want. Outcome document of the Rio + 20 conference. UN-General Assembly Resolution A/RES/66/288

United Nations (2015) Transforming our world: the 2030 Agenda for sustainable development. UN-General Assembly Resolution A/RES/70/1

Wikipedia (2016) Common good (economics). https://en.wikipedia.org/wiki/Common_good_%28economics%29. Accessed 29 April 2016

Worldbank (2016) World DataBank. http://databank.worldbank.org/data/home.aspx. Accessed 1 April 2016

Land Degradation Neutrality and the UNCCD: From Political Vision to Measurable Targets

Sara Minelli, Alexander Erlewein, and Victor Castillo

1 Establishing Land Degradation Neutrality as a Political Goal

For several years the concept of Land Degradation Neutrality (LDN) has been discussed as a promising new strategy that might have the potential to revitalize the international agenda to fight desertification/land degradation and drought (DLDD).[1] Policies and programmes to halt and reverse DLDD, as well as the corresponding multilateral environmental agreement, the UNCCD, have long suffered from the absence of a clear overarching goal and quantitative targets that could guide action and make progress measurable. LDN, which was initially discussed under the title "Zero Net Land Degradation", offered a way forward by suggesting the goal of "no net loss".[2] From the very beginning, it was evident that a major challenge in pursuing a neutrality approach lies in quantifying land degradation. Given the multiple dimensions of land degradation and its high dependency on local circumstances, there is no commonly agreed upon standard for measuring land degradation. Instead, a variety of different methods and approaches compete with each other, leading to varying and partially contradicting results that make unambiguous conclusions difficult.[3] Against this background, the concept of LDN is also seen as

[1]In the context of the UNCCD, "Land degradation means reduction or loss, in arid, semi-arid and dry sub-humid areas, of the biological or economic productivity and complexity of rainfed cropland, irrigated cropland, or range, pasture, forest and woodlands" (UNCCD 1994, article 1(f)).

[2]UNCCD (2012).

[3]Caspari et al. (2015).

S. Minelli (✉) • A. Erlewein • V. Castillo
Science, Technology and Implementation Unit, UNCCD Secretariat, Platz der Vereinten Nationen 1, 53113 Bonn, Germany
e-mail: sminelli@unccd.int; AErlewein@unccd.int; VCastillo@unccd.int

H. Ginzky et al. (eds.), *International Yearbook of Soil Law and Policy 2016*, International Yearbook of Soil Law and Policy,
DOI 10.1007/978-3-319-42508-5_9

an opportunity to harmonize assessment approaches and achieve a minimum consensus on how to quantify land degradation.

The concept of "Zero Net Land Degradation" was first introduced to a wider international policy audience at the United Nations Conference on Sustainable Development (Rio + 20) in 2012. Parties at the conference acknowledged that addressing land degradation is a key element in pursuing a sustainable development pathway and concluded in the outcome document "The Future We Want": "we will strive to achieve a land degradation neutral world in the context of sustainable development".[4] The report was approved as a resolution by the United Nations General Assembly (UNGA) and served as a basis for defining the Sustainable Development Goals (SDG), thus paving the way for including LDN in the SDG negotiation process. This process was, to a large extent, dominated by the question of how wide-ranging the catalogue of SDGs should be. While some Parties argued in favour of a short, prioritized list of SDGs, the ultimate consensus resulted in a rather comprehensive set of 17 SDGs, accompanied by 169 targets that define more specific and concrete objectives.[5]

This approach made it possible that DLDD was not only addressed in the more general form of an SDG, but that the specific target of LDN was included when the 2030 Agenda for Sustainable Development was eventually approved by the UNGA in September 2015. Therein, SDG 15 calls to: "Protect, restore and promote sustainable use of terrestrial ecosystems, sustainably manage forests, combat desertification, and halt and reverse land degradation and halt biodiversity loss", while target 15.3 makes particular reference to LDN: "By 2030, combat desertification, restore degraded land and soil, including land affected by desertification, drought and floods, and strive to achieve a land degradation-neutral world".[6]

In response to this recognition and to maintain the political momentum, the twelfth session of the Conference of the Parties to the UNCCD (COP 12), only 1 month later in October 2015, embraced SDG target 15.3 by "recognizing the unique role of the UNCCD in addressing DLDD in affected areas" and stating that "the Convention will make a significant contribution to achieving LDN".[7] Going one step further, COP 12 decided to integrate LDN in the UNCCD process and stipulated that "striving to achieve SDG target 15.3 is a strong vehicle for driving implementation of the UNCCD, within the scope of the Convention".[8] In this respect COP 12 also explicitly tasked the UNCCD to "take the initiative and invite other relevant agencies and stakeholders [...] to seek cooperation to achieve SDG target 15.3".[9]

---

[4]UNGA (2012), p. 40.

[5]UNGA (2015).

[6]ibid.

[7]UNCCD (2015a), decision 3/COP.12.

[8]ibid.

[9]ibid.

While the same decision recalls "that affected areas under article 1, paragraph (h), of the Convention means: arid, semi-arid and/or dry sub-humid areas affected or threatened by desertification",[10] decision 8/COP.12 notes "that a significant proportion of land degradation occurs beyond arid, semi-arid and dry sub-humid areas".[11] In the latter sense, it was recognized in general terms that "Parties may use the UNCCD to guide their policies relating to DLDD and voluntary targets when striving to achieve LDN at national and subnational levels".[12] These decisions can be seen as re-orienting the Convention's implementation process towards LDN and making it a guiding principle for addressing land degradation in all countries.

The question of how a global SDG target on LDN can be translated into national plans and policies, and ultimately trigger local action, was debated within the UNCCD ever since the LDN concept emerged. A first step in this direction was seen in the formulation of national voluntary LDN targets that would contribute to approaching the overarching global target of LDN in 2030. A pilot project, spearheaded by the UNCCD secretariat, was conducted in 2014 and 2015 and supported 14 countries to define baselines and tentatively set LDN targets based on the UNCCD monitoring and reporting indicator framework.[13] Based on the experience of the LDN pilot project, COP 12 agreed on officially inviting "Parties to formulate voluntary [national] targets to achieve LDN in accordance with their specific national circumstances and development priorities".[14] This invitation was not restricted to affected country Parties.

Consistent with this call for voluntary target setting, the COP also invited affected country Parties to integrate LDN in their national action programmes (NAP) which are the main instruments for implementing the Convention's objectives within member countries. Decision 2/COP.12 "Invites affected country Parties [...] to establish baselines and national-level voluntary land degradation neutrality (LDN) targets within their NAPs [...] [and] to develop and implement strategies through their NAPs to achieve the objectives of the Convention in the light of target 15.3 of the 2030 Agenda for Sustainable Development".[15]

At the same time, COP 12 requested the secretariat and Convention bodies to support the integration of LDN in the UNCCD processes and to "develop guidance for formulating national LDN targets and initiatives".[16] A particular need for support was seen with regard to monitoring LDN. Parties tasked the secretariat to "Further develop, keep under review and facilitate, including through pilot projects,

[10]ibid.

[11]UNCCD (2015a), decision 8/COP.12.

[12]ibid.

[13]The 14 countries which participated in the LDN pilot project were Algeria, Armenia, Belarus, Bhutan, Chad, Chile, Costa Rica, Ethiopia, Grenada, Indonesia, Italy, Namibia, Senegal and Turkey.

[14]UNCCD (2015a), decision 3/COP.12.

[15]UNCCD (2015a), decision 2/COP.12.

[16]UNCCD (2015a), decision 3/COP.12.

the use of the UNCCD indicator framework as a contribution to the monitoring, evaluation and communication of progress towards the national LDN targets".[17] In response to this decision, the Global Mechanism (GM) of the UNCCD has established the "LDN Target Setting Programme" (LDN-TSP) at the end of 2015.

The main objective of the LDN-TSP is to enable country Parties to define national baselines and measures to achieve voluntary targets for LDN. The LDN-TSP provides data and information on methodologies and approaches related to the assessment of the LDN baseline at country level, including the identification of priority areas for LDN implementation. The programme also supports countries to gather and process LDN related datasets, and to identify suitable (sub)national data sources, as appropriate. Based on this assessment, the LDN-TSP supports countries in defining LDN targets and identifying measures to address the drivers of land degradation and approach the targets set. Furthermore, the LDN-TSP aims to facilitate large-scale LDN project opportunities and engage multiple stakeholders in LDN implementation. By April 2016, around 80 countries expressed their interest to take part in LDN-TSP and a number of regional inception workshops will be held throughout 2016 to start the LDN target setting process in interested countries. It is expected that the first voluntary national LDN targets will be set until the next UNCCD COP in late 2017.[18]

2 Conceptualizing Land Degradation Neutrality

While the political process of establishing LDN as a goal in the arena of international development and environmental politics advanced significantly in 2015, uncertainties remained as to what implementing LDN means exactly. This need for clarification was already evident in 2013 when COP 11 decided to form an Intergovernmental Working Group (IWG) in order "to establish a science-based definition of land degradation neutrality in arid, semi-arid and dry sub-humid areas; [and] develop options [. . .] that Parties might consider should they strive to achieve land degradation neutrality".[19] The LDN definition was eventually endorsed by COP 12 as follows: "Land degradation neutrality is a state whereby the amount and quality of land resources necessary to support ecosystem functions and services and enhance food security remain stable or increase within specified temporal and spatial scales and ecosystems".[20] In its report, the IWG clarified that, in line with the request by COP 11, this definition was elaborated with regard to "LDN in arid,

[17]ibid.

[18]http://www.global-mechanism.org/content/supporting-countries-set-land-degradation-neutrality-targets.

[19]UNCCD (2013a), decision 8/COP. 11.

[20]UNCCD (2015a), decision 3/COP.12.

semi-arid and dry sub-humid areas" but that "this definition is suggested by the IWG as equally applicable to other areas".[21]

While fully acknowledging the contextuality of land degradation in terms of local ecological and socio-economic conditions and the need to thoroughly identify the drivers of land degradation, the IWG identified two basic pathways for implementing LDN: "(a) Prevent, avoid or minimize land degradation [...] through, inter alia: (i) national and local land use planning that fully accounts for the potential and resilience of land resources; (ii) the adoption of sustainable land management practices"; and as the second basic pathway: "(b) Rehabilitate or restore degraded land [...] by reducing the drivers and impacts of current land degradation processes and by the implementation of projects and other measures for rehabilitation and recovery".[22] Moreover, the IWG report highlighted the need for applying a consistent monitoring approach to allow for regularly assessing the status of land degradation and track progress towards LDN.

The guidance by the IWG and its endorsement at COP 12 was a major step forward in achieving consensus on the meaning of LDN across Parties to the UNCCD. However, at COP 12 several Parties and other stakeholders expressed the need to further elaborate the concept of LDN and options for its operationalization. The Science-Policy Interface (SPI) of the UNCCD was tasked to "Provide scientific guidance to the operationalization of the voluntary land degradation neutrality (LDN) target".[23] The SPI responded to this request by convening, in February 2016, an interdisciplinary workshop on LDN in collaboration with external experts. The workshop triggered the development of a conceptual framework for LDN with the aim of providing a scientifically-sound basis for understanding and defining LDN and to inform practical guidance for implementing LDN.[24]

The conceptual framework, which was still in its infancy at the time of writing, builds on the approved definition of LDN by reconfirming its ecosystem-based approach, while putting particular emphasis on provisioning ecosystem services in the sense of land productivity. Land degradation causes a decline in the multiple goods and services provided by ecosystems, including food and fibre, carbon sequestration potential, air quality, hydrological function and cultural values. Ultimately, long-term food productivity and food security are threatened by land degradation. LDN, as an aspirational goal, seeks to sustain the natural capital of the land, to support ecosystem functions and services, and thus meet the needs of the current and future human population. The framework furthermore describes how and to what extent this objective and the associated land-based ecosystem services are reflected and covered by the indicator and associated sub-indicators suggested for monitoring LDN (see Sect. 3.3).

---

[21]UNCCD (2015b), p. 9.

[22]ibid.

[23]UNCCD (2015a, b), decision 21/COP.12.

[24]UNCCD (2016) Scientific Conceptual Framework for Land Degradation Neutrality. A Report of the Science-Policy Interface. Barron J. Orr, Annette L. Cowie, et al. Manuscript in preparation.

In view of operationalizing LDN for national implementation, the framework intends to advance the understanding of a number of key aspects related to LDN. One of the biggest challenges lies in clarifying the implications of neutrality.[25,26,27] In the sense of a no-net-loss approach, LDN implies a commitment to avoid the net deterioration of a given status quo in land degradation and to achieve the maintenance, or improvement, of the reference state. In strict technical terms, this implies that the target (2030) should, at least, be equivalent to the baseline (see Sect. 3.3). However, in this context it is also important to note that the 2030 Agenda for Sustainable Development stipulates that "targets are defined as aspirational and global, with each Government setting its own national targets guided by the global level of ambition but taking into account national circumstances".[28]

Inherent to the notion of a no-net-loss approach is the assumption that losses and gains can be balanced. As such, the LDN concept acknowledges that land degradation cannot be stopped completely everywhere, while suggesting that a balance can be reached through the restoration and rehabilitation of degraded areas. It is obvious that the instruments and approaches which may be used to achieve such a balance are diverse and will depend on country circumstances and preferences, in line with the above mentioned provision of national sovereignty. Against this background, the conceptual framework suggests a number of principles that should be considered when counterbalancing losses with gains in order to prevent unintended outcomes.

As a basic principle, the framework recommends to follow a "response hierarchy" when making use of counterbalancing options under an LDN approach. As it is known in many other fields of environmental management,[29] the response hierarchy prioritizes *avoiding* degradation, followed by *reducing* the negative impacts of ongoing degradation as far as possible. Only once the possibilities for avoiding and reducing land degradation have been fully used, *reversing* degradation through restoration and rehabilitation of already degraded land should be an option to compensate for the remaining part of what might be termed "unavoidable degradation". However, the framework also clarifies that restoration activities that are not intended to counterbalance the degrading use of other areas are obviously not a "last resort" option, but a direct contribution towards approaching LDN.

In line with the recommendations made by the IWG, the framework emphasizes the importance of land use planning in approaching LDN targets. It suggests mainstreaming LDN into land use planning by making it a "planning principle", whereby land use planning decisions not only consider threats of serious or irreversible damage within a particular site, but also the net contribution of each of those decisions, positive or negative, to the goal of neutrality.

[25]Welton et al. (2014).

[26]Altvater et al. (2015).

[27]Tal (2015).

[28]UNGA (2015), p. 13.

[29]E.g. Ramsar Convention on Wetlands (2012).

Without doubt, the biggest challenge in counterbalancing land degradation lies in ensuring equivalence between losses and gains. Given the vast heterogeneity of land and its associated ecosystem services, it is a key principle to counterbalance "like with like" not only in terms of quantity, but also quality. Operationalizing the latter poses an enormous and partially insurmountable challenge which requires detailed analysis at local level. As a basic principle to limit the comparison of unequal lands (in terms of ecosystem services), the framework, in line with the definition of LDN, recommends that counterbalancing should only take place between land units belonging to the same ecosystem and having the same long-term potential to sustainably generate ecosystem services.

Another basic consideration in this context refers to the different temporal scales of land degradation and restoration. While healthy land can become completely degraded within very short timeframes, in particularly as a consequence of land use change, restoration activities take a long time to become effective. The framework calls for taking such time-lags fully into consideration.

The framework emphasizes that any kind of LDN action will have to be based on a thorough analysis of the drivers of land degradation processes which not only can be found in poor land use practices and management, but are often related to indirect socio-economic and political causes.

3 Measuring, Monitoring and Reporting Land Degradation Neutrality

3.1 The Sustainable Development Goals Indicator Framework

The adoption of the SDGs has launched a process for the identification or development of appropriate indicators for measuring global and national progress towards the achievement of its ambitious targets. A solid framework of indicators and data to establish a baseline, monitor progress, inform policy and ensure accountability of all stakeholders is essential for a robust review of implementation mechanism. At its 47th session in March 2016, the United Nations Statistical Commission (UNSC) endorsed a global indicator framework for the 2030 Agenda to be further adopted by the Economic and Social Council and the General Assembly. The global indicator framework includes a total of 231 indicators.[30] In line with the provisions related to national sovereignty embedded in the 2030 Agenda, the global indicator framework will be complemented by indicators at the national level to be developed by Member States and by indicators for thematic monitoring.

[30]United Nations (2016a).

Global indicators form the core of all other sets of indicators and are expected to fulfil a number of criteria that can be summarized as follows[31]:

- *Constructed from comparable and standardized national data*: Global monitoring and reporting should be based, to the largest possible extent, on comparable and standardized national data, obtained through well-established reporting mechanisms from countries to the international statistical system, and aggregated to the regional and global level.
- *Consensus based, in line with international standards*: Global indicators should be underpinned by a broad international consensus on their measurement and be based on international standards to facilitate international comparison.
- *Disaggregated:* In order to track inequalities in SDG achievement, global indicators should lend themselves to disaggregation where relevant.
- *Allow for high frequency monitoring*: Global indicators should lend themselves to annual production since an annual SDG progress report will be prepared by the United Nations Secretary-General. For a number of indicators, including the one proposed for SDG 15.3, annual monitoring may be impossible or inadvisable; in such cases producing data every 2–3 years and doing robust projections, extrapolations or modelled estimates may be sufficient.

A successful follow-up and review of progress for the implementation of the new 2030 Agenda will depend upon the availability of adequate data. Efforts will have to be made to fill data gaps, establish baselines where these are not available, and improve international comparability by adopting internationally agreed standards at the national level.

3.2 *Opportunities and Challenges of Monitoring Land Degradation Neutrality*

By promoting the recognition of land degradation as a global threat, the adoption of LDN as a target for sustainable development is generating momentum for initiating monitoring and reporting actions at the national level, while boosting efforts for the global assessment of land degradation. A global partnership will be needed to guide the collection of required data and the establishment of a baseline in a universally agreed, harmonized approach. Indeed, COP 12 directed "the secretariat of the UNCCD, as the lead organization for DLDD, to take the initiative and invite other relevant agencies and stakeholders [...] to seek cooperation to achieve SDG target 15.3".[32] It is in this spirit, that the UNCCD secretariat together with the Convention on Biological Diversity, the Food and Agriculture Organization (FAO) of the United Nations, and the Scientific and Technical Advisory Panel of the Global Environment Facility, hosted an expert meeting on a land degradation

[31] United Nations (2016b).

[32] UNCCD (2015a, b), decision 3/COP.12.

indicator for SDG target 15.3 in February 2016. The objective of the meeting was twofold: to help bring about a consensus on a standard approach to monitor and report on the status and trends in land degradation and restoration; and to lay the foundation for establishing a global partnership to assist countries with regular reporting on progress made towards the SDG target 15.3, but also towards the objectives of the Rio Conventions and other relevant targets, commitments and initiatives.

In particular the meeting explored the contribution to be made by a wide range of data, including Earth observation and geospatial information, while ensuring national ownership in tracking progress. Remote sensing offers objective, comparable and disaggregated data over large areas on a near real time basis. Therefore, when complemented by bottom-up ground surveys, it may currently provide one of the most viable, low-cost means of compliance verification.

A number of attempts have been made in recent decades to map the extent of land degradation globally.[33] Recently, the Intergovernmental Science-Policy Platform on Biodiversity and Ecosystem Services (IPBES) has launched an assessment of land degradation and restoration. This assessment is expected to cover the global status of and trends in land degradation, the effect of degradation on biodiversity values, ecosystem services and human well-being, as well as the state of knowledge of ecosystem restoration extent and options. Although mapping and monitoring capabilities are progressing rapidly, such high-level assessments face constraints due to a number of challenges inherent in the land degradation concept. Land degradation is a complex and multi-faceted phenomenon triggered by a range of man-induced and climatic processes.[34,35] It is a global phenomenon, but highly contextual and spatially heterogeneous. Therefore, it cannot be judged independently of its spatial, temporal, economic, environmental and cultural context.[36] Consequently, its monitoring must effectively address complex human-environment interactions, requires measurement over large areas coupled with bottom-up technique that starts at the local level and entails continuous measurement scales and grades.[37] Ultimately, a comprehensive assessment of land degradation status and trends requires an analysis of the following four dimensions: spatial extent of degradation, degree of degradation, type of degradation and causes of degradation (its direct and indirect drivers).[38] As a result of these inherent difficulties, global estimates of land degradation range from a low of 15 % to a high of 63 % of total land area.[39]

Thus, monitoring LDN will be challenging, owing to the need of establishing a scientifically robust and consistent baseline against which to measure future

[33]Caspari et al. (2015).

[34]Geist and Lambin (2004).

[35]Reynolds et al. (2007).

[36]Warren (2002).

[37]Grainger (2015).

[38]Yengoh et al. (2016).

[39]Safriel (2007).

changes in the rates of land degradation and improvement. As suggested by the adopted LDN definition, monitoring the pace towards achieving LDN targets requires means of assessing where the amount and quality of land resources remain stable or increase. A pre-requisite to LDN monitoring is therefore the technical ability to assess levels of land degradation at different spatial and temporal scales; that is, the ability to distinguish between already degraded land, land under degrading use and land under non-degrading use. This aspect is particularly important not only to quantify LDN, but to support the selection of the appropriate response options, from resilience-conferring actions and SLM for avoiding and reducing land degradation, to rehabilitation and restoration interventions for already degraded lands.[40,41] Furthermore, monitoring and assessment must be set up in such a way that addresses the temporal scale disparity between "losses" and "gains". While degradation can happen quite rapidly, the impacts and sustainability of gains are frequently not obtained in the short term.[42] Attribution is also challenging because of the sheer range of social, economic and environmental factors that can influence long-term outcomes, many of which are external to specific SLM and restoration policies or actions.

The most difficult part of methodological considerations is to agree on a measurable indicator to be able to assess the state and trend of the system in question. Given the complexity of the land degradation phenomenon, there is no single indicator which can unambiguously track trends in land degradation. The challenge is therefore to identify a few sub-indicators that in combination can enable regular reporting on the status and trends in land degradation and restoration. In addition to fulfilling the generic criteria for SDG's global indicators (see Sect. 3.1), such a restricted set of sub-indicators needs to indicate the essence of degradation processes, be sufficiently sensitive, draw on existing data and be cost-effective to monitor.[43] A balance needs to be struck between comparability across countries and the contextual nature of land degradation.

To this end, it is important to reflect on the difference between monitoring and reporting. Monitoring and reporting are distinct yet closely linked processes. Monitoring refers to the continuous process of observing a state and its changes and might also include examining the context, the local environment, as well as direct and indirect drivers. Reporting refers to how monitoring information is communicated, often across governance scales, and might provide an overview of progress, for instance at the global level.[44] Therefore, while reporting generally requires a few comparable indicators, national-level monitoring systems benefit from being flexible and pragmatic, using additional indicators, data and methods that are appropriate to the national context.

---

[40]Chasek et al. (2015).

[41]UNCCD (2016) Scientific Conceptual Framework for Land Degradation Neutrality. A Report of the Science-Policy Interface. Barron J. Orr, Annette L. Cowie, et al. Manuscript in preparation.

[42]ibid.

[43]Caspari et al. (2015).

[44]European Environment Agency (2015).

3.3 *An Indicator for Reporting on Land Degradation Neutrality*

"Proportion of land that is degraded over total land area" is the global indicator adopted by the UNSC to measure progress towards the SDG target 15.3. This indicator, originally named "Trends in land degradation", was proposed to the Inter-Agency and Expert Group on Sustainable Development Goal Indicators (IAEG-SDGs) by the UNCCD secretariat and the FAO. The proposed method of computation is based on the combined use of three sub-indicators, namely land cover, land productivity and carbon stocks above and below ground, including soil organic carbon, in conjunction with other nationally relevant indicators and contextualized with information at the national and sub-national level (see Fig. 1).

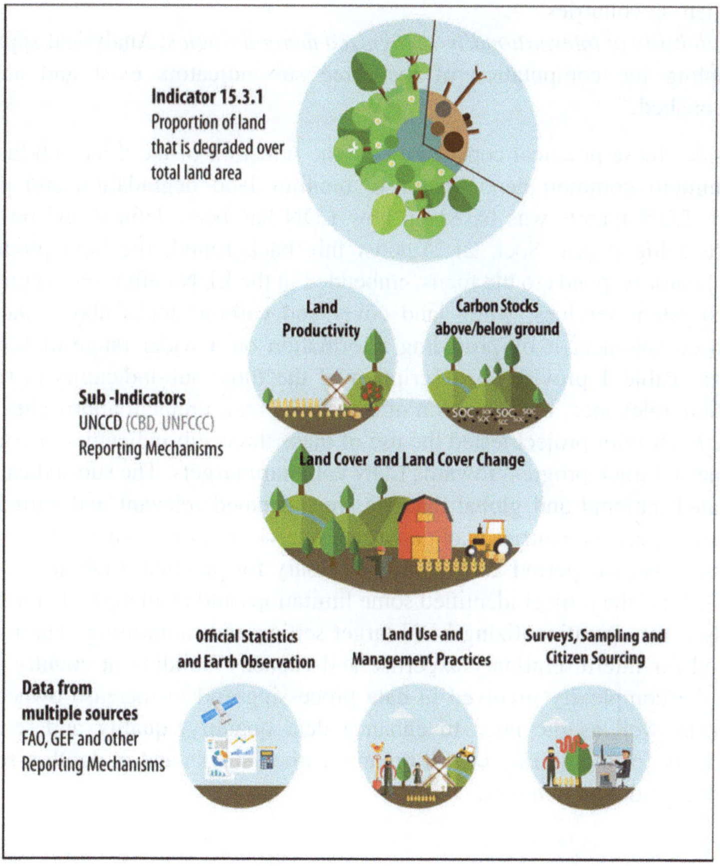

Fig. 1 Indicator framework for monitoring and reporting on SDG target 15.3. Adapted from UNCCD, CBD, FAO, STAP (2016) Framework and Guiding Principles for a Land Degradation Indicator [Outcomes of the Expert Meeting], manuscript in preparation

The indicator proposal was shaped by a number of practical considerations:

- *Alignment with the existing UNCCD monitoring and reporting indicator framework:* The three sub-indicators recommended for tracking land degradation are part of a set of six indicators adopted by the UNCCD to track progress in the implementation of the Convention through national reporting.[45] The existence of a well-established reporting mechanism from countries under the UNCCD represents a clear advantage and ensures the availability of a channel to derive harmonized national data on a regular basis.
- *Availability of global data sources to complement and support national reporting:* In the absence of suitable national datasets, global data sources exist for the three sub-indicators and can be used as default data for validation by countries. While in the long term all countries should be enabled to independently perform relevant data collection, analysis, and report on land degradation, global data sources can bridge data gaps and ultimately decrease the reporting burden on countries.
- *Availability of internationally-recognized methodologies*: Analytical approaches enabling the computation of the three sub-indicators exist and are well-established.[46]

Besides these practical considerations, the selection of the three sub-indicators as minimum common denominator to monitor land degradation and progress towards LDN targets was based on how LDN has been defined and on what it aims to achieve (see Sect. 2). Against this background, the land productivity sub-indicator responds to the focus, embedded in the LDN definition, on provisioning ecosystem services, while land cover and carbon stocks above and below ground complement it by providing information on a wider range of ecosystem services. Table 1 provides a description of the three sub-indicators in terms of definition, relevance, interpretation of trends and measurement approaches.

The LDN pilot project tested the use of these three sub-indicators to establish a baseline and track progress towards LDN voluntary targets. The sub-indicators and associated national and global datasets were deemed relevant and sufficient for most participating countries to obtain a synoptic view of on-going land degradation processes over the period 2000–2010, necessary for practical LDN target setting. Nevertheless, the project identified some limitations and challenges that need to be addressed for operationalizing LDN target setting and monitoring. These include the need for interdisciplinary expertise and capacity building at country level to tackle the complexity involved in data processing and to increase ownership of results, as well as the need to enhance data quantity, quality and resolution, particularly for monitoring of soil organic carbon stocks and in-depth assessment of land degradation hotspots.

---

[45]UNCCD (2013a), decision 22/COP. 11.

[46]Yengoh et al. (2016).

Table 1 Description of sub-indicators for computing the indicator "Proportion of land that is degraded over total land area" for the SDG 15.3 (Adapted from UNCCD, CBD, FAO, STAP (2016) Framework and Guiding Principles for a Land Degradation Indicator [Outcomes of the Expert Meeting], manuscript in preparation)

Sub-indicator	Description	
Land cover	Definition	Land cover refers to the observed physical cover of the Earth's surface (IPCC 2003).
	Relevance	Land cover is a fundamental land surface parameter that assists with the interpretation and stratification of the other two sub-indicators. Changes in land cover is also an important sub-indicator in its own right as it provides a first indication of a reduction or increase in vegetation, habitat fragmentation and land conversion.
	Interpreting trends	Changes in land cover may be characterized as positive or negative when contextualized with national or local information.
	Measurement	Most often derived from Earth observation, the sub-indicator requires geospatial mapping of land cover classes using comparable methodologies at regular time intervals. Standardized land cover classification systems exist (e.g. the Land Cover Meta Language of the Food and Agriculture Organization (FAO)); their use should be encouraged among countries to enable global comparisons.
Land productivity	Definition	Land productivity refers to the total above-ground net primary productivity (NPP) defined as the energy fixed by plants minus their respiration (Millennium Ecosystem Assessment 2005).
	Relevance	Land productivity refers to the biological productive capacity of the land, the source of all the food, fibre, and fuel that sustains humans (i.e. provisioning ecosystem services). Maintaining and enhancing the productivity of agro-ecosystems in a sustainable manner reduces the pressure for expansion and thus minimizes the loss and degradation of natural ecosystems.
	Interpreting trends	Areas with increasing NPP can generally be interpreted as improving, unless assessed otherwise at country level (for example, bush and tree encroachment (i.e., land cover change from grassland to shrub dominated) in drylands is considered to be land degradation, even though the NPP and soil organic carbon (SOC) may both be increasing).
	Measurement	The sub-indicator can be calculated across large areas from Earth observation data on NPP. Proxies to measure NPP, such as vegetation indices, are influenced in the short-term by crop phenology, rainfall, nutrient fertilization and other variables which must be corrected for to accurately interpret trends.

(continued)

Table 1 (continued)

Sub-indicator	Description	
Carbon stocks above and below ground	Definition	Carbon stock is the quantity of carbon in a pool (i.e. a system which has the capacity to accumulate or release carbon). Carbon pools are biomass (above-ground biomass and below-ground biomass); dead organic matter (dead wood and litter); and soil (soil organic matter) (IPCC 2003).
	Relevance	Carbon stocks reflect the integration of multiple processes affecting plant growth and the losses/gains from terrestrial organic matter pools. These changes also reflect trends in soil health and ecosystem functioning as well as land use and management practices.
	Interpreting trends	Areas with increasing carbon stocks can generally be interpreted as improving, unless assessed otherwise at country level (see example of bush and tree encroachment above).
	Measurement	Carbon stocks can be estimated by applying carbon density values from ground data or national inventories across land cover maps obtained by remotely-sensed data. The IPCC provides a systematic approach for estimating carbon stock changes from biomass, dead organic matter, and soils.

The expert meeting held in February 2016 came to a consensus on the use of these three sub-indicators as the most practical approach available to date for assessing the status and trends in land degradation and clarified how they can be used in combination to estimate the "Proportion of land that is degraded over total land area" and track progress towards LDN. Discussions at the expert meeting were fuelled by the preliminary outcomes of the interdisciplinary workshop convened by the SPI to develop a scientific conceptual framework for LDN. The main preliminary conclusions of the two meetings relevant to estimating the proportion of degraded land and tracking progress towards LDN can be summarized as follows:

- *Setting a baseline:* A first necessary step is setting a baseline, intended as an initial state which every future state gets compared to. LDN, in the sense of a no net loss approach, implies a commitment to maintain or enhance the land based natural capital relative to the baseline. The baseline can be expressed as the initial estimated value of each of the three sub-indicators and LDN is achieved if all three sub-indicators remain stable or improve. The assumption is that for most countries that baseline will be tied to current conditions. However, in order to minimize the effects of seasonal and inter-annual climate variability, the baseline should be calculated for each indicator as an average across at least 5 years.
- *Detecting changes:* Detecting 'changes' from the baseline should be the focus of efforts at the national and global levels, rather than the determination and accuracy of the baseline per se. An assessment of current trends in the three

sub-indicators can be used to analyse where interventions need to occur (e.g., identification of hotspots). Areas with increasing land productivity and carbon stocks can generally be interpreted as improving (i.e. gains), while areas with decreasing land productivity or carbon stocks can be interpreted as deteriorating (i.e. losses), unless assessed otherwise at country level. For example, often bush and tree encroachment (i.e., land cover change from grassland to shrub dominated) in the drylands leads to the loss of natural capital; with less forage for grazing animals and wildlife and is therefore considered to be land degradation, even though land productivity and carbon stocks may both be increasing. Such an "exception" assessment (i.e., "false positive") requires local knowledge, and interpretation according to local context, and should therefore be made at the national or sub-national level. In the case of land cover, changes may be characterized as positive or negative when contextualized with national or local information. The interpretation of changes in land cover is therefore ultimately up to national and local authorities who should provide explanations as on why changes are evaluated to be positive (gains) or negative (losses) in the given context.

- *Deriving the indicator "Proportion of land that is degraded over total land area"*: the proportion of degraded land can be derived by summing all those areas subject to changes that are considered negative by national authorities in one or more of the three sub-indicators (i.e., land degradation) and dividing by total land area. Attention should be paid to avoid double counting those areas that are subject to negative trends in more than one sub-indicator.
- *Using additional national indicators and contextual information*: Recognizing that the three sub-indicators do not capture the entirety of land degradation aspects, countries should evaluate whether these three sub-indicators should be supplemented with additional national or sub-national indicators. These national indicators would be used to interpret the changes detected, assess their cause, and devise management interventions to address land degradation.

3.4 Embedding the LDN Indicator in a Broader Monitoring Approach

The expert meeting held in February 2016 concluded that the indicator "Proportion of land that is degraded over total land area" and its associated sub-indicators, while suitable for reporting purposes, do not comprehensively address all quantitative and qualitative aspects of land degradation. Recognizing the limits of a "one-size-fits-all" approach, participants suggested that complementary indicators at national and sub-national scales could assist in monitoring issues relevant to specific national contexts. This is in line with the principles of the 2030 Agenda for Sustainable Development and with the idea of complementing the SDG global indicator framework with indicators at the national level to be developed by Member States.

Nested indicator systems capable of catering for the information needs of diverse groups of users, describing national/local issues and integrating with information derived from regional to global scales need to be developed.[47,48,49]

Given the links between target 15.3 and other SDGs,[50] indicators reported under other SDG targets could also contribute to interpret changes detected, assess causes, and devise management interventions to address land degradation. For instance, the indicator "Proportion of agricultural area under productive and sustainable agriculture" for target 2.4, and "Progress towards sustainable forest management" for target 15.2, complement the suggested indicator for target 15.3 and can help to provide a more comprehensive monitoring and evaluation framework for this target.[51] Being activity-based, these indicators have the potential to show changes sooner than sub-indicators such as land productivity and carbon stocks, provided that clear methodologies and agreed international standard to measure them are available. Other socio-economic indicators, such as those reported under SDG 1 "no poverty", 2 "zero hunger" and 6 "clean water" can help tracking the impact of land degradation on human wellbeing, while noting that distinguishing the contributive influence of land degradation and its remedies on socio-economic conditions is, at least at present, difficult to achieve.[52]

As suggested by the adopted LDN definition and the use of the words "amount and quality", monitoring the pace towards achieving the LDN targets also requires a blending of qualitative and quantitative measures and indicators. A mixed-methods approach to monitoring and reporting, which combines multiple sources of information, provides a strong basis for assessing progress.[53] Quantitative indicators are seen as attractive and practical, as they can communicate clear messages that are easy for policymakers to grasp. However, used in isolation, indicators are not effective in revealing the narrative behind complex issues. Qualitative information, drawn from local/national storylines, complement quantitative approaches and reveal critical contextual information that can help to explain the narrative behind the numbers.[54] A mixed-methods approach can produce a more comprehensive assessment of the relationships between policy actions and outcomes, and therefore help to address the attribution challenge (see Sect. 3.2).

Participatory processes can prevent an over-reliance on indicators and are perceived to play an important role in interpreting changes detected through

[47]Sommer et al. (2011).

[48]Verstraete et al. (2011).

[49]UNCCD (2013b).

[50]Akhtar-Schuster et al. (2016).

[51]UNCCD, CBD, FAO, STAP (2016) Framework and Guiding Principles for a Land Degradation Indicator, manuscript in preparation.

[52]Orr (2011).

[53]European Environment Agency (2015).

[54]UNCCD (2013b).

quantitative indicators, assess causes, and devise management interventions. A range of methods exists to gather stakeholder perspectives including surveys, workshops, in-depth interviews, consultations and the establishment of expert panels. These methods are often based on the principles of expert elicitation (i.e., the synthesis of opinions of technical and scientific experts). Ultimately, it is likely to be the task of the national authority that coordinates measurement, monitoring and reporting at national level to interpret what this combination of quantitative and qualitative data is saying regarding key land degradation and rehabilitation trends.

4 Conclusions

Over the last 5 years, and particularly in 2015, the concept of LDN went from being considered merely as an ambitious proposal to becoming firmly established as a political goal in the international development arena. The recognition of LDN at the Rio + 20 Conference paved the way for its incorporation in the 2030 Agenda for Sustainable Development as SDG target 15.3. Shortly after, COP 12 integrated LDN in the UNCCD process and decided that "striving to achieve SDG target 15.3 is a strong vehicle for driving the implementation of the UNCCD",[55] thus making LDN a kind of guiding principle for the Convention. Equally important, COP 12 invited all Parties to define national voluntary targets for achieving LDN, a process which is now being supported by the LDN target setting programme. Over 20 years after the Convention was first adopted, the confirmation by COP 12 of the UNCCD's key position in achieving LDN in the future means that the Convention now has a much clearer overarching goal and quantitative targets that are expected to guide action and make progress measurable.

Following the successful establishment of LDN in the SDG and UNCCD process, the challenge is now one of operationalization. LDN is expected to draw the attention of the international community on the significance of land, foster action on the ground and harmonize approaches to quantify and monitor land degradation. As such, it requires a scientifically-sound understanding of the LDN concept and what it seeks to achieve. The LDN definition, as endorsed by COP 12, and the development of a LDN conceptual framework by the SPI are significant advances in this regard. LDN is now understood as an aspirational goal that seeks to sustain the natural capital of the land, to support ecosystem functions and services, and thus meet the needs of the current and future human population. Accepting that some land degradation is inevitable, LDN seeks to restore and rehabilitate degraded land in order to counterbalance and reverse negative trends and achieve a no net loss outcome.

Operationalizing LDN also requires establishing a scientifically robust and consistent baseline against which to measure future changes in the rates of land

[55] UNCCD (2015a), decision 3/COP.12.

degradation. The indicator adopted for SDG target 15.3 and its associated sub-indicators provide a promising way forward in this regard. They constitute a practical approach for assessing the status and trends in land degradation, highlighting areas of potential concern and tracking progress towards LDN. Recognizing that land cover, land productivity and carbon stocks above and below ground are the absolute minimum requirement and on their own may be inadequate for a full assessment of land degradation, the indicator adopted to measure target 15.3 is considered the cornerstone of a broader monitoring framework that will include complementary indicators at the national level, as well as a blending of quantitative and qualitative information for a more rigorous analysis of the actual driving forces, state of the environment, and impacts of changes.

Earth observation, complemented by bottom-up ground surveys, may currently provide one of the viable, low-cost sources of data for reporting on a number of SDG indicators and targets, including SDG target 15.3. Capacity building will be essential for national statistical systems to meet the demands of the 2030 Agenda for Sustainable Development. Particularly important will be to look at the institutional arrangements and transform statistical systems to ensure they respond to the new monitoring requirements.

Finally, an enabling environment must be put in place to transform LDN from a global aspirational goal to a reality in national policies and in terms of environmental benefits. Key actors for creating an enabling environment for LDN implementation at the (sub-)national level are planning authorities, land user associations, civil society organizations, the private sector and development partners. The extent to which they take up the LDN concept and mainstream it in their activities will decide whether or not LDN makes a substantial difference on the ground.

LDN is still in its infancy. While considerable progress has been made to promote its political uptake and advance its understanding and monitoring, there are no final answers yet. LDN is a living concept that will continue evolving and adapting through experience.

References

Akhtar Schuster M, Stringer LC, Erlewein A, Metternicht G, Minelli S, Safriel U, Sommer S (2016) Unpacking the concept of land degradation neutrality and addressing its operation through the Rio conventions. Manuscript submitted for publication

Altvater S, Dooley E, Roberts E (2015) Legal Instruments to implement the objective "Land Degradation Neutral World" in International Law. Federal Environment Agency (Germany). Texte 19/2015. http://www.umweltbundesamt.de/publikationen/legal-instruments-to-implement-the-objective-land. Accessed 3 Mar 2016

Caspari T, van Lynden GWJ, Bai ZG (2015) Land Degradation Neutrality: an evaluation of methods. Report commissioned by German Federal Environment Agency (UBA), p 57. https://www.umweltbundesamt.de/publikationen/land-degradation-neutrality-an-evaluation-of. Accessed 3 Mar 2016

Chasek P, Safriel U, Shikongo S, Fuhrman VF (2015) Operationalizing zero net land degradation: the next stage in international efforts to combat desertification? J Arid Environ 112:5–13. doi:10.1016/j.jaridenv.2014.05.020

Intergovernmental Panel on Climate Change (2003) Glossary in good practice guidance for land use, land-use change and forestry. http://www.ipcc-nggip.iges.or.jp/public/gpglulucf/gpglulucf_files/Glossary_Acronyms_BasicInfo/Glossary.pdf. Accessed 3 Mar 2016

European Environment Agency (2015) National monitoring, reporting and evaluation of climate change adaptation in Europe. Technical report No 20/2015. http://www.eea.europa.eu/publications/national-monitoring-reporting-and-evaluation. Accessed 3 Mar 2016

Geist HJ, Lambin EF (2004) Dynamic causal patterns of desertification. BioScience 54 (9):817–829. doi:10.1641/0006-3568(2004)054[0817:DCPOD]2.0.CO

Grainger A (2015) Is land degradation neutrality feasible in dry areas? J Arid Environ 112:14–24. doi:10.1016/j.jaridenv.2014.05.014

Millennium Ecosystem Assessment (2005) Glossary in ecosystems and human well-being: current state and trends, Volume 1. http://www.millenniumassessment.org/documents/document.767.aspx.pdf. Accessed 3 Mar 2016

Orr (2011) Scientific review of the UNCCD provisionally accepted set of impact indicators to measure the implementation of strategic objectives 1, 2 and 3. White Paper Version 1. Report commissioned by the UNCCD, p 145. http://www.unccd.int/en/programmes/Science/Monitoring-Assessment/Documents/White%20paper_Scientific%20review%20set%20of%20indicators_Ver1.pdf. Accessed 3 Mar 2016

Ramsar Convention on Wetland (2012) An Integrated Framework and guidelines for avoiding, mitigating and compensating for wetland losses. Ramsar COP 11 Resolution XI.9. http://www.ramsar.org/sites/default/files/documents/library/cop11-res09-e.pdf. Accessed 3 Mar 2016

Reynolds JF, Stafford-Smith DM, Lambin EF, Turner BL, Mortimore M, Batterbury SPJ, Downing TE, Dowlatabadi H, Fernandez RJ, Hernick JE, Huber-Sannwald E, Jiang H, Leemans R, Lynam T, Maestre F, Ayarza M, Walker B (2007) Global desertification: building a science for dryland development. Science 316:847–851. doi:10.1126/science.1131634

Safriel UN (2007) The assessment of global trends in land degradation. In: Sivakumar MVK, Ndiaugui N (eds) Climate and land degradation. Springer, Berlin, pp 1–38

Sommer S, Zucca C, Grainger A, Cherlet M, Zougmore R, Sokona Y, Hill J, Della Peruta R, Roehrig J, Wang G (2011) Application of indicator systems for monitoring and assessment of desertification from national to global scales. Land Degrad Dev 22:184–197. doi:10.1002/ldr.1084

Tal A (2015) The implications of environmental trading mechanisms on a future Zero Net Land Degradation protocol. J Arid Environ 112:25–32. doi:10.1016/j.jaridenv.2014.05.013

UNCCD (2012) Zero Net Land Degradation: Sustainable Development Goal for Rio + 20. http://www.unccd.int/Lists/SiteDocumentLibrary/Publications/UNCCD_PolicyBrief_ZeroNetLandDegradation.pdf. Accessed 3 Mar 2016

UNCCD (2013a) Report of the conference of the parties on its eleventh session, held in Windhoek from 16 to 27 September 2013. Part two: Action taken by the Conference of the Parties at its eleventh session. Addendum. ICCD/COP11/23/Add.1. http://www.unccd.int/Lists/OfficialDocuments/cop11/23add1eng.pdf. Accessed 3 Mar 2016

UNCCD (2013b) Refinement of the set of impact indicators on strategic objectives 1, 2 and 3. Recommendations of the ad hoc advisory group of technical experts. ICCD/COP(11)/CST/2. http://www.unccd.int/Lists/OfficialDocuments/cop11/cst2eng.pdf. Accessed 3 Mar 2016

UNCCD (2015a) Report of the conference of the parties on its twelfth session, held in Ankara from 12 to 23 October 2015. Part two: Action taken by the Conference of the Parties at its twelfth session. Addendum. ICCD/COP12/20/Add.1. http://www.unccd.int/en/about-the-convention/official-documents/Pages/SymbolDetail.aspx?k=ICCD/COP%2812%29/20/Add.1&ctx=COP%2812 %29. Accessed 3 Mar 2016

UNCCD (2015b) Integration of the sustainable development goals and targets into the implementation of the United Nations Convention to Combat Desertification and the report of the

Intergovernmental Working Group on Land Degradation Neutrality. ICCD/COP12/4. http://www.unccd.int/Lists/OfficialDocuments/cop12/4eng.pdf. Accessed 3 Mar 2016

United Nations (2016a) Report of forty-seventh session of the United Nations Statistical Commission. http://unstats.un.org/unsd/statcom/47th-session/documents/Report-on-the-47th-session-of-the-statistical-commission-E.pdf. Accessed 28 Apr 2016

United Nations (2016b) Report of the Inter-Agency and Expert Group on Sustainable Development Goal Indicators. http://unstats.un.org/unsd/statcom/47th-session/documents/2016-2-IAEG-SDGs-E-Revised.pdf. Accessed 3 Mar 2016

United Nations General Assembly (UNGA) (1994) Elaboration of an international convention to combat desertification in countries experiencing serious drought and/or desertification, particularly in Africa. Final text of the Convention. A/AC.241/27. http://www.unccd.int/Lists/SiteDocumentLibrary/conventionText/conv-eng.pdf. Accessed 3 Mar 2016

United Nations General Assembly (UNGA) (2012) The future we want. Resolution adopted by the General Assembly on 27 July 2012. A/RES/66/288. https://documents-dds-ny.un.org/doc/UNDOC/GEN/N11/476/10/PDF/N1147610.pdf?OpenElement. Accessed 3 Mar 2016

United Nations General Assembly (UNGA) (2015) Transforming our world: the 2030 agenda for sustainable development. Resolution adopted by the General Assembly on 25 September 2015. A/RES/70/1. http://www.un.org/ga/search/view_doc.asp?symbol=A/RES/70/1&Lang=E. Accessed 3 Mar 2016

Verstraete MM, Hutchinson CF, Grainger A, Stafford Smith M, Scholes RJ, Reynolds JF, Barbosa P, Léon A, Mbow C (2011) Towards a global drylands observing system: observational requirements and institutional solutions. Land Degrad Dev 22:198–213. doi:10.1002/ldr.1046

Warren A (2002) Land degradation is contextual. Land Degrad Dev 13:449–459. doi:10.1002/ldr.532

Welton S, Biasutti M, Gerrard MB (2014) Legal and scientific integrity in advancing a "Land Degradation Neutral World". Sabin Center for Climate Change Law. Columbia Law School. https://web.law.columbia.edu/sites/default/files/microsites/climate-change/welton_-_land_degradation.pdf. Accessed 3 Mar 2016

Yengoh GT, Dent D, Olsson L, Tengberg AE, Tucker III CJ (2016) Use of the Normalized Difference Vegetation Index (NDVI) to assess land degradation at multiple scales. Current status, future trends, and practical considerations. Springer International Publishing

The Role of Soils in International Climate Change Policy

Charlotte Streck and Agustina Gay

1 Introduction

Soils play an important role in the provision of vital ecosystem services such as food production, biomass, carbon storage, storage and supply of water, biodiversity, and added cultural and social services.[1] In addition, due to their specific features soils both contribute to and are affected by climate change. On one hand, soils under use release large amounts of greenhouse gases (GHG), such as carbon dioxide (CO_2), methane (CH_4) or nitrous oxide (N_2O), into the atmosphere; on the other changing climate affects soils, in particular in agriculture, where a change in temperature or precipitation can affect crop production, and the increase in catastrophic weather events can wipe out harvests on large areas of land.

Despite their fundamental importance for the survival of humankind, soils receive little attention in international climate negotiations. This may not be surprising as the negotiations convened by the UN Framework Convention on Climate Change (UNFCCC or the Convention) are hardly featuring discussions on natural bodies, such as oceans, rivers, or soils—forests being a notable exception. Instead climate negotiations discuss targets and efforts to address the adverse effects of climate change in rather general terms focusing on the results of action rather than any particular action per se. While they receive little mentioning in international climate policy, the management and protection of soils are important strategies to address climate change. This is recognized, albeit indirectly, in the GHG accounting frameworks that support international climate policy, and the

[1]FAO (2011), p. 25.

C. Streck (✉) • A. Gay
Climate Focus, Berlin, Germany
e-mail: c.streck@climatefocus.com; agustinagay@gmail.com

H. Ginzky et al. (eds.), *International Yearbook of Soil Law and Policy 2016*,
International Yearbook of Soil Law and Policy,
DOI 10.1007/978-3-319-42508-5_10

discussions of land-based mitigation and adaptation, whether in the context of agriculture, forestry or the protection of livelihoods and ecosystems.

The objective of this article is to discuss how soils feature in international climate policy as well as opportunities that the UNFCCC, including the recently adopted Paris Agreement, offers for soil conservation. We are well aware that the limitation of our analysis to the climate change regime—that is the set of rules and procedures adopted under the UNFCCC—will result in an incomplete picture of relevant policies as it leaves out important international instruments, such as the United Nations Convention to Combat Desertification (UNCCD) or the 2015 Sustainable Development Goals (see the chapters of Ehlers and Castillo & Erlewein in this volume). Considering, however, that the climate change regime provides the overarching framework for national and international collaborative efforts to address climate change, the proposed limitations seem acceptable.

This chapter is organized as follows: After this brief introduction, Sect. 2 discusses the role of soils and their importance for climate change. Section 3 provides an introduction into the climate regime while Sect. 4 is dedicated to the role of soils in international agreements, with particular attention paid to the recently adopted Paris Agreement. Section 5 concludes the paper.

2 Soils and Climate Change

Soils are the second largest reservoir of carbon after oceans and when not managed properly, they can become an enormous emitter. On the other hand, soils are affected by climate change where changed rainfall patterns lead to soil structure degradation and decreased porosity as well as increased erosion and runoff, less availability of organic matter supply, water saturation and chemical depletion.[2] As consequence soils are depleted from their essential components and lose their vital functions.

2.1 *Soils and Climate Change Mitigation*

The most significant source of soil emissions are associated with agricultural practices, such as methane emissions from rice fields, from drained peatlands and manure on fields, as well as with nitrous oxide emissions from the application of fertilizer. Consequently, the largest potential to reduce emissions from soils relates to the management of croplands (including fertilizer management) and the restoration of degraded and cultivated organic soils (including peat soils) present the greatest potential for carbon dioxide emission reduction, while livestock and rice field management present the highest potential for methane reduction.[3] The most

[2]IPCC (2001).

[3]IPCC (2014), p. 29.

relevant direct and indirect mitigation strategies include the restoration of cultivated peaty soils and degraded lands, the enhancement of rice cultivation methods, better livestock and manure management practices, the development of nitrogen fertilizer application techniques reducing nitrous oxide emissions.[4] In the aggregate, the emissions reduction potential of the agricultural sector through the management of soils can be as high as 2 Gt CO_2e per year by 2030.[5] In most cases, these interventions would yield productivity gains and ought to be in the best interest of farmers and governments.

Additional emissions come from the loss of carbon stored in soils.[6] Among the most important human-induced reasons of soil carbon loss are deforestation, biomass burning, plowing, resource-based and subsistence agriculture, and practices that mine soil fertility and deplete the soil organic C [carbon] pool.[7] It has been calculated that we lose around 24 billion tonnes of fertile soil per year due to misuse[8] and conversion of land for agricultural purposes.[9] Soils can, however, be restored. Restoration of soils does not only improve the quality and productivity of soils but also rehabilitates their organic carbon pool. Due to fluctuations and the absence of data, the mitigation potential of carbon sequestration in agricultural soils and above-ground biomass shows large uncertainties, with a global carbon sequestration potential of between 700 and 1600 Mt CO_2e per year by 2030.[10] The mitigation, yield, and economic impacts of sequestration are not well understood for all practices, and there are complicating factors such as the impermanent nature of carbon stocks.[11] However, long-term management and preservation of soil carbon is critical for agricultural productivity because it increases soil fertility, reduces erosion, and increases moisture retention.

2.2 Soils and Climate Change Adaptation

The loss of carbon from soils often leads to soil degradation, including accelerated erosion and mineralization, and land use change.[12] One of the most urging risks for soils associated with climate change is desertification, defined as 'land degradation in arid, semiarid, and dry sub-humid lands' (Art. 1 UNCCD). Degraded soils lead to

[4]IPCC (2014), p. 29.

[5]Dickie et al. (2014), p. 14.

[6]Lal (2004a), p. 2.

[7]Lal (2004a), p. 1.

[8]Bartz (2015), p. 6.

[9]Lal (2004b), p. 3.

[10]Dickie et al. (2014), p. 4.

[11]As soil can experience unexpected events that release or remove the carbon, for example floods, fires, droughts.

[12]FAO (2008), p. 2.

reduced aptness for agriculture that leads to an overall loss of fertile land and an enhanced risk of lost or insufficient harvests. Degraded soils are also more vulnerable to extreme weather conditions like heavy rainfall, droughts or extreme wind. The rising temperatures accelerate the loss of organic matter and more intense rainfall increases the exposure of the soil to erosion and degradation.

Appropriate soil management is therefore essential for climate change adaptation. Adaptation strategies for the agricultural sector include the adjustment of planting dates and crop variety, crop relocation and improvement of land management.[13] Soil carbon sequestration can increase the ability of soils to hold moisture, withstand erosion and enrich the biodiversity of the ecosystem, thus helping cropping systems to better adapt to the increase in droughts and floods.[14] The peculiar ability of soils to seize carbon from the atmosphere and store it—the carbon sequestration process—means that soil restoration combines adaptation with mitigation benefits.

3 The Climate Change Regime

On 12 December 2015, 196 Parties to the UN Framework Convention on Climate Change (UNFCCC) adopted the Paris Agreement, a new legally-binding framework for an internationally coordinated effort to tackle climate change. The Agreement represents the culmination of almost a decade of international climate change negotiations under the auspices of the UNFCCC, and was reached under intense international pressure to avoid a repeat failure of the Copenhagen conference in 2009. The Paris Agreement is the latest and most important addition to the global climate change regime that can be traced back to the mid-1980s when climate change started to become a political issue and a growing global concern that led to the adopting of the UNFCCC in 1992.[15] Before we describe those elements of the regime that are relevant to soils, we propose a quick recapitulation of the path to and the current status of international climate policy.

3.1 Quick History of International Climate Negotiations

The development of the regime can be described in four phases that led from the UNFCCC to the Paris Agreement.[16] The first phase ranges from the adoption of the UNFCCC in 1992 up to the adoption of the Berlin mandate in 1995. During these

[13]IPCC (2014), p. 27.

[14]FAO (2015), p. 4.

[15]Bodansky (2001), pp. 23, 26.

[16]Brenton (2013), pp. 541–546.

early years of climate policy, developing countries' role in the negotiations was limited and they were excluded from any regulation and constraints. At the time the U.S. was already reluctant to subscribe to a binary agreement dividing Parties into lists fixed in Annexes, but eventually went along and supported a Convention that introduced separate provisions for developed and developing countries, but failed to formulate targets.

The second phase describes the negotiations that led to the adoption of the Kyoto Protocol, and the creation of binding emission reduction targets for developed countries. The U.S. position had changed compared to Rio 1992 and a negotiation team led by Vice-President Al Gore was open to discuss the adoption of mitigation commitments.[17] In the meantime, China and other emerging economies which had become responsible for a sizable share in global annual emissions (in absolute, if not per-capita terms) were unwilling to commit to pre-defined action and ultimately joined the other developing countries to remain, as a block, outside of the mitigation framework of the Protocol. This created a serious weakness of the Kyoto Protocol in terms of country coverage and effectiveness, as well as political acceptability of the Protocol over time.

The implementation of the Kyoto Protocol had hardly begun in 2005 when new efforts and partnerships emerged with the goal to bring all major emitters to the table, marking a third phase of negotiations but neither the cooperation of the Parties in the context of the UNFCCC or the Kyoto Protocol nor the various discussions outside the U.N. and its climate regime helped to make the 15th session of the Conference of the Parties (COP-15) that was scheduled to produce a new and more comprehensive climate treaty a success. In fact, COP-15 set to adopt a post-2012 climate treaty was by all measure a spectacular failure as it concluded with nothing but a non-binding political agreement involving developed and developing countries.

And still, despite the shortcomings of the Copenhagen conference, the formulation of the 'Copenhagen Accord'[18] laid the seed for the Paris Agreement. This fourth phase was characterized by a more horizontal, inclusive and balanced approach. The broadening of the consulted Parties over the past 6 years made the negotiations significantly more complex, to the extent that they stood at the brink of failure more than once. It is the merit of the Presidencies of the COP since Copenhagen (Mexico, South Africa, Qatar, Poland and, in particular, France) combined with an increasingly committed U.S. leadership that prepared the road to Paris with bilateral agreements and consultations, and an extraordinary effort of the UNFCCC Secretary General that contributed to the adoption of the Paris Agreement. The achievement of the Paris Agreement is that it provides a common framework within which individual countries (or alliances of countries) are invited to define action plans or 'nationally determined contributions' (NDCs) taking into

[17]Brenton (2013), p. 544.

[18]Copenhagen Accord, FCCC/CP/2009/L.7, 18 December 2009.

account the overall goal of the Convention and the Agreement as well as their own capacities.

3.2 At a Glance: The UNFCCC

As its name indicates, the UNFCCC is a framework treaty, designed to lay down the foundational architecture of the international climate regime; as such, when formulating Party obligations, the Convention, rather than focusing on specific mandatory obligations and targets, focuses on general measures. Adopted in 1992, the UNFCCC's ultimate goal—'the stabilization of greenhouse gas concentrations in the atmosphere at a level that would prevent dangerous anthropogenic interference with the climate system' (Art. 2 UNFCCC) remains valid till today. This goal has been the reference point of international climate negotiations and referenced over and over again, in particular in the many times when the breakdown of international climate negotiations threatened was imminent.

Without digressing too much from the main topic of this article, it is worth to highlight three features of the UNFCCC are relevant for the climate change regime until today:

(i) The leadership in mitigation and finance by developed countries. The primary responsibility for climate change mitigation sits with developed countries in accordance with the 'principle of common but differentiated responsibilities' (CBDR) (Art. 3.1 UNFCCC). Without committing to targets, developed countries are bound to adopt policies and measures that reduce GHG emissions and demonstrate their leadership in bending the curve of global emissions (Art. 4.2 UNFCCC).

(ii) Financial transfers to support developing country action. The demand for 'new and additional' financial resources to help developing countries to implement the Convention led to the designation of a financial mechanism for 'the provision of financial resources on a grant or concessional basis' (Art. 11.1 UNFCCC).

(iii) The establishment of a framework for measuring and reporting of emissions and emission reductions. All countries should report on the steps they are taking to implement the Convention (Articles 4.1 and 12 UNFCCC) and developed countries have to prepare national GHG inventories.

The UNFCCC's governing body, the Conference of the Parties, meets annually to review the status of the implementation and take decisions to support the further elaboration and development of the international climate change regime. It is in this context that both the 1997 Kyoto Protocol and the 2015 Paris Agreement were negotiated.

3.3 At a Glance: The Kyoto Protocol

The Kyoto Protocol formulates the binding emission reductions and limitations targets that the UNFCCC failed to include. These targets are however limited to developed countries which it was consonant with the CBDR principle and, in particular, with the notion of developed-country leadership on climate action. The Kyoto Protocol allows developed countries some flexibility in meeting the established targets thereby reducing the costs of compliance. The Protocol formulated three 'flexible mechanisms', including emissions trading and Joint Implementation (JI) among industrialized countries, as well as a Clean Development Mechanism (CDM) for emission reduction projects in developing countries (Art. 6, 12, and 17 KP). These mechanisms are linked with new sources of finance, e.g. from the auctioning of a mandatory 'share of proceeds' related to the transaction of CDM credits. The Kyoto Protocol also tightens the measurement, reporting and verification (MRV) for developed countries and establishes a comprehensive accounting and tracking framework for emission allowances.

3.4 At a Glance: The Paris Agreement

Different from the Kyoto Protocol, which will phase out when the new treaty is to enter into force in 2020, the Paris Agreement is not based on binding, top-down, GHG emission reduction commitments for developed countries. Rather, it relies on countries making non-binding dynamic, bottom up, and ever-more ambitious mitigation pledges to hold the increase in the global average temperature to well below 2 °C. Through a global process of engagement, follow-up (Art. 4 PA), regular stock-take exercises (Art. 14 PA) and cooperative action (Art. 6 PA)—including on the provision of funding for developing countries from a floor of 100 billion USD per year—the Paris Agreement offers the best chance in decades to overcome the many divisions that had marked the Kyoto area: between developed and developing countries, between industrialized nations inside the Protocol and those outside, between those supportive of market mechanisms and those that vehemently opposed them.

At the same time, the Paris Agreement remains within the framework of the UNFCCC, and there is institutional and conceptual continuity, including with respect to developed country leadership and the principle of CBDR expressed through the adoption of economy-wide absolute emission reduction targets and the provision of finance, technology and capacity building. The core instruments of the Paris Agreement are NDCs, which have to be formulated by all Parties and which are to become more ambitious over time. In the run-up to Paris, Parties had already communicated their 'intended NDCs' to the UNFCCC Secretariat which will have to be reconfirmed before the commitments of the Paris Agreement replace the Kyoto Protocol targets in 2020, and be strengthened every 5 years.

The Agreement is based on the recognition that developing countries need support in the implementation of their NDCs and that such support will allow for more ambitious mitigation action (Art. 4.5). More specifically, the Paris Agreement mandates that developed countries provide financial resources (Art. 9), notes the importance of technology transfer (Art. 10), and calls for Parties to cooperate and enhance capacities (Art. 11).

Finally, the Paris Agreement lays the groundwork for installing a robust communication and accounting framework. Parties must provide their NDCs with 'necessary' levels of 'clarity, transparency and understanding' (Art. 14.8). Furthermore, all Parties shall account for their actions and results achieved in implementing the NDCs (Art. 14.13). In accounting for emissions and removals 'corresponding to' their NDCs, Parties shall 'promote environmental integrity, transparency, accuracy, completeness, comparability and consistency, and ensure the avoidance of double-counting', in accordance with guidance to be adopted (ibid).

The hope is that with increased transparency and an ambitious overall target, countries will step forward with ambitious national plans. The risk however, is that individual country contributions fall short of the overall goal and that the Paris Agreement remains a shell without sufficient action and support, unable to address the collective action problem of climate change.

4 Soils and the Climate Change Regime

Neither the UNFCCC, nor the Kyoto Protocol or the Paris Agreement deal with soils explicitly, instead they set a framework for national climate action, which mandates ratifying countries—the Parties to the respective legal instrument—to adopt policies and measures that reduce GHG emissions and increase the resilience of our livelihoods, economy, and ecosystems. As we have seen in Sect. 2 above, soils are relevant for climate action in mainly two aspects:

(i) the restoration and protection of soils to ensure the continuous delivery of ecosystem functions in a changing climate, and the sequestration of carbon;
(ii) the use of land for crops and livestock results in high levels of emissions, in particular in the form of methane and nitrous oxide.

Both aspects are closely related to the economic sectors of agriculture and forestry and most mitigation and adaptation activities will pertain to these two sectors. We will therefore briefly summarize the way how the UNFCCC, the Kyoto Protocol, and the Paris Agreement address agriculture and forestry as well as land-based emission sources more generally. We will organize our arguments along three topics that we mentioned in the previous section as cornerstone themes of the UNFCCC and the climate change regime:

(i) Climate action, covering both activities that reduce GHG emissions (mitigation) and those that increase the adaptive capacities of our human and natural environments (adaptation);
(ii) Climate finance, that is transfers from developed to developing countries to support them to take the climate action necessary to contribute to mitigation and enable adaptation;
(iii) MRV and accounting for emissions and emission reductions, that provide the basis for measuring the contribution of a Party—and activity to the problem of climate change as well as the impact of mitigation measures at the Party—or activity—level.

4.1 Climate Action: Mitigation and Adaptation

The UNFCCC recognizes the importance of agriculture (Art. 2 and 4.1) and forestry (Art. 4.1 and 4.8), calls for mitigation in both sectors, and mandates the preparation of adaptation strategies that consider the needs of the agriculture, forestry, and other land uses (Art. 4.1 and 4.2). Under Article 4 of the Convention, Parties shall

- 'promote sustainable management, and promote and cooperate in the conservation and enhancement, as appropriate, of sinks and reservoirs of all greenhouse gases [. . .], including biomass, forests and oceans as well as other terrestrial, coastal and marine ecosystems' (Art. 4 (1)(d) UNFCCC); and
- 'cooperate in preparing for adaptation to the impact of climate change; develop and elaborate appropriate and integrated plans for coastal zone management, water resources and agriculture, and for the protection and rehabilitation of areas, particularly in Africa, affected by drought and desertification, as well as floods' (Art. 4 (1)(e) UNFCCC), and developed countries have the specific obligation to
- 'adopt national policies and take corresponding measures on the mitigation of climate change by protecting and enhancing its greenhouse gas sinks and reservoirs' (Art. 4 (2)(a) UNFCCC), in accordance with the CBDR principle.

However despite the explicit mentioning of the need to address sector-specific emissions, the climate regime fails to formulate a coherent vision or set of incentives for mitigation in the land sectors. While the Convention is demonstrably concerned about land use emissions, the Kyoto Protocol did not single out any sectoral action. The Protocol lists sustainable promotion of sustainable forest management practices, afforestation and reforestation and sustainable forms of agriculture, as a way for developed countries to achieve their emissions reductions and limitations.[19]

[19] KP (1997), Art. 2 (1)(a)(ii) and (iii).

Such call for sustainable land-use policies was uncontroversial in the negotiations that led to the Kyoto Protocol. Far more contentious was the treatment of land use emissions in the context of these commitments. Negotiators recognized that a decision on whether to include land use emissions into baseline emissions of developed countries and how to account for them in meeting commitments could dramatically alter the scope and stringency of Parties' targets. Among the contentious rule choices was whether to measure net human-induced land use emissions in absolute terms or relative to a base year. This issue is substantially more complex in land use, land-use change and forestry (LULUCF, as it is referred to in the context of the Kyoto Protocol) that in other sectors due to the frequently long time-scales between land use emissions or removals and the activities from which they arise, as well as the sometimes complex relationship between human-induced and non-human induced emissions or removals, including from soils. The Kyoto Protocol established that emissions for grassland management, revegetation, and cropland management (including emissions from soil management) are subject to a 'net-net' accounting approach that compares net emissions to those in the reference year.[20] Emissions for deforestation, afforestation, reforestation, and forest management, meanwhile, are subject to a 'gross-net' approach that measures changes in carbon stocks during the commitment period without referring to a reference year.[21] As with other emissions under the Protocol, LULUCF emissions reductions can be used to meet the emission limitation targets of developed countries. Article 3.7 of the Protocol, however, constrains which developed country Parties can use LULUCUF emissions reductions to this end, noting that only: 'Developed country Parties for whom land-use change and forestry constituted a net source of greenhouse gas emissions in 1990 shall include in their 1990 emissions base year or period the aggregate anthropogenic carbon dioxide equivalent emissions by sources minus removals by sinks in 1990 from land-use change for the purposes of calculating their assigned amount'.[22] As a result of this constraint, for the large majority of developed countries,[23] LULUCF emissions are not included in their baseline emissions. Contrastingly, emissions from agricultural soils (CH_4 and NO_2) are accounted for as baseline emissions and account against the Protocol's targets. With its focus on developed countries, the Kyoto Protocol does not encourage agricultural emissions by developing countries, neither does it formulate priority actions for adaptation.

The accounting framework for LULUCF of the Kyoto Protocol was constructed in such a way that land use emissions were kept separate from other sources of GHG emissions. The Kyoto Protocol with its focus on a small group of countries, failed

[20]KP (1997), Art. 3.3.

[21]KP (1997), Art. 3.4.

[22]KP (1997), Art. 3.7.

[23]The main beneficiary of this clause was Australia, as it allowed to include the net emissions from LULUCF in 1990 to include deforestation emissions in the base year emissions, thereby inflating Australia's base year emissions, effectively converting Australia's 108 % target into the equivalent of a 142 % increase on 1990 levels using the standard accounting rules. Source: Macintosh (n. d.), p. 2.

also to set incentives for large emission sources in developing countries, most notably emissions from the conversion of land from forests to agricultural land. Finally, it did not do much to spur investments into the restoration and protection of soils that would have increased their climate resilience while contributing to climate mitigation.

Over the last decade, climate negotiators went back to the more holistic approach promoted to by the Convention when it comes to considering the land sectors. In 2009 the Bali Action Plan introduced sector-specific approaches—including in relation to agriculture—as well as a framework to encourage emissions from deforestation within the post-2012 negotiating framework.[24] From 2010 to 2012, agriculture was negotiated under the agenda item 'cooperative sectoral approaches and sector-specific actions' in the context of the Ad-Hoc Working Group on Long-Term Cooperative Action. For the first time agriculture was discussed as a sector that required special attention. Additionally, a working group on agriculture was created in June 2013, raising the profile of mitigation in agriculture.

Many commentators supported the development of a work program for agriculture[25] which would identify data gaps, and prioritize adaptation and mitigation action. While the Convention calls for action on both agricultural mitigation, developing countries have been wary of the fact that developed countries may use cooperative sectoral approaches and sector-specific actions in the agriculture sector to gain a competitive advantage in the sector and establish trade barriers for developing country agricultural commodities. In 2011, the COP stresses the importance sustainable economic growth and development of all Parties, and that measures that address climate change should not serve as 'a means of arbitrary or unjustifiable discrimination or a disguised restriction on international trade or used to gain competitiveness on agriculture trade'.[26] Parties were also concerned that mitigation action in agriculture could threaten food security and smallholder agricultural systems, which result in well justified 'survival emissions' by small farmers. The majority of developing countries therefore supports the view that discussions on agriculture in the context of international climate negotiations should not 'lead to new commitments by developing countries on mitigation, but rather step up action on how to safeguard our agriculture sector against the worst effects of climate change'.[27]

Countries that confirm the need for mitigation in the agricultural sector, stress the critical link between adaptation and mitigation actions, including the fact that many mitigation actions (e.g. improved protection and enhancement of soil organic matter, restoring degraded land, improving efficiency and productivity of agricultural systems in a sustainable manner)[28] can at the same time build resilience to

[24]Decision 1/CP.17, para 1 (b)(iv).

[25]FAO (2010), p. 3.

[26]UNFCCC 1/CP.16. para 90.

[27]Gambia (2012), p. 15.

[28]EU (2012), pp. 11, 12.

climate change impacts. The term 'climate-smart agriculture' has become the shorthand for activities that yield adaptation and mitigation benefits while increasing productivity (see Box 1).

Box 1: Climate-Smart Agriculture as an Integrative Approach

Climate-smart agriculture (CSA) describes an approach for developing agricultural strategies to secure sustainable food security under climate change. CSA reduces/removes GHGs (mitigation), and enhances achievement of national food security, climate resilience and development goals. CSA activities include better management of farms, crops, livestock for improved productivity while increasing resilience; conservation ecosystem and landscape management and services for farmers and land managers to help them implement the changes.

Source: FAO (n.d.) Climate-Smart Agriculture. http://www.fao.org/climate-smart-agriculture/en/. Accessed March 29 2016

In contrast to agriculture, developing countries were fundamental in developing a framework for 'reduced emissions from deforestation and forest degradation, the role of conservation, enhancement of forest carbon stocks, and sustainable management of forests' (REDD+) in the context of international climate negotiations. Since REDD+ got fully integrated into the global climate agenda at COP-13 in Bali, it has come to be regarded as potentially one of the most effective and efficient mitigation strategies by developing countries. REDD+ is supported by a broad agreement between developed and developing countries with the latter developing mitigation plans and robust measurement systems and the former providing financial incentives.

The Paris Agreement recognizes the progress that has been made on REDD+ in its Article 5 which is dedicated to land use, with a focus on forests. The main purpose of that article is to anchor existing forest-related provisions, frameworks and decisions, including on REDD+, in the new Agreement. Paragraph 1 of the Article refers to the UNFCCC and the mandate to sustainably manage, conserve and enhance biological carbon reservoirs, a reference that includes the management and conservation of soils in developed and developing countries.

Parties to the Paris Agreement will have to formulate mitigation and adaptation action in all relevant sectors. For many developing countries, the land sector is one of the largest contributors to national GHG, and many have already pledged to reduce emissions from agriculture. Around 80 % of the intended NDCs include references to agriculture and carbon sequestration. Some also include references to soil enhancement and to the necessity of better soil management practices (see Table 1 for examples).

NDCs are not limited to formulate mitigation goals. Most countries have used the opportunity and include adaptation priorities in their intended NDCs. The Paris

Table 1 Examples for references to soil management in intended NDCs

Country	Approach	Intended nationally determined contribution
Nigeria	Crops and livestock	Adopt improved agricultural systems for both crops and livestock (for example, diversify livestock and improve range management; increase access to drought resistant crops and livestock feeds; adopt better soil management practices; and provide early warning/meteorological forecasts and related information).
Botswana	CSA as part of Adaptation Plan	Botswana is developing a National Adaptation Plan (NAP) and Action Plan which will highlight all the priority areas including Climate Smart Agriculture which include techniques such as low to zero tillage, multi-cropping to increase mulching which reduce evapotranspiration and soil erosion.
Zambia	Focus on food security	Guaranteed food security through diversification and promotion of climate-smart agricultural practices for crop, livestock and fisheries production including conservation of germplasm for land races and their wild relatives.
Uruguay	Protection of croplands	Development of soil use and management plans to reduce erosion and preservation of organic matter in croplands.

Agreement also creates a global goal on adaptation that had been absent from previous UNFCCC agreements, aiming to enhance 'adaptive capacity, strengthening resilience and reducing vulnerability to climate change'.[29] The Agreement determines that countries should put more emphasis on adaptation planning, and based on this planning Parties should strengthen their cooperation, including through the transfer of funds.[30] The adequacy of action and support will be reviewed as part of the global stocktake.

Summarizing the main points relating to mitigation and adaptation action as mandated by the climate regime and relevant to the priority action for soils (cropland and pasture management, soil protection and restoration), the following should be noted:

(i) The Convention mentions both agriculture and forestry as important sectors for adaptation and mitigation. The Kyoto Protocol in contrast is less concerned about sectoral action and more worried that erroneous accounting for land use emission reductions and removals may lead to a watering-down of developed country climate goals. The Protocol, consequently, does not count land use emissions as baseline emissions for the majority of developed countries. Non-carbon gases from agriculture are however fully included in the Protocol's mitigation and accounting system, which is however limited to developed countries. The Paris Agreement does neither judge nor prioritize economic sectors for mitigation and adaptation action. With its bottom-up approach it leaves it to the Parties to decide on which sectors to target for climate change

[29]PA (2015) Art. 7.1.

[30]PA (2015) Art. 7.

action. Considering the importance that land use emissions have in many developing countries, it is expected that most developing countries (and developed countries) will include land use action in their NDCs.

(ii) Developing countries have expressed concerns with respect to a focus on agricultural mitigation. They have made it unambiguously clear that adaptation takes priority over mitigation. Many concrete activities, such as pasture management, agroforestry, and sustainable intensification combine adaptation and mitigation benefits while increasing yields. Such 'triple-win' activities are considered as 'climate-smart' agricultural practices.

4.2 Climate Finance

In many countries, climate change has already impacted the local food production and the priority of the discussion of agriculture for these countries is on receiving support for the transition to a climate resilient agriculture. Investments in improved agricultural practices will be essential to address the intertwined challenges of climate change mitigation, adaptation and food security. Appropriate incentives, governance, institutions, and funding mechanisms will be needed for an integrated approach that delivers on the multiple goals of agriculture—and as important element healthy soils—which include climate mitigation and adaptation, safeguarding livelihoods, generation of food, fuel and fiber. A mix of instruments and governance arrangements that include positive incentives, safeguards, and regulations will be needed. Enabling means such as finance, technology, capacity building and measurement of results are all important to achieving the various objectives expected from agriculture.

Early in the negotiations of the UNFCCC, developing countries left no doubt that they require international assistance to support adaptation (a demand that has been reflected in Articles 4.4, 4.8 and 4.9 UNFCCC) and that mitigation engagement was dependent on support. Article 4(7) UNFCCC makes developing country action conditional of the effective implementation of commitments under the UNFCCC and the transfer of resources and technologies.[31] This clause is a response to possible shortfalls of funding. The Convention also includes an explicit commitment to assist countries particularly vulnerable to the effects of climate change and to promote technology transfer. Funding for developing countries under the Convention is channeled through the Global Environment Facility and the Green Climate Fund.

[31]"The extent to which developing country Parties will effectively implement their commitments under the Convention will depend on the effective implementation by developed Country Parties of their commitment under the Convention relating to financial resources and transfer of technology." The UN Convention on Biological Diversity contains a similar provision in Article 20 (4). 31. I.L.M. 818.

Complementing the notion of fund-based financial transfers to developing countries, the Kyoto Protocol introduced the notion of market-based climate finance to the climate regime. With the goal to add institutional flexibility and mobilize private sector investment, the Protocol largely relied on market forces for organizing financial transfers to developing countries it created three 'flexible mechanisms' which might be used to supplement domestic mitigation action. Two of these instruments are available only to countries with qualified targets: Joint Implementation set forth in Article 6 and Emission Trading set forth in Article 17 of the Kyoto Protocol. In addition, the Kyoto mechanisms also include in Article 12 a Clean Development Mechanism (CDM), which aims to enhance co-operation among industrialized and developing countries to achieve sustainable development and reduce emissions. The project-based mechanism CDM allows developing countries to earn Certified Emission Reduction credits through emission reduction projects. These credits can be later traded, and eventually used by developed countries to meet a part of their reduction targets. The CDM recognizes only afforestation and reforestation as mitigating land use activities, but overlooks the carbon sequestration potential from sustainable agricultural land management activities, agroforestry systems and avoided deforestation through REDD+. Despite its limited scope, the CDM has funded a number of fertilizer production and manure management projects, as well as afforestation projects, some of which explicitly stated the objective of restoring degraded land (see Box 2). The accounting for soil carbon is however still excluded under the CDM,[32] although there are efforts to expand the CDM to allow 'agriculture CDM' projects to get credits for soil carbon sequestration.[33]

Box 2: Soil Conservation Under the Clean Development Mechanism (CDM)

The CDM accredited Soil Conservation Project in Moldova aims to: 'to achieve multiple objectives of restoring productivity of degraded lands, enhancing forest product supplies to local communities and promoting actual net GHG removals by sinks.' The project area covers degraded lands in the northern, central and southern regions of the country. The financing of the project includes the sale of Certified Emission Reductions from afforestation/reforestation activities under the CDM. The amount of reductions projected is: 179,242 metric tonnes CO_2e per annum. Other similar projects include a large number of manure management and avoidance projects around the globe, and afforestation project on degraded extensive grazing land.

(continued)

---

[32]FAO (2013), p. 2.

[33]UNFCCC (2015).

Source: CDM-Executive Board (2008) PROJECT DESIGN DOCUMENT FORM FOR AFFORESTATION AND REFORESTATION PROJECT ACTIVITIES (CDM-AR-PDD)—Version 04.

Climate finance was also a major and most controversial issue in the negotiations up to and in Paris. Financial and non-financial support to developing countries remains an essential element of the international climate architecture, and the Paris Agreement is based on the recognition that developing countries need support in the implementation of their NDCs and that such support will allow for more ambitious mitigation action (Art. 4.5). More specifically, the Agreement mandates that developed countries provide financial resources (Art. 9), notes the importance of technology transfer (Art. 10), and calls for Parties to cooperate and enhance capacities (Art. 11). The language on finance neither provides concrete figures for climate finance nor a timetable for disbursement, though it does note the 'significant role of public funds' in climate finance (Art. 9.3). Developed countries are asked 'to take the lead in mobilizing climate finance' (Art. 9.3), but all Parties are 'encouraged to provide or continue to provide such support voluntarily' (Art. 9.2).

The lack of a concrete, time-bound commitment on climate finance in the Paris Agreement was a major concession by developing countries for whom greater levels of climate finance was a key demand. To partly rectify this, the COP clarified that the 'existing mobilization goal' (i.e. USD 100 billion per year from 2020) formulated in Cancún (COP-16) will continue through 2025, and that from 2025, 'Parties to the Paris Agreement shall set a new collective quantified goal from a floor of USD 100 billion per year'.[34] These pledges build on the financial commitments include in the 2009 Copenhagen Accord, where developed countries had agreed to provide USD 30 billion 'fast start finance' during 2010–2012, with balanced allocation between adaptation and mitigation, prioritizing the most vulnerable and least developed countries.

While generally there is a lack of transparency with respect to the allocation of climate finance, it is clear that finance is biased towards sectors or emission sources where quantification and monitoring of the mitigation benefits is relatively easy and where the economic and political risks are lower. Agriculture has received only a small share in mitigation finance, with a higher percentage being dedicated to adaptation and food security with concrete however figures lacking.

There are no disaggregated figures about climate finance dedicated to agriculture, forestry or other land use. It is clear however that the majority of fast start finance to date has been destined for mitigation in industry and energy sectors. Of the USD 35 billion fast-start finance disbursed by mid 2014, agriculture (adaptation and mitigation) accounted for USD 0.75 billion, equivalent to 2.1 %.[35] This is

[34]Para. 54 Decision 1/CP-21).

[35]Hoogzaad et al. (2014), p. 2.

modest given the contribution of agriculture to GHG emissions, estimated at around 14–24 %. To put these figures into perspective, official development assistance (ODA) for agriculture was estimated at USD 6.4 billion in 2012, and the level of total accumulated investment by farmers worldwide, as measured by the value of agricultural capital stock, at USD 5 trillion in 2007.[36]

REDD+ has received both more attention and finance. Between 2008 and 2015, almost USD 2.9 billion has been pledged to multilateral funds that support the reduction of deforestation emissions.[37] Additional finance has been made available through bilateral agencies and funds. Most of these resources have however been dedicated to governance reform, data collection and other efforts that support the ability of countries to reduce land use emissions. We assume that very little of the climate funds pledged via REDD+ have—so far—benefits soils.

4.3 MRV and Accounting

The measurement of Parties' emissions and emission reductions has been essential to inform Parties and stakeholders of the effectiveness and comparability of mitigation actions. MRV is a precondition for the planning of mitigation and adaptation measures as well as for the relative and absolute success of these measures. The UNFCCC and the Kyoto Protocol have set up a system of national communications and national inventory reports which are to be compiled by Parties and published by the UNFCCC Secretariat. In Bali COP-13 took these elements and with a new emphasis on transparency set the development of a comprehensive transparency framework in action which is now firmly anchored in the Paris Agreement.

The UNFCCC establishes a dedicated reporting framework for land use, forestry and agriculture. Article 12 UNFCCC requires both developed and developing country Parties to submit national GHG inventories that include sources and sinks. The requirements for each grouping are substantially different in terms of frequency and detail, while developing country Parties are additionally entitled to receive technical and financial support to assist in inventory preparation. Developed country Parties are required to submit a detailed national inventory each year,[38] which is subject to an in-depth review by expert panels.[39] Developed country Parties, by contrast, have until recently been required only to include inventory

[36]FAO (2012), p. 11.

[37]ODI (2015), p. 1.

[38]Decision 11/CP.4, para 2 (b); Updated UNFCCC reporting guidelines on annual inventories following incorporation of the provisions of decision 14/CP.11, UN Doc FCCC/SBSTA/2006/9; Decision 1/CP.16, para 40 (a).

[39]Decisions 2/CP.1 and 6/CP.3.

reports as part of national communications and in accordance with less stringent guidelines.[40]

Reporting of land use emissions under the UNFCCC follows a land-based approach, that is, the reporting of all land-use change and forestry emissions and removals.[41] This contrasts with the activity-based approach employed under the Kyoto Protocol. Land-based accounting reports the total carbon stock change in applicable carbon pools on particular land units (e.g. a Party's territory). Activity-based accounting accounts for the carbon stock change attributable to designated land-use activities.

Land-use reporting follows six main land-use categories: Forest land, cropland, grassland, wetlands, settlements, other land, and five different carbon pools: above-ground biomass, below ground biomass, dead wood, litter, soil organic carbon.[42] Land use, land-use change and forestry (mainly to CO_2 emissions and removal from LULUCF activities) and agriculture (mainly CH_4 and NO_2 emissions) emissions are reported separately in national inventory reports.[43] Parties can apply different estimation methods ("tiers") depending on available data and capabilities. Higher tiers should be applied to key categories with higher climate impact.

Under the Kyoto Protocol, the measurement and reporting of emissions became more than merely informative as it led to the conversion of measured data into emission reduction credits that could be used to meet targets adopted under the Kyoto Protocol. As we have seen above, during the negotiations and implementation of the Protocol accounting rules turned into a deeply political issue as they had a direct impact on the ability of Parties to meet their emission reduction and limitation targets.

In the context of their Kyoto targets, Parties must account for emissions and removals from deforestation, afforestation and reforestation, while accounting for forest management, cropland management, grazing land management and revegetation is optional. In order to ensure any subsequent changes in emissions or removals are accounted for, any optional activities included must also be included in subsequent commitment periods. For the second commitment period of the Kyoto Protocol (2012–2020), Parties have agreed to make accounting for forest management and harvested wood products (not previously included) mandatory, while wetland drainage and rewetting has been added as an additional optional category. For agriculture this means that emissions resulting from land management (crop or grazing land) are optional in their accounting while emissions resulting from industrial processes supporting agriculture (N_2O from fertilizer production and use) and livestock (CH_4) are not considered under the land use accounting rules and fall into the mandatory emissions accounting of industrial emissions. The CDM, as noted above, allows a limited number of options for

[40]Decision 10/CP.2 and Decision 17/CP.8.

[41]Prag et al. (2013), p. 19.

[42]IPCC (2003), p. 4.

[43]UNFCCC (2012).

agricultural projects. Difficulties in the measurement of emissions in agricultural settings and high uncertainties are one of the reason for the exclusion of soil carbon projects from the CDM. There are fewer existing local/regional studies about GHG emissions levels in the agricultural field which prevents further simplification and standardization of emission factors.[44]

The absence of binding emission targets in the Paris Agreement makes transparency and accountability even more essential for the Parties' accountability to assess individual and cumulative progress in meeting the stated climate goals. Consequently, transparency and accountability on the steps countries are taking toward voluntary goals included in NDCs are essential elements of the Agreement. All Parties have to account for their NDCs and in doing so, Parties should take into account existing methods and guidance under the Convention.[45] They also have to report their emissions, since only if there is transparent tracking of progress will it be possible to adjust and sufficiently strengthen the ambition of NDCs. The Agreement therefore foresees a process that evaluates the progress of individual Parties in meeting their NDCs, and another mechanism that looks at the overall accumulated progress in avoiding dangerous climate change. Article 13 provides an 'enhanced transparency framework for action and support' that aims at providing a clear understanding of mitigation action and available climate finance. Parties will have to collect and make available information necessary to track progress made in implementing and achieving its NDC and keep track of their emissions in national inventory reports.

When it comes to achieving high levels of transparency, robust and harmonized estimation of emissions and removals in agriculture and forestry have to be established, something that requires investment in monitoring and reporting capacity. Even among developed countries that are Parties to the Kyoto Protocol and report annual emissions to the UNFCCC significant data exists, especially on agricultural soils.

This is particularly true for carbon storied in soils. Of the many debates in the agricultural-climate world, perhaps none has endured as many fluctuations in recent years as the discussion surrounding the role of carbon sequestration in agricultural soils and above-ground biomass. Estimates of global mitigation potential vary greatly and have often not been generated in ways that reflect actual uses of land and crop residues resulting in large uncertainties. However, long-term management and preservation of soil carbon is critical for agricultural productivity because it increases soil fertility, reduces erosion, and increases moisture retention. Maintaining soil organic matter is vital for farmers and ranchers everywhere, regardless of the potential to measure or monetize sequestration.

---

[44]UNFCCC (2015), p. 4.

[45]PA (2015), Art. 4.13.

5 Conclusions and Outlook

Soils have received little attention in international climate policy. They are our silent and often unnoticed allies[46] in producing food and supporting our ecosystems. They come in countless forms and constitutions, and despite their profound importance for our survival, they are among the least considered topics in the international environmental discourse. This is not different for the climate regime. As we have seen, incentives for better cropland and pasture management were for the longest time limited to developed countries, and even there soil carbon was often not appropriately considered in national inventories. The role of developing countries was limited, as were the resources made available to support climate resilience of soils. Table 2 summarizes the aspects where climate action and soils intersect with the three thematic areas intersect with the three main thematic areas of the climate regime.

The good news for those concerned about healthy soils and the ecosystems and livelihoods that depend on them is that activities around adaptation and mitigation action in rural landscapes have accelerated. A number of stakeholder initiatives inside and outside of the UNFCCC negotiations have formed alliances to close data gaps and facilitate much needed investments. The importance of soils to ensure food security, protect biodiversity and watersheds, and combat climate change led also to the anchoring of their protection in the Sustainable Development Goals. Under Target 3 of Goal 15 the international community promises to 'combat desertification, restore degraded land and soil, including land affected by desertification, drought and floods, and strive to achieve a land degradation-neutral world'[47] by 2030. In Paris, a group of 50 governments launched the "4 Per 1000 Initiative: Soils for Food Security and Climate" declaration, which emphasizes the important role of soils and soil carbon as a mitigation strategy.[48] The goal of the initiative is to engage stakeholders in a transition towards a productive, resilient agriculture, based on a sustainable soil management and generating jobs and incomes, hence ensuring sustainable development. The Government of France has committed to ensure that at least 50 % of French agricultural holdings adopt an agro-ecological approach by 2020, while Australia has pledged to build the resilience of 300 million hectares of Australia's inland and northern rangelands in the front line and at risk of degradation.[49]

These accelerated voluntary efforts correspond to the bottom-up approach of the Paris Agreement. The Agreement provides new political impetus by obligating all countries to contribute to climate change mitigation and help to achieve the below

[46]The multiple roles of soils often go unnoticed. Soils don't have a voice, and few people speak out for them. They are our silent ally in food production. Mr. José Graziano da Silva, FAO Director-General at the occasion of the launch of the International Year of Soils.

[47]United Nations (2015).

[48]http://4p1000.org.

[49]ibid.

Table 2 Main considerations for soils under the UNFCCC climate regime

	Climate action (mitigation and adaptation)	Climate finance	MRV and accounting
Reduction of GHG emissions from croplands and pasture	– Most significant source of emissions pertaining to agricultural soils. – Included in developed country action, but without systematic consideration in developing countries. – Likely to receive more attention in the context of the Paris Agreement, both in the context of mitigation and adaptation.	– There are significant investment needs in the agricultural sector in developing countries. – Until recently incentives for cropland and pasture management in developing countries was limited to CDM projects (manure management, fertilizer production, afforestation). – Climate-smart agricultural finance provides an opportunity for integrated finance that considers the tree objectives of agriculture (productivity, mitigation, adaptation). – The pledges of finance of a minimum of USD100 billion/year holds significant opportunities for enhanced climate finance in the sector.	– Emissions from the management of agricultural soil (N_2O, CH_4) form part of baseline emissions of developed countries under the Kyoto Protocol. – Data hold uncertainties (in particular in relation to peat soils) but less so than the MRV of soil carbon.
Restoration and protection of soil carbon pools	– Maintaining and increasing soil carbon contents is essential for soil fertility and the climate resilience of agricultural systems. Soil carbon restoration is also an important mitigation strategy. – REDD+ provides significant opportunities for the protection of soil carbon reservoirs in forested landscapes. Supporting an integrated approach can promote adaptation and mitigation across landscapes.	– Due to large insecurities and lacking incentives, there has been limited financing been available for soil restoration. – Soil carbon accounting is too uncertain and benefits too limited at the farm level to make worth accounting in the context of carbon markets. – Additional funding for MRV and the closing of data gaps is necessary. – In the meanwhile support for sustainable agriculture and soil protection and restoration in the context of REDD+	– Soil carbon accounting is unsatisfactory both in developed and developing countries. – Soil carbon has been excluded from the CDM. – While MRV of soil carbon has to improve generally, major capacity building and technological efforts are needed to establish reliable soil carbon data bases.

(continued)

Table 2 (continued)

	Climate action (mitigation and adaptation)	Climate finance	MRV and accounting
		provides significant combined adaptation and mitigation opportunities that should be prioritized for funding.	

2 °C goal of the agreement. To achieve this goal, reducing land use emissions, including emissions from soil management, and enhancing carbon sequestration, in above and below ground biomass, is important in the near-term and maintaining natural ecosystems is critical for the future. For many developing countries, forests and land use comprise a large part of their emissions profile, and reducing these emissions provides an opportunity to engage in climate mitigation while strengthening the climate resilience of agricultural systems. Adaptation action will also have to target soil degradation and erosion.

It remains to be seen, however, whether the Paris Agreement indeed triggers meaningful action, including action that helps to restores and protects soils. Developing countries require substantial support to make the investments required to meet agricultural development needs, and climate finance is a potential source of investment to support adoption of agricultural practices with adaptation, mitigation and food security benefits. A UNFCCC assessment suggests that the additional costs of adapting to climate change in agriculture in developing countries will amount to more than USD 3 billion by 2030.[50] While climate finance is falling short of the needs, and of the overall amount of climate finance, less than one per cent of climate finance is invested in the agriculture sector.[51] This will have to change for soils to form part of meaningful climate action.

One barrier to mitigation action in agriculture is, among others, the lack of data and—where data are available—the large levels of insecurity. Here too, the Paris Agreement provides an opportunity. It is likely that additional funds become available that help developing countries to improve their capabilities to measure soil carbon and crop-related emissions.

In sum, the Paris Agreement provides a common framework within which individual countries are invited to define climate goals taking into account the overall goal of the Convention and the Agreement as well as their own capacities. The hope is that with increased transparency and an ambitious overall target, countries will step forward with ambitious national plans, which for most countries have to cover the land sector, including agriculture and forestry. The risk however, is that individual country contributions fall short of the overall goal and that the

[50]UNFCCC (n.d.), p. 93.

[51]Hodas (2012), p. 14.

Paris Agreement remains a shell without sufficient action and support, unable to address the collective action problem of climate change. Addressing this risk will require a new international effort to form coalitions and agreements with which to populate the Paris Agreement in the months and years ahead.

References

Bartz D (2015) Soil Atlas. Heinrich Böll Foundation and IASS Potsdam, Berlin

Bodansky D (2001) The history of the global climate change regime. In: Luterbacher U, Sprinz DF (eds) International relations and global climate change. Massachusetts Institute of Technology, Cambridge, pp 23–40

Brenton A (2013) Great Powers' in climate politics. Climate Policy 13(5):541–546

Dickie A, Streck C, Roe S, Zurek M, Haupt F, Dolginow A (2014) Strategies for mitigating climate change in agriculture: recommendations for philanthropy. Climate Focus and California Environmental Associates, prepared with the support of the Climate and Land Use Alliance

EU (2012) Paper no. 4: Denmark and the European Commission on behalf of the European Union and its member States, Views on issues relating to agriculture, submission of Parties, published on 27 March 2012, pp 11–15. http://unfccc.int/resource/docs/2012/sbsta/eng/misc06.pdf. Accessed 21 Mar 2016

FAO (2008) The carbon sequestration potential in agricultural soils. Submission by Food and Agriculture Organization of the United Nations. 3rd Session of the Ad Hoc Working Group on Long-term Cooperative Action under the Convention (AWG-LCA3), Accra

FAO (2010) Agriculture, food security and climate change in post-Copenhagen processes. http://foris.fao.org/static/data/nrc/InfoNote_PostCOP15_FAO.pdf. Accessed 21 Mar 2016

FAO (2011) The state of the world's land and water resources for food and agriculture. Managing systems at risk, Rome

FAO (2012) The State of Food and Agriculture 2012 - investing in agriculture for a better future. Food and Agriculture Organization of the United Nations, Rome, http://www.fao.org/docrep/017/i3028e/i3028e.pdf. Accessed 21 Mar 2016

FAO (2013) Agriculture, forestry and other land use mitigation project database. Second assessment of the current status of land-based sectors in the carbon market. http://www.fao.org/docrep/017/i3176e/i3176e.pdf. Accessed 21 Mar 2016

FAO (2015) Soils help to combat and adapt to climate change by playing a key role in the carbon cycle. http://www.fao.org/3/a-i4737e.pdf. Accessed 21 Mar 2016

Gambia (2012) Submission of Gambia on behalf of the least developed countries, views on issues relating to agriculture, submission of Parties, published on 27 March 2012, pp 15–18. http://unfccc.int/resource/docs/2012/sbsta/eng/misc06.pdf. Accessed 21 Mar 2016

Hodas R (2012) Bilateral Finance Institutions and Climate Change: a mapping of 2012 climate financial flows to developing countries. UN Environment Programme, Nairobi

Hoogzaad J, Hoberg J, Haupt F (2014) The geographical distribution of climate finance for agriculture, analysis prepared for CCAFS, June 2014. http://www.climatefocus.com/sites/default/files/the_geographical_distribution_of_climate_finance_for_agriculture_0.pdf. Accessed 21 Mar 2016

IPCC (2001) Climate Change 2001: the scientific basis. In: Houghton JT, Ding Y, Griggs DJ, Noguer M, van der Linden PJ, Dai X, Maskell K, Johnson CA (eds) Contribution of Working Group I to the Third Assessment Report of the Intergovernmental Panel on Climate Change. Cambridge University Press, Cambridge, UK/New York, NY, USA, p 881

IPCC (2003) Good practice guidance for land use. Land-use change and forestry. Vol. Intergovernmental panel on climate change, Hayama, Japan

IPCC (2014) Climate Change 2014. AR4 Synthesis Report. Summary for Policymakers. https://goo.gl/NVotzQ. Accessed 21 Mar 2016

KP (1997) Kyoto Protocol to the United Nations Framework Convention on Climate Change adopted at COP3 in Kyoto, Japan, 11 Dec 1997

Lal R (2004a) Agricultural activities and the global carbon cycle. Nutr Cycling Agroecosyst 70:103–116

Lal R (2004b) Soil carbon sequestration impacts on global climate change and food security. Science 304:1623

Macintosh A. (n.d.) LULUCF offsets and Australia's 2020 abatement task. ANU Center for Law and Policy. http://goo.gl/GwG8ua. Accessed 21 Mar 2016

ODI (2015) Climate finance thematic briefing: REDD+ finance. http://www.odi.org/sites/odi.org.uk/files/odi-assets/publications-opinion-files/10054.pdf. Accessed 21 Mar 2016

Paris Agreement (2015) Adoption of the Paris Agreement. FCCC/CP/2015/L.9/Rev.1

Prag A, Hood C, Martins Barata P (2013) Made to measure: options for emissions accounting under the UNFCCC, International Energy Agency Environment Directorate COM/ENV/EPOC/IEA/SLT. 1, 41

UNFCCC (1992) United Nations Framework Convention on Climate Change adopted in Rio de Janeiro, Brazil, 12 June 1992

UNFCCC (2012). Guidelines for the preparation of national communications by Parties included in Annex I to the Convention, Part I: UNFCCC reporting guidelines on annual greenhouse gas inventories, Annex to Decision 15/CP.17

UNFCCC (2015) Concept note: exploration of methodological options for developing 'agriculture CDM'. CDM Document CDM-EB87-AA-A10, https://cdm.unfccc.int/public_inputs/2015/mp69_01/index.html. Accessed 21 Mar 2016

United Nations (2015) General Assembly Resolution 70/1. Transforming our world: the 2030 agenda for sustainable development from September 2015. Available at: https://sustainabledevelopment.un.org/sdg15. Accessed 21 Mar 2016

The UN Convention on Biological Diversity and Soils: Status and Future Options

Franziska Wolff and Timo Kaphengst

1 Introduction[1]

The global loss of fertile soils is widely recognized as one of the most pressing environmental and social problems of the next decades: land and soil degradation impact the food supply and livelihoods for a growing world population, as well as the protection of water, biodiversity and ecosystem services. It is thus a genuine "nexus" resource.[2] In spite of this, the issue has been strongly neglected in the past. Soils are under pressure: worldwide, crop- and grazing lands expand at the cost of forests and natural areas; urbanization and built-up areas encroach on fertile land; land degradation, the loss of fertile soil and terrestrial biodiversity, continues unabated.[3] The European Commission has identified eight "soil threats": erosion,

[1]This article draws substantially on an earlier discussion paper (Wolff and Kaphengst 2015), developed within the GLOBALANDS project (www.globalands.de) which ran from 2011 to 2015, funded by the German Federal Environmental Ministry and Federal Environmental Protection Agency. The earlier paper, however, had a broader scope: sustainable *land* use. Land is frequently defined as a "physical entity in terms of its topography and spatial nature; a broader integrative view also includes natural resources: the soils, minerals, water and biota that the land comprises" (Agenda 21, Para 10.1). The focus here is on soils.

[2]Weigelt et al. (2014).

[3]See FAO and ITPS (2015), FAO (2010), SCBD (2014), UNEP (2014), and ISRIC (2008).

F. Wolff (✉)
Environmental Law & Governance Division, Oeko-Institute, Freiburg, Darmstadt, Berlin, Germany
e-mail: F.Wolff@oeko.de

T. Kaphengst
Ecologic Institute, Berlin, Germany
e-mail: Timo.Kaphengst@ecologic.eu

H. Ginzky et al. (eds.), *International Yearbook of Soil Law and Policy 2016*,
International Yearbook of Soil Law and Policy,
DOI 10.1007/978-3-319-42508-5_11

organic matter decline, salinization, compaction, landslides, contamination, sealing and the decline in soil biodiversity.[4]

Against this backdrop, political action is urgently needed to promote the sustainable use and management of soil,[5] both domestically and at the international level. Soil management is defined by the World Soil Charter (2015) as sustainable "if the supporting, provisioning, regulating, and cultural services provided by soil are maintained or enhanced without significantly impairing either the soil functions that enable those services or biodiversity".[6] Soil management merits being addressed by international policies because it has international implications—despite being often classified as a "local" (i.e., non-transboundary) issue.[7] For instance, the storage of carbon in soils has an immense effect on the global climate. The pool of organic and inorganic carbon in soils is 3.3 times the size of atmospheric carbon and 4.5 times the size of the biotic pool.[8] The way soils are managed strongly determines the content of soil organic matter and therefore the rate of carbon sequestration for mitigating GHG emissions. Land use changes such as deforestation release large amounts of carbon, not only by removing vegetation but also through the associated degradation of soils. Soil use is thus linked to the stabilization of the climate, a global common good—with land use-related emissions accounting for about a quarter of all GHG emissions.[9] Another international task of paramount importance is the promotion of food security, which is tightly interlinked with the management of soils. Finally, international rule making can strengthen domestic soil use regimes and foster learning processes with positive effects on sustainable development both nationally and internationally.

---

[4]European Commission (2006).

[5]In the literature, the distinction between soil management, soil use and land use is often unclear and blurred. We employ the terms "soil use" and "soil management" largely interchangeably, with a tendency to use "soil use" in a broader sense (referring to uses such as plant production, grassland management, soil use for residential settlements vs. industrial sites etc.) and "soil management" in a narrower sense (referring to more specific practices (mostly) to prevent soil degradation, like crop rotation, soil cover, tillage practices, afforestation, nutrient management, water retention etc.). By "land use" we mean broad categories such as agriculture, forestry, building land, mineral extraction etc.

[6]More specifically, FAO and ITPS (2015) set the following criteria for sustainably managed soils: leakage of nutrients is low; biological production is high relative to the potential limits set by climate and water availability; levels of biodiversity within and above the soil are relatively high; rainfall is efficiently captured and held within the root zone; rates of soil erosion and deposition are low, with only small quantities being transferred out of the system; contaminants are not introduced into the landscape and existing contaminants are not concentrated to levels that cause harm; systems for producing food and fibre for human consumption do not rely on large net inputs of energy; and net emissions of greenhouse gases are zero or less.

[7]Ginzky (2015).

[8]Lal (2004).

[9]IPCC (2014).

To some extent, land and soil degradation have already gained political momentum in international discussions in the past years.[10] This is reflected by the fact that the 5th of December was recently declared by the United Nations (UN) General Assembly as "World Soil Day", while 2015 was designated "International Year of Soils" under the framework of the Global Soil Partnership, established in 2011. In the Rio + 20 process, land and soil degradation became an issue of major concern. The idea of a land degradation-neutral world, originally promoted by the UNCCD, found its way in the Rio outcome document, which states in section 206: "We recognize the need for urgent action to reverse land degradation. In view of this, we will strive to achieve a land-degradation-neutral world in the context of sustainable development." Despite various uncertainties about the concept of a land degradation-neutral world[11], it was also incorporated into the Sustainable Development Goals (SGDs). In target 15.3, it says: "By 2030, combat desertification, restore degraded land and soil, including land affected by desertification, drought and floods, and strive to achieve a land degradation-neutral world". The protection of soils also appears in other SDG goals and targets.

As a key multilateral environmental agreement (MEA) within international environmental law, this paper analyses how the UN Convention on Biological Diversity (CBD) contributes to the international governance of sustainable soil use. The CBD promotes the conservation and sustainable use of biodiversity, and this includes soil biodiversity. Soil biodiversity encompasses the variability among living organisms in the soil, such as microbes, bacteria, fungi as well as macrofauna like earthworms and termites.[12] Conserving and sustainably using soil biodiversity is a crucial facet of sustainable soil management (SSM). This is because soils are ecosystems that perform numerous crucial functions and services, ranging from the provision of food and cleaning of water to archiving cultural heritage.[13] Moreover, "[s]oil organisms contribute a wide range of essential services to the sustainable function of all ecosystems, by acting as the primary driving agents of nutrient cycling, regulating the dynamics of soil organic matter, soil carbon sequestration and greenhouse gas emission; modifying soil physical structure and water regimes, enhancing the amount and efficiency of nutrient acquisition by the vegetation and enhancing plant health."[14] Soil biodiversity is thus essential both for the health of natural ecosystems and the productivity of agricultural systems.

The paper consists of five sections. Firstly, we give a brief overview of the international governance of soil sustainability to date, locating the role of the CBD in this context. Secondly, we elaborate on how the CBD deals with soil (biodiversity) issues—e.g., in its Strategic Plan and Aichi Targets, through principles for the sustainable use of biodiversity, the CBD's Ecosystem Approach as well as the

[10] Cf. the chapter by Boer et al. (2016).

[11] Altvater et al. (2015).

[12] UNEP/CBD/SBSTTA/7/INF/11.

[13] Jeffery et al. (2010), p. 8.

[14] Ibid.

International Soil Biodiversity Initiative. In the discussion, we highlight the value and shortcomings of the soil topic under the CBD. We also develop options on how sustainable soil management could be strengthened within the CBD and through the CBD. We conclude that the CBD can indeed contribute to the political promotion of sustainable soil use but that presently there are few incentives for CBD parties to push the process forward. Also, to be effective, the CBD is dependent on meaningful progress in international politics on the broader topic of sustainable *land* use.

2 Sustainable Soil Use and Management: International Governance to Date

A great number of international policies directly or indirectly affect the sustainable use and management of soils, both positively and negatively. This can be concluded from an extensive survey of international policies relevant for sustainable land-use conducted by Wunder et al. (2013). Various international policies that do not focus on land or soils (such as trade and investment policies) can cause substantial (often negative) side-effects on soils and the sustainability of soil use. Most international policies affecting soils are sectoral approaches such as agricultural, forestry, biodiversity, climate, resource or development policies and programmes.[15] To date, no binding agreement exists that explicitly aims at promoting sustainable soil use at a truly international scale.

The first attempt to put soils on the international agenda was the World Soil Charter (WSC), a set of principles and guidelines to counter soil degradation adopted by the UN Food and Agriculture Organisation (FAO) in 1982. The document was non-binding and has not led to significant impacts on the ground.[16] Soil-related recommendations in the Brundtland Report (1987) and Chapter 12 of the Agenda 21 (1992), too, had only an appellative character. A draft 'Convention on the Sustainable Use of Soils' (Tutzing Proposal)[17] developed by civil society in 1997 and intended to be binding was widely discussed but ultimately did not find sufficient governmental support. The failure so far to introduce a generic environmental liability regime at the international level also hampers soil protection.

Against this backdrop, the question is how best to strengthen the objective of sustainable soil use in international policies? Various binding and non-binding instruments and processes exist that address individual aspects of sustainable soil use and can potentially be utilised for this purpose. Among these, we will focus on the UN Convention on Biological Diversity (CBD). The CBD has various thematic points of entry for sustainable soil management (see Sect. 3). It is a legally binding

[15]Wunder et al. (2013).

[16]Montanarella and Vargas (2012), p. 1.

[17]Held and Kümmerer (1997).

agreement with almost universal membership and has been politically strengthened in recent years—both through its Strategic Plan and subsequently increased funding, and through the establishment in 2012 of the Intergovernmental Science-Policy Platform on Biodiversity and Ecosystem Services (IPBES). One of the thematic assessments to be conducted by IPBES in coming years will cover land degradation and restoration (in close cooperation with the UN Convention to Combat Desertification, UNCCD).[18]

Promoting sustainable soil management through and mainstreaming it into other suitable international policies, processes and fora will also be necessary. To some extent, this does already take place. For instance, the FAO's Global Soil Partnership (GSP), launched in 2011, has become an important policy arena for SSM. The multistakeholder forum aims at addressing all dimensions of soils including soil biodiversity, soil fertility and soil carbon. It is based on five "pillars of action", i.e. the promotion of: sustainable management of soil; investment, policy, and awareness in soil; targeted soil research; more and better soil data; and harmonization of methods. The GSP's scientific advice body, the Intergovernmental Technical Panel on Soils, has contributed to generating some international movement by revising the World Soil Charter (2013) and collating the first 'Status of the World's Soil Resources' report.[19] It has now embarked on developing 'Voluntary Guidelines for Sustainable Soil Management' to technically operationalize the Soil Charter. As a public private network with a mere technical (not political) intergovernmental mechanism, however, the GSP does not have the same political clout as an international agreement and is unlikely to provide any binding provisions.

Another international instrument which has a strong thematic overlap with sustainable soil use is the UNCCD. It addresses desertification and the degradation of ecosystems in arid, semi-arid and dry sub-humid countries (with an emphasis on Africa). Its geographical focus and a certain political inertia hamper the UNCCD in serving as a global 'vehicle' to promote sustainable soil use.[20] The UN Framework Convention on Climate Change (UNFCCC) affects soil management as well—both by trying to mitigate climate change and by improving adaptation to its effects. Politically, it creates incentives related to soil carbon management, land use change and forestry activities (LULUCF), biomass production and the reduction of deforestation (REDD+). Finally, the UN's Agenda 2030 for Sustainable Development provides some entry points for SSM, with individual targets relating to soil quality and land degradation neutrality. This includes, in particular, Target 15.3 which states: "By 2030, combat desertification, restore degraded land and soil, including

[18]The assessment which is due in 2018 will cover the global status of and trends in land degradation and ecosystem restoration (by region and land cover type). It will also identify the effects of degradation on biodiversity (including soil biodiversity), on ecosystem services (among others, soil formation, soil fertility, nutrient cycling, erosion prevention) and on human well-being.

[19]FAO and ITPS (2015).

[20]Ginzky (2015). In contrast, other authors highlight opportunities related to the UNCCD, e.g. Dooley et al. (2015).

land affected by desertification, drought and floods, and strive to achieve a land degradation-neutral world", but also Targets 2.4, 3.9 and 12.4. This is particularly valuable for boosting soil-related efforts in the operationalisation and implementation of the SDGs.

In the following, we identify in some more detail how the CBD can contribute to strengthening sustainable soil management.

3 Thematic Cornerstones of Sustainable Soil Management in the CBD

At first sight, it does not seem very obvious that the CBD could function as a new trigger of sustainable soil management in international policies. The terms "sustainable soil use" or "soil management" do not appear very prominently in CBD documents. This might be rather for political than for substantive reasons. Soil management is strongly related to land use such as agriculture and forestry, which in turn are highly contested issues in international policy, and most countries see their national sovereignty at risk, should these issues be internationally regulated. Unlike global commons such as the atmosphere or the seas, soil management and land use have a long tradition of being governed at the domestic level, frequently based on private ownership. As a consequence, vested interests exist and policy makers are usually reluctant to deal with soil-related issues in international negotiations. Furthermore, biodiversity is only a sub-aspect of sustainable soil management, which also encompasses water availability and quality, soil erosion and degradation, nutrient balances, etc.

However, soil biodiversity was identified as an area requiring particular attention soon after the adoption of the Convention. Within the CBD, soil biodiversity is described as the "many thousands species of animals and micro-organisms living in soils contributing to many essential ecosystem services. Soil biota ranges from the almost invisible microbiota (e.g. bacteria, fungi and protozoa) to the more conspicuous macrofauna and megafauna (e.g. earthworms, termites, millipedes, moles and rats)".[21]

In the following, we highlight the most relevant elements of the CBD that can serve as a basis for strengthening sustainable soil management in the CBD. The selected elements exist at different conceptual levels of the CBD: objectives, principles, programmes of work and a specific cross-cutting initiative. A key element is the CBD understanding and objective of "sustainable use of biodiversity and its components", which has some strong conceptual links to sustainable soil management. The sustainable use objective is operationalised by the CBD's "Addis Ababa Principles" and the "Ecosystem Approach". Furthermore, the Convention's programme of work on sustainable agriculture and the International Initiative on the

---

[21]SCBD (2016a).

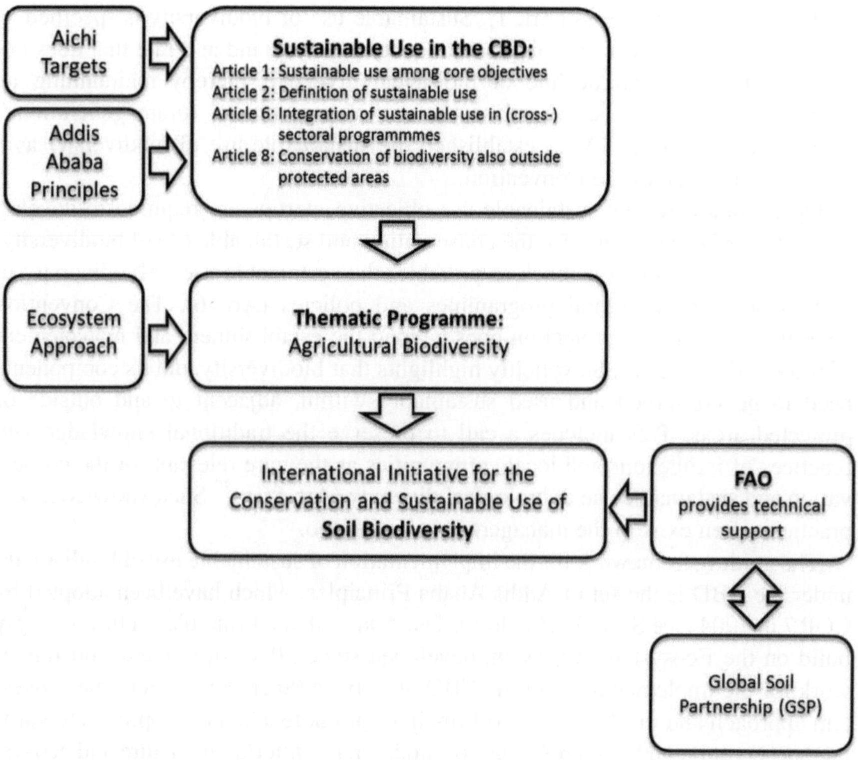

Fig. 1 Thematic and institutional links between the CBD and sustainable soil management. *Source*: own

Conservation and Use of Soil Biodiversity are relevant for soil management (for an overview see Fig. 1).

3.1 *Objectives, Targets and Principles: The Role of Sustainable Use*

Biodiversity has been defined by the CBD to encompass genes, species and ecosystems (Art. 2). Including the concept of "ecosystems" widens the understanding of biodiversity to a dynamic system (and a functional entity), of which human beings and their activities constitute an integral part. Here we can already trace a conceptual link to the management and protection of soils: soil management constitutes an important part of the management of ecosystems, be it in agro-ecosystems, forests, coasts, savannahs or others.

Another important 'anchoring point' for soil management in the CBD is the Convention's definition of sustainable use of all components of biodiversity as one of its three core objectives (Art. 1). Sustainable use of biodiversity is specified as "the use of components of biological diversity in a way and at a rate that does not lead to the long-term decline of biological diversity, thereby maintaining its potential to meet the needs and aspirations of present and future generations" (Art. 2). COP Decision V/25 established the sustainable use of biodiversity as a cross-cutting issue of the Convention.

Operationalizing the sustainable use objective, parties are required to develop strategies and programmes for the conservation and sustainable use of biodiversity, as well as to integrate (as much as possible) the sustainable use of biodiversity in sectoral and cross-sectoral programmes and policies (Art. 6). The Convention underlines that nature protection goes beyond the establishment and management of protected areas: Art. 8 explicitly highlights that biodiversity and its components need to be conserved and used sustainably within, adjacent to and outside of protected areas. This includes a call to preserve the traditional knowledge and practices of indigenous and local communities as they are relevant for the conservation and sustainable use of biological diversity (Art. 8(j)).[22] Such knowledge and practices often exist in the management of soils, too.

The guiding framework for the implementation of sustainable use of biodiversity under the CBD is the set of Addis Ababa Principles, which have been adopted by COP 7 in 2004 (see Sect. 3.1.1 below). The Addis Ababa Principles in turn strongly build on the Ecosystem Approach, developed since 2000—a fundamental framework for the implementation of the CBD objectives (Sect. 3.1.2). Both the ecosystem approach and the Addis Ababa Principles promote a holistic approach towards the conservation and sustainable use of biodiversity, interfacing nature and ecosystems with the human sphere. The more recent Aichi Targets (adopted in 2010) furnish both the principles on sustainable use and the Ecosystem Approach with a framework of targets, timelines and indicators (Sect. 3.1.3).

3.1.1 The Addis Ababa Principles on Sustainable Use of Biodiversity

In its Decision VII/12, the COP adopted the Addis Ababa Principles on Sustainable Use of Biodiversity as "an important tool to achieve the 2010 target endorsed by the World Summit on Sustainable Development, the Millennium Development Goals and the three objectives of the Convention". The Addis Ababa Principles are not prescriptive but rather provide a governance framework for the sustainable use of components of biodiversity. Their guidelines and instruments can be used by governments, resource managers, indigenous and local communities, the private sector and other stakeholders,[23] who not only deal with the protection of biodiversity, but are also confronted with the need to manage land and soils in a sustainable way.

[22]Note that the provisions of Art. 8(j) are subject to national legislation.

[23]SCBD (2016b).

Although "soil use" is not explicitly mentioned in the Principles, a strong conceptual link to soil management can be found in Principle 5, which says: "Sustainable use management goals and practices should avoid or minimize adverse impacts on ecosystem services, structure and functions as well as other components of ecosystems." Further, the rationale of the principle explains "for use of any resource there is a need to take into account the functions that resource may fulfill within the ecosystem in which it occurs, and that use must not adversely affect ecosystem functions."

Soils are an obvious biotic and abiotic component of (terrestrial) ecosystems. Without soils, no vegetation could exist; no water, nutrients, carbon and other substances could be stored and made available for all kinds of organisms. Furthermore, as outlined in the introduction to this paper, the core objective of sustainable soil management is to maintain and enhance soil functions and related ecosystem services. Ecosystems services are defined as "the benefits people obtain from ecosystems"[24] and soil change is regarded as one main factor directly affecting the provisioning of ecosystem services.[25] Sustainable soil management can thus be seen as a key component for achieving the objective of a sustainable use of biodiversity under the CBD.

The Addis Ababa Principles include 14 principles in total, which stress a wide range of (governance) issues that can be relevant for the sustainable use of soils, although they widen the scope beyond the biophysical aspects of sustainable soil management, for example:

- Multi-level governance and scaling of management to the ecological and socio-economic needs of resource use (Principle 1 and 7)
- Adaptive management (Principle 4)
- International cooperation (Principle 8)
- Rights-based and participatory approach (Principle 2, 9 and 12)
- Education, awareness raising and communication (Principle 14)

3.1.2 The Ecosystem Approach

The "Ecosystem Approach" is the CBD's primary framework for action, adopted by the parties in 2000 (Decision V/6) and then further developed (Decision VII/11). The ecosystem approach is a strategy for the integrated management of land, water and living resources that promotes conservation and sustainable use in an equitable way. The approach requires adaptive management to deal with the complex and dynamic nature of ecosystems, as well as precautionary action in the absence of complete knowledge or understanding of ecosystem functioning. It plays a particular conceptual role in the design and implementation of the programme of work on

[24] MA (2005), p. v.

[25] FAO (2015).

agricultural biodiversity and others and recognizes that humans and their activities are elementary parts of an ecosystem.

Principle 5 of the Addis Ababa Principles (see above) is strongly related to the Principles 3, 5 and 6 of the ecosystem approach, which are:

3. Ecosystem managers should consider the effects (actual or potential) of their activities on adjacent and other ecosystems;
5. Conservation of ecosystem structure and functioning, in order to maintain ecosystem services, should be a priority target of the ecosystem approach;
6. Ecosystems must be managed within the limits of their functioning.

Again, the terms (ecosystem) "functions" and "functioning" appear prominently, representing the most obvious linkage to sustainable soil management. Like the Addis Ababa Principles, the ecosystem approach depicts other relevant aspects, which can potentially widen the perspective of soil management to more social and political aspects such as decentralized management (Principle 2), recognition of the economic context (6), and consideration of different forms of knowledge and practices (11).

3.1.3 The Strategic Plan for Biodiversity and the Aichi Targets

The CBD's most recently agreed Strategic Plan for Biodiversity (2011–2020) takes a broad perspective on what is needed to preserve biodiversity and to use ecosystems more sustainably. It does so by focusing on food security, human health, local livelihoods, clean air, water and so on.

The "Aichi Biodiversity Targets" (which form part of the Strategic Plan) aim at achieving five strategic goals.[26] The targets are mostly set for 2020 (partly for 2015), and they are accompanied by a vision for 2050. Soil health and soil biodiversity are cross-cutting amongst these targets. For instance, they are indirectly addressed in the call to keep the impacts of natural resource use well within safe ecological limits (Target 4); to reduce the rate of loss of all natural habitats and to reduce degradation and fragmentation significantly; to manage agriculture sustainably (Target 7); to reduce pollution, including from excess nutrients (Target 8); to restore ecosystem services (Target 14); and to enhance ecosystem resilience and the contribution of biodiversity to carbon storage, including through the restoration of degraded ecosystems (Target 15) (Table 1).

To measure progress towards the Aichi Targets, the CBD's flagship publication—the Global Biodiversity Outlook (GBO) 4—uses indicators which are partly

[26]These are: addressing the underlying causes of biodiversity loss by mainstreaming biodiversity across government and society (Strategic Goal A); reducing the direct pressures on biodiversity and promote sustainable use (B); improving the status of biodiversity by safeguarding ecosystems, species and genetic diversity (C); enhancing the benefits to all from biodiversity and ecosystem services (D); and enhancing implementation through participatory planning, knowledge management and capacity building (E).

Table 1 Indicators of Aichi Targets with relevance for sustainable soil management

No.	Aichi target	Selected GBO-4 indicators[a]
4	By 2020, … keep the impacts of use of natural resources well within safe ecological limits	Ecological footprint
5	By 2020, … the rate of loss of all natural habitats is at least halved or close to zero, and degradation and fragmentation is significantly reduced	Extent of forests & forest types; areas of forest under sustainable management—degradation & deforestation
7	By 2020, areas under agriculture, aquaculture and forestry are managed sustainably, ensuring conservation of biodiversity	Area of forest under sustainable management (FSC, PEFC-certified); area under organic agriculture; area under conservation agriculture
8	By 2020, pollution, including from excess nutrients, has been brought to levels that are not detrimental to ecosystem function and biodiversity	Loss of reactive nitrogen to the environment; global surplus of nitrogen
14	By 2020, ecosystems … are restored and safeguarded …	Wetland extent
15	By 2020, ecosystem resilience and the contribution of biodiversity to carbon stocks has been enhanced, through conservation and restoration, including restoration of at least 15 % of degraded ecosystems …	[no indicator available]

Source: own
[a]UNEP/CBD/SBSTTA/20/13 (Annex); Leadley et al. (2014); http://www.bipindicators.net/globalindicators

also interesting as proxies for sustainable soil management. These include, for instance, areas of agricultural land under organic production and agricultural land under conservation agriculture (Target 7); as well as loss of reactive nitrogen to the environment and global surplus of nitrogen (Target 8).[27] Missing, so far, are indicators for soil organic carbon (for Target 15) as well as for soil organic matter content, soil productivity, biodiversity of soil organisms or soil water storage capacity (Target 7).[28] For the next GBO, which seeks to align its indicators with those of other Rio Conventions and those under discussion for biodiversity-relevant SDGs, some new soil and land-related indicators are being considered. These include the ratio of land consumption rate to population growth rate (Target 4) and the proportion of land that is degraded over total land area (Target 5).[29]

With regard to one of the Aichi Targets' overriding strategic goals—mainstreaming biodiversity concerns into relevant sectors of the economy (including those which heavily utilize soil)—the Subsidiary Body on Scientific, Technical and Technological Advice (SBSTTA) of the CBD made recommendations in its

[27]Leadley et al. (2014).

[28]European Commission, IASS & Umweltbundesamt (2013).

[29]UNEP/CBD/SBSTTA/20/13.

2016 meeting that have positive implications for soil management. Parties and other governments are encouraged, for example, to develop clear legal or policy frameworks for land use to enhance biodiversity-related ecosystem services from agricultural land, such as pollination, pest control, water provision and erosion control.[30]

3.2 *Specific Soil-Related Activities*

Besides the conceptual linkage between its objectives and sustainable soil management, the CBD addresses soils directly, focussing on soil biodiversity. The importance of soil biodiversity was first recognized formally in 1996 (at CBD COP3) when the CBD's programme of work on agricultural biodiversity was adopted (Decision III/11). During programme development, the role of soil and other below-ground biodiversity in supporting agricultural production, especially in nutrient cycling, was considered in depth.[31]

In 2002, the CBD parties agreed to establish an International Initiative for the Conservation and Sustainable Use of Soil Biodiversity (in short: International Soil Biodiversity Initiative) as part of the Programme of Work on Agricultural Biodiversity (Decision VI/5). A preparatory study on the subject was carried out by FAO. Based on this input, SBSTTA expertise and a technical workshop held in Brazil in 2002, the International Soil Biodiversity Initiative was developed. The CBD parties adopted the initiative in 2006 and established soil biodiversity as a "cross-cutting issue" (CBD Decision IIX/23).

3.2.1 The Programme of Work on Agricultural Biodiversity

Over the years, the COP has endorsed seven thematic programmes of work that reflect the major biomes of the world and provide concrete guidance by describing principles, key issues, outputs and timetables. With regard to soil management, the most relevant programme of work is that on agricultural biodiversity.[32] One of the two components of the work programme addresses the impacts of agricultural systems and practices on biodiversity in different ecosystems. It thus has a clear link to soil management.

Specifically, the work programme stipulates that the biological diversity of soils is responsible for nutrient circulation and fertility within agricultural ecosystems. Decision III/11 of the COP entails an indicative list of thematic areas which highlights soil erosion under the topic of "land resources". Other thematic areas

[30]UNEP/CBD/SBSTTA/15.

[31]UNEP/CBD/SBSTTA/7/INF/11, p. 2.

[32]Jeffery et al. (2010), pp. 82–83. Links (though weaker ones) also exist to the programmes of work on forest biodiversity and on dry and sub-humid lands.

such as sustainable tillage, integrated land and resource management, restoration of degraded landscapes and irrigation management are key issues, which are currently also discussed in the debate on sustainable soil use. Furthermore, the work programme gives soil micro-organisms priority for protection. Annex 3 of the programme identifies the measurement and monitoring of symbiotic micro-organisms, in particular nitrogen-fixing bacteria and mycorrhizal fungi, as an initial issue of conducting case studies within the programme of work. Finally, the work programme calls for the identification and promotion of sustainable land-use practices to enhance micro-organisms.

3.2.2 The International Initiative for the Conservation and Sustainable Use of Soil Biodiversity

The International Soil Biodiversity Initiative (CBD Decision IIX/23, 2006) calls on governments "to integrate soil biodiversity conservation and sustainable use into their national strategies and action plans and to put in place multi-sectoral programmes and initiatives for the conservation and sustainable use of soil biodiversity". The principles on which the initiative is based are mostly derived from the Ecosystem Approach and the Addis Ababa Principles outlined above. Its main goals are[33]:

1. **Promote awareness-raising, knowledge and understanding** of key roles, environmental services, functional groups and impacts of diverse soil management practices, including those performed by indigenous and local communities, in different farming systems and agro-ecological and socio-economic contexts.
2. **Increase understanding** of the role of soil biodiversity in agricultural production, traditionally applied land management practices and ecosystem and environmental health.
3. Promote the understanding of the impacts, ownership, and adaptation of all land use and soil-management practices as an integral part of agricultural and **sustainable livelihood** strategies.
4. Promote the **mainstreaming** of soil biodiversity conservation into land and soil-management practices.

The initiative is managed by FAO in cooperation with various international partners. FAO is also responsible for carrying out an "expanded progress report" of the Initiative's implementation (CBD Decision X/34). Generally, the FAO ensures an important institutional link to the more agriculture-driven (rather than conservation-driven) initiatives on soil protection; most importantly, the Global Soil Partnership (see Sect. 2).[34]

---

[33]For an overview of the Initiative's activities, see Dias and Coates (2012), pp. 9–10.

[34]It also acts as a connecting entity to other relevant initiatives on soil biodiversity, such as the science-driven Global Soil Biodiversity Initiative (GSBI).

4 Discussion

We have shown that the CBD rules and activities are conducive to sustainable soil management. Soils—which are both ecosystems themselves and a key component of most terrestrial ecosystems—are managed sustainably if, inter alia, levels of biodiversity within and above the soil are relatively high. Also, if (soil-related) ecosystem functions and services are maintained and enhanced, this contributes decisively to sustainable soil management as defined by the revised World Soil Charter (see Sect. 1). Both high levels of biodiversity and abundant ecosystem services are core objectives of the CBD and particularly with regards to soils, of the International Soil Biodiversity Initiative. More specifically, from a substantive perspective, the CBD's objective to promote the sustainable use of biodiversity, its concrete Aichi Targets, supportive management concepts (Addis Ababa Principles, Ecosystem Approach), the programme of work on agriculture and the Soil Biodiversity Initiative can all be considered as conducive for sustainable soil management, though they are not yet sufficiently targeted towards soils and soil biodiversity.

However, while conducive to sustainable soil management, the Convention does not yet deal with soils in a comprehensive way. With the exception of the International Soil Biodiversity Initiative, soil is addressed mostly *indirectly* by the various targets, guidelines and initiatives within the CBD. There is also an implementation deficit: while the International Soil Biodiversity Initiative "calls upon" parties to integrate the conservation and sustainable use of soil biodiversity into their national biodiversity *strategies*, they have, for the most part, failed to follow through.[35] Nor do soil biodiversity or related issues (such as pollution from excess nutrients) figure prominently in the national *reports* that parties are obliged to submit to the CBD.[36] This may have to do with the general political neglect of soil biodiversity. However, it may also result from the fact that there is no self-standing work programme on "soils" on the implementation of which they would need to report. Also, newer reports focus on parties' performance with regard to the Aichi Targets (or to their respective national equivalents), which do not directly relate to soils (see Sect. 3.1.1). Summing up, there is potential for an increased contribution of the CBD to sustainable soils. In the following paragraphs, we discuss how the CBD could support the global governance of sustainable soil use more strongly in the future.

From a formal perspective, CBD provisions that form part of the Convention have full legal status under international law, though many of them are open-ended and can be balanced against economic and social conditions in the implementa-

[35]Pisupati and Prip (2015).

[36]See in detail at https://www.cbd.int/reports/analyzer.shtml or the syntheses UNEP/CBD/COP/10/INF/2 (2010), UNEP/CBD/WG-RI/2/INF/1 (2007) and UNEP/CBD/WG-RI/2/INF/1/Add.3 (2007). The syntheses hardly mention soils, except for the last one which focuses on the implementation of thematic programmes of work. This includes the programme on agricultural biodiversity. In this context the synthesis points out that "a few parties" in their reports mentioned assessment of soil biodiversity (ibid, para 61).

tion.[37] The legal status of decisions of the Conference of Parties (adopted by consensus and without the need for domestic ratification) is more ambiguous. Generally, COP decisions (including CBD work programmes and initiatives) do not have a clearly binding character.[38] However, as 'soft law' they form an important part of treaty development and guide treaty interpretation. Summing up, while the CBD is a legally binding international treaty, not all of its provisions create sufficient obligations for states to act accordingly.[39]

As one step, the topic of soils could be anchored more directly within and mainstreamed through the CBD. This includes the development of soil-specific implementation or capacity building actions as well as an integration of soil-related provisions in the CBD text itself, to increase the formality of responsibilities under the Convention. A low-threshold activity in the context of implementation and capacity-building activities could be the production or commissioning of a guidance document by the CBD Secretariat within the CBD's Technical Series,[40] addressing the topic of "sustainable use of soils" under the CBD. As a minimum, this document would collate (handbook-style) all CBD objectives, principles, rules, decisions etc. with relevance for the sustainable use of soil, and it would describe their respective soil implications. A CBD publication could also review soil (biodiversity)-related ecosystem services and/ or highlight their role for achieving the SDGs, so as to strengthen awareness of the benefits of SSM. In this context, the economic costs resulting from the loss of soil biodiversity could further be demonstrated, too.[41] Alternatively, a CBD publication might present management practices and policy instruments that successfully promote soil biodiversity, so as to disseminate best practices. Workshops, technical expert groups, side events etc. could complement such efforts at awareness-raising and ultimately contribute to political agenda-setting. The pending FAO assessment of the International Soil Biodiversity Initiative could be used to launch a targeted initiative for scaling up the Initiative's implementation at country level. This would include pushing for a better integration of soil biodiversity into national biodiversity strategies and action plans, and into sectoral policies and national reports.

A more structural way to attain similar outcomes would be to work towards integrating sustainable soil use into the next revision and update of the Strategic Plan for Biodiversity (2021–2030). As some of the Aichi Targets 'expire' in 2020, it seems likely that they will either be updated or replaced by new targets. This could be a window of opportunity to introduce a soil-specific target or at least include soil

---

[37]See Jóhannsdóttir et al. (2010), p. 143. Cf. qualifying wording in the Convention text, such as the provision that parties shall "integrate, *as far as possible and as appropriate*, the conservation and sustainable use of biological diversity into relevant sectoral or cross-sectoral plans, programmes and policies" (Art. 6(b) CBD).

[38]E.g., Brunnée (2002); Gehring (2007), p. 491.

[39]Jóhannsdóttir et al. (2010), p. 143.

[40]See publications at https://www.cbd.int/ts/.

[41]Cf. ELD Initiative (2015).

biodiversity in existing Aichi targets (see potential starting points in Sect. 3.1.3). Doing so would be a more powerful pathway than an implementation initiative for better integrating soils into the CBD: there would be increased pressure on CBD parties to comply with and report towards this target, which could in turn trigger national soil protection activities. Moreover, targets at such a high strategic level typically are reflected in bilateral as well as multilateral funding instruments (e.g., the Global Environmental Facility). As a consequence, funding for projects through such instruments would explicitly have to consider soil biodiversity.

Integrating a soil focus in the CBD text itself would require an amendment to the Convention and thus a decision of the COP (Art. 29, CBD) which would de facto needed to be taken by consensus. Hence, amendments are cumbersome procedures which furthermore require, once they are adopted by the COP, that each party ratifies them (Art. 29.4). Also, explicitly mentioning soil (biodiversity) in the CBD would constitute a precedence, as no other types of biodiversity (e.g., forest, marine, mountain, island etc. biodiversity) are currently highlighted in the Convention text itself. Overall, the value-added of integrating "soils" in the Convention text might also be limited as a political dynamic might result to incorporate references to other types of biodiversity.

The most far-reaching option to better integrate soils into the CBD would be the development of a "Soil Protocol" under the Convention. The protocol could both address the conservation and sustainable use of soil biodiversity through specific mechanisms, and it could promote sustainable soil management as a strategy to bolster terrestrial biodiversity. Unlike the envisaged FAO Voluntary Guidelines on Sustainable Soil Management, a CBD Protocol would constitute a binding international agreement with some political clout. Formally, a protocol would require a COP decision (Art. 28 CBD) and subsequent ratification by sufficient parties.[42] Politically, at the moment it seems unlikely that the CBD parties will engage in developing a new protocol. The process of negotiating protocols has proven to be time- and resource-intensive, and the tough negotiations of the CBD's Nagoya Protocol—adopted only in 2010—still linger in the memory of delegates. The negotiation histories of the CBD's Cartagena Protocol (on biosafety) as well as its Nagoya Protocol (on access and benefit-sharing relating to genetic resources) also show that significant political momentum—including public pressure—around the issues and a critical mass of promoters are necessary to trigger protocol negotiations. In both cases, a substantial number of developing countries had called for a protocol, fearing economic injustice and/ or ecological risk from the lack of international regulation. In other words, a perception had emerged among a sufficient number of countries that they would gain from introducing international regulation. Also, both Protocols' topics (biosafety and access & benefit-sharing) had been "novel" issues, for which no national regulations or property rights had pre-existed; this could have kept opposition to an international regulation of the

[42]The necessary number of signatories is determined in the Protocol; the Cartagena and Nagoya Protocols each specify the need for 50 parties to ratify, accept, approve or accede the Protocol.

issues comparatively low. Finally, the ultimate adoption of both protocols can be attributed to issue linkages and/or compensatory arrangements. Most of these conditions are not (likely to be) prevalent when it comes to addressing 'sustainable soil management'. The topic of sustainable soils is not (yet) sufficiently politicized at the international level. Among the CBD parties, only a number of CBD parties have made steps towards a greater attention to soils. These include a number of EU Member States[43] and countries committed to the concept of a 'land degradation-neutral world' in the context of the UNCCD.[44] A range of both developing and developed country parties to the CBD, on the other hand, have rejected international 'interventions' with regard to their domestic soil and land use regimes in the past (most notably, with regard to forestry and agriculture).[45] To date, no group of states regards themselves as potential 'winners' from the international regulation of sustainable soil management, and thus there is little incentive for governments to become leaders and push the process forward. Pondering a Soil Protocol under the CBD, therefore, can only be a medium- to long-term perspective when all other options have been pursued.

How else can the CBD support the global governance of sustainable soil use? A different approach (with more short-term prospects of success) is the use and expansion of existing channels of institutional coordination between the CBD and other fora with regard to soil issues. The cooperation with IPBES could thus be used to promote a spotlight on soil, most importantly in conjunction with the currently ongoing assessment on land degradation. With regard to cooperation among the Rio Conventions (CBD, UNCCD, UNFCCC),[46] efforts are ongoing to harmonize among the three conventions joint land-based indicators and monitoring systems towards the achievement of the SDGs. For instance, one proposed indicator for SDG target 15.3—"proportion of land that is degraded over total land area"—could also be used as an indicator for Aichi Target 5 ("At least halving the rate of loss of all natural habitats"). Beyond this initiative, the CBD and UNCCD should be closely involved in the GSP's development of the Voluntary Guidelines on Sustainable Soil Management. Looking to the future, all Rio Conventions will need to strengthen their material provisions on sustainable soil management. Among the three multilateral agreements, the CBD is less likely to take the lead on soil issues

[43]Namely, those 23 member states willing to commit themselves to an EU Soil Framework Directive (which was ultimately blocked by a minority of five EU countries, leading to the withdrawal of the draft directive in 2014).

[44]The so-called "Group of Friends (GoF) of Desertification, Land Degradation and Drought". It includes Australia, Burkina Faso, Iceland, Lesotho, Namibia, Qatar, the Republic of Korea and Turkey.

[45]With regard to forestry these have been, among others, Brazil, Indonesia, Malaysia, but also Russia, Finland and Sweden. With regard to agriculture, these have included the "Miami Group" of agricultural exporters (US, Canada, Argentina etc.). In Europe, Austria, France, Germany, the Netherlands and the UK successfully blocked the draft EU Soil Protection Directive.

[46]Cf. Böhringer (2014).

than the UNCCD with its soil-related mandate. However, but it can and should still play a relevant role.

5 Conclusions

In this paper, we argued that soil management practices crucially affect sustainable development worldwide and that the political promotion of sustainable soil use is urgently needed. The CBD can contribute to this: it deals with aspects of sustainable soil use, under the umbrella of "sustainable use of biodiversity", though thus far mostly indirectly. We have outlined several options for highlighting soil management within the CBD and how the CBD subsequently, with the help of its institutional architecture and well-tried principles, could contribute to strengthening the international governance of sustainable soil use. A Soil Protocol under the CBD need seems unlikely in the medium term.

Strategically, a political 'entrepreneur' is needed push the soil agenda forward. So far, few countries seem to recognize the inherent value of sustainable soil use. Promoting sustainable soil governance may therefore be more successful if benefits other than direct pay-offs are stressed. Building consensus within the international community would certainly profit from reframing the issue of 'global sustainable soil use' in a way that creates winners. A success example is the reframing of international forest protection through a climate protection mechanism called "REDD+"[47]: after years of stalemate in the global forest negotiations, the perspective of tying forest protection to international payments under the climate regime created potential winners and an immense political dynamic.[48] Is a promising reframing possible in international soil politics, too? Some strategic brainstorming may be needed to explore possible options here (e.g., concrete instruments[49] for effective and equitable solutions as well as opportunities with regard to issue linkage, forum shopping, promising 'narratives', etc.).

Lastly, to be effective, the CBD is dependent on meaningful progress in international politics on sustainable *land* use.[50] The sustainability of land use depends, among other things, on access to land, secure tenure rights, participation and other social preconditions. Moreover, the sustainable management of soil—as *the*

[47]"Reducing Emissions from Deforestation and forest Degradation in developing countries". The mechanism created under the UNFCCC will compensate forest-rich developing countries for not cutting down their forests.

[48]Wolff (2011).

[49]Introducing an international payment scheme *analogous* to REDD+ is unlikely to work in the context of international soil policy, where the incentive for states is low to compensate other countries for sustainably managing their soils. We put the example of REDD+ forward here to emphasize potential political dynamics linked to reframing an issue, not for modeling a soil instrument.

[50]See Wolff and Kaphengst (2015) and Wunder et al. (2013).

important subsystem to land—will play a significant role. The SDG targets on biodiversity and ecosystem services thus must be strongly aligned with those addressing land and soil degradation (for example, Target 15.3 striving for a land degradation-neutral world). This way, the implementation of the SDGs imposes a need for closer cooperation between the three Rio Conventions, and the CBD needs to be prepared for this.

References

Altvater S, Dooley E, Roberts E (2015). Legal Instruments to implement the objective "Land Degradation Neutral World" in International Law. UBA-Texte 19/2015, Dessau

Boer B, Ginzky H, Heuser IL (2016) International soil protection law – history, concepts and latest developments. In: Ginzky H, Heuser IL, Qin T, Ruppel OC, Wegerdt P (eds) International yearbook of soil law and policy 2016. Springer, Heidelberg (in this volume). doi:10.1007/978-3-319-42508-5

Böhringer A-M (2014) Die Kooperationsvereinbarungen der Sekretariate multilateraler Umweltschutzübereinkommen. Mohr Siebeck, Tübingen

Brunnée J (2002) COPing with consent: law-making under multilateral environmental agreements. Leiden J Int Law 15(1):1–52

de Souza Dias BF, Coates D (2012) Soil biodiversity, functions, ecosystem services and the international sustainable development agenda. Issue Paper for the Global Soil Week 2012, Berlin

Dooley E, Roberts E, Wunder S (2015) Land degradation neutrality under the SDGs: National and international implementation of the land degradation neutral world target. Elni Rev 1+2/2015

European Commission (2006). Communication from the Commission to the Council, the European Parliament, the European Economic and Social Committee and the Committee of the Regions – Thematic Strategy for Soil Protection, COM/2006/0231 final

European Commission, IASS & Umweltbundesamt (2013) Soils and land in the SDGs and the post-2015 development agenda: a proposal for a goal to achieve a Land Degradation Neutral World in the context of sustainable development. Brussels/Potsdam/Dessau

FAO (2010) Global Forest Resources Assessment 2010: Main report. Rome

FAO, ITPS (2015) Status of the World's Soil Resources (SWSR) – technical summary. Food and Agriculture Organization of the United Nations and Intergovernmental Technical Panel on Soils, Rome, Italy

Gehring T (2007) Treaty-making and treaty evolution. In: Bodansky D, Brunnée J, Hey E (eds) The Oxford handbook of international law. Oxford, pp 467–497

Ginzky H (2015) Bodenschutz weltweit – Konzeptionelle Überlegungen für ein internationales Regime. Zeitschrift für Umweltrecht 26(4):199–208

Held M, Kümmerer K (1997) Preserving soils for life: proposal for an international soil convention. The Tutzing Project "Time Ecology". Gaia 6(3):212–216

ELD Initiative (2015) The value of land. Prosperous lands and positive rewards through sustainable land management. Bonn

IPCC (ed) (2014) Climate Change 2014 – Synthesis Report: Fifth Assessment Report (AR5)

ISRIC (ed) (2008) Global Assessment of Land Degradation and Improvement (GLADA). Wageningen

Jeffery S, Gardi C, Jones A, Montanarella L, Marmo L, Miko L, Ritz K, Peres G, Römbke J, van der Putten WH (eds) (2010) European atlas of soil biodiversity. European Commission, Publications Office of the European Union, Luxembourg

Jóhannsdóttir A, Cresswell I, Bridgewater P (2010) The current framework for international governance of biodiversity: is it doing more harm than good? RECIEL 19(2):139–149
Lal R (2004) Soil carbon sequestration impacts on global climate change and food security. Science 304(1623):1623–1627
Leadley PW, Krug CB, Alkemade R, Pereira HM, Sumaila UR, Walpole M, Marques A, Newbold T, Teh LSL, van Kolck J, Bellard C, Januchowski-Hartley SR, Mumby PJ (2014) Progress towards the Aichi Biodiversity Targets: an assessment of biodiversity trends, policy scenarios and key actions. Secretariat of the Convention on Biological Diversity, Montreal, Canada. CBD Technical Series No. 78
MA (2005) Millennium ecosystem assessment. Ecosystems and human well-being: synthesis. Island Press, Washington, DC
Montanarella L, Vargas R (2012) Global governance of soil resources as a necessary condition for sustainable development. Curr Opin Environ Sustain 4(5):559–564
Pisupati B, Prip C (2015) Interim Assessment of Revised National Biodiversity Strategies and Action Plans (NBSAPs). UNEP WCMC, Cambridge, UK and Fridtjof Nansen institute, Lysaker, Norway
SCBD (ed) (2014) Global Biodiversity Outlook 4. A mid-term assessment of progress towards the implementation of the Strategic Plan for Biodiversity 2011-2020. Secretariat of the Convention on Biological Diversity, Montréal
SCBD (2016a) Key issues. Secretariat of the Convention on Biological Diversity, https://www.cbd.int/agro/soilkeyissues.shtml (last accessed 04/04/2016)
SCBD (2016b). Addis Ababa principles and guidelines. Secretariat of the Convention on Biological Diversity, https://www.cbd.int/sustainable/addis.shtml (last accessed 04/04/2016)
UNEP (ed) (2014) Assessing global land use: balancing consumption with sustainable supply. A Report of the Working Group on Land and Soils of the International Resource Panel. Unter Mitarbeit von Stefan Bringezu, H. Schütz, W. Pengue, M. O'Brien, F. Garcia, R. Sims et al. United Nations Environment Programme, Nairobi
Weigelt J, Müller A, Beckh C, Töpfer K (eds) (2014) Soils in the nexus – a crucial resource for water, energy and food security, München
Wolff F (2011) Explaining the construction of global carbon markets: REDD+ as a test case? Int J Glob Energy Issues 35(2/3/4):255–274
Wolff F, Kaphengst T (2015) Exploring options for strengthening sustainable land use within the UN Convention on Biological Diversity. GLOBALANDS Discussion Paper, Berlin
Wunder S, Hermann A, Heyen DA, Kaphengst T, Smith L, von der Weppen J, Wolff F (2013) Governance screening of global land use. Discussion Paper prepared by GLOBALANDS Project. Ecologic Institute & Oeko-Institute, Berlin

The Alpine Convention's Soil Conservation Protocol: A Model Regime?

Till Markus

1 Introduction

Like many other natural areas, the Alpine region is increasingly being exploited for economic gain. The steady growth of tourism (e.g., construction of ski runs, lifts and hotels), intensified agricultural use, the expansion of (transit) traffic, energy projects (including power lines, dams, hydroelectric plants and reservoirs), settlement development and human-induced climate change are placing a growing burden on the environment. The soil in the Alps is not exempt from this development. Land utilisation and anthropogenic climate change result in severe sealing, erosion, degradation and destabilisation effects amongst other things.[1] The Member States of the Alpine Convention (AC) thus adopted the *Protocol on the implementation of the Alpine Convention of 1991 in the field of soil conservation—Soil Conservation Protocol* (SoilProt) at the 5th Alpine Conference in 1998. To this day the SoilProt remains the only binding instrument under international law to deal specifically and directly with soil conservation in a particular region.[2] In light of this special role, description and analysis of the SoilProt can lead to new findings for current regional and global efforts focused on internationally coordinated and legally binding soil conservation. To this end, the ecological and socio-economic Alpine region is outlined briefly and the background, nature and extent of soil impact here is highlighted (B. and C.). In addition, the Soil Conservation Protocol

[1]See, for example, IPCC, Climate Change: Impacts, Adaptation, and Vulnerability. Part B (2014), p. 1281.

[2]For details on the development status of international soil conservation law, see Boer and Hannam (2014); for earlier developments see: Wyatt (2008), p. 165; Lee (2006).

T. Markus (✉)
University of Bremen, Bremen, Germany
e-mail: t.markus@uni-bremen.de

H. Ginzky et al. (eds.), *International Yearbook of Soil Law and Policy 2016*,
International Yearbook of Soil Law and Policy,
DOI 10.1007/978-3-319-42508-5_12

in particular is described and analysed as an important component of the Alpine Convention (D.). Finally, the added value of an international approach with respect to soil conservation compared to the national measures of the Contracting Parties is discussed along with the question of to what extent the solutions it contains could be used as an expedient model for similar problems in other regions (E). A final review concludes these considerations (F.).

2 The Ecological and Socio-Economic Alpine Region[3]

The natural Alpine region is one of the largest ecosystems in Europe and home to some 13,000 plant species and 30,000 species of animal. Transboundary rivers such as the Rhine, the Rhône, the Danube and the Po are predominantly fed by Alpine waters. From a socio-cultural and socio-economic perspective, the Alps form a highly heterogeneous habitat. The Alpine region extends across eight states, is a permanent home to around 14 million people and is visited by approx. 120 million guests each year (number of overnight stays by tourists). Around 190 million tons of goods are transported each year by road and rail in this region. There is a high concentration of settlements, utilisation and traffic in some areas, whilst others remain sparsely populated and less economically developed. The Alpine region is subject to structural and climatic changes. Traditionally, agriculture in this region was characterised by small rural farms, yet these are increasingly being replaced by modern agricultural structures as well as industrial and service economies. Many young people are moving permanently away from the country into larger towns. Moreover, climate change is exercising a growing influence on the development of the Alpine environment and in particular its glaciers and rivers.

3 Important Impacts on Alpine Soils

For some time now the Alpine region has also been affected by the loss of ecologically and biologically intact soil. The reasons for this are many and varied.[4] One of the most important impacts is agriculture, in particular intensive grazing, which has resulted in soil erosion in many areas. In addition, the construction of hotels, holiday homes, roads and hydroelectric power plants results in the long-term sealing of soil. Industry and mining activities also have a variety of impacts on the soil (e.g., via the input of harmful substances). In particular, the construction of ski

[3]The information in the following paragraph is a summary from: Permanent Secretariat of the Alpine (2010); Veit (2002); Also see Norer (2002), p. 4 ff.

[4]See the following overview from Meusburger and Alewell (2014); also see Poulenard and Podwojewski (2006), p. 75; Körner (1999).

runs leads to the almost complete destruction of the ecological functions of mountain soils in the affected area. It is forecasted that the consequences of human-induced climate change, e.g., increased rainfall and snowmelt as well as rising temperatures will serve to accelerate erosion and the natural weathering process of Alpine soil.[5] Generally, given their location at great altitudes and them being usually subject to low temperatures, biological activities in mountain soils are limited.[6] Accordingly, mountain soils develop and regenerate especially slowly, meaning that any kind of impact and destruction has a very long-lasting effect.

4 The Alpine Convention and Its Soil Conservation Protocol

The Alpine Convention is unique in two respects. For one, it is the first international convention to deal with the protection and sustainable use of a mountain region. Thus, it could be regarded as the starting point for the development of a special "international mountain law" or "mountain conservation law".[7] For two, it also resulted in the adoption of the SoilProt, which remains to this day the only binding international convention to deal specifically with soil conservation.

4.1 The Alpine Convention as a Framework Convention of the Soil Conservation Protocol

The Alpine Convention (AC) was signed by Austria, Switzerland, Germany, France, Liechtenstein, Italy and the European Community in Salzburg on 7.11.1991. Slovenia and Monaco have subsequently signed this convention. The AC came into force in 1995 following ratification by Austria, Liechtenstein and Germany.[8] Today, all Parties have ratified the Convention.[9] It is designed as a binding framework convention under international law. It contains general objectives and obligations as well as institutional and procedural provisions. In particular Art. 2 Para. 3 of the AC expressly stipulates that the objectives and specific policies

---

[5]European Environment Agency (2009); Steiniger and Weck-Hannemann (2002).

[6]FAO, Understanding Mountain Soils (2015), p. 3.

[7]See: Villeneuve et al. (2002), p. 3; for the particular challenges for the protection of mountainous regions see: Quillacq and Onida (2011).

[8]Germany ratified the Convention dated 7 November 1991 for the protection of the Alps (Alpine convention) with the Act on the Convention dated 7.11.1991 for the protection of the Alps (Alpine convention), Federal Law Gazette II 1994, p. 2538.

[9]See overview of ratifications under http://www.alpconv.org/de/convention/ratifications/default.html (last accessed on 18 Feb 2016).

of the AC are to be developed by means of protocols "laying down the details for implementing the Convention". As per Art. 11 and Art. 6 Para. 2 of the AC, the protocols are to be adopted by the Conference of Contracting Parties (Alpine Conference). The protocols constitute legally independent agreements under international law and are only binding for the parties who have signed and ratified them.[10]

As per Art. 1 Para. 1 of the AC, the convention covers the Alpine region as described and depicted in the Annex to the AC. The objectives of the AC are stated in Art. 2 Para. 1 of the AC. In accordance with these, the Contracting Parties "shall pursue a comprehensive policy for the preservation and protection of the Alps by applying the principles of prevention, payment by the polluter (the 'polluter pays principle') and cooperation, after careful consideration of the interests of all the Alpine States [. . .] and through the prudent and sustained use of resources." To this end, "appropriate measures" have to be taken in a variety of areas. These areas comprise, as per Art. 2 Para. 2 (a–l) AC, the issues of population and culture (a), spatial planning (b), prevention of air pollution (c), soil conservation (d), water management (e), conservation of nature and the countryside (f), mountain farming (g), mountain forests (h), tourism and recreation (i), transport (j), energy (k) as well as waste management (l). Art. 3 and Art. 4 of the AC contain further general cooperation obligations (research and systematic monitoring; legal, scientific, economic and technical cooperation). The AC thus pursues a comprehensive, trans-sectoral conservation concept and targets different economic, political and social areas along with their interdependencies.[11]

From an institutional perspective, the AC has established three bodies: The Alpine Conference, the Standing Committee, and the Secretariat. The Alpine Conference is the decision-making body of the AC.[12] As a rule it comprises the responsible ministers from the Contracting Parties. It meets every 2 years and usually resolves changes to the AC, its protocols and financial decisions unanimously.[13] Germany took over chairmanship from Italy at the XIIIth Alpine Conference on 21st November 2014 in Turin (term ends 13th October 2016).[14] The Standing Committee for its part is made up of public official delegates from the Contracting Parties. It essentially analyses all the relevant information for implementation of the AC and the protocols, prepares the Alpine Conferences and appoints Working Groups in order, for example, to formulate protocols.[15] The

[10]See Art. 11 Para. 2 of the AC as well as, e.g., Art. 26 of the Soil Conservation Protocol. Not all protocols have as yet been signed and ratified by all states. For the status of ratification see: http://www.alpconv.org/de/convention/ratifications/default.html (18 Feb 2016).

[11]See Schroeder (2006), p. 134; Norer (2002), p. 9.

[12]Art. 6 of the AC.

[13]Art. 5 Para. 2 and Art. 7 Para. 1 of the AC.

[14]Information on the German chairmanship is available at http://www.bmub.bund.de/themen/europa-international/int-umweltpolitik/alpenkonvention (last visited on 13.01.2015).

[15]Art. 8 Para. 1, Para. 6 (a–g) of the AC.

administrative tasks in connection with the Alpine Conference and the Standing Committee are handled by the Secretariat.[16] The latter is also responsible for technical tasks, including the coordination of Alpine research projects. In addition to the established institutions, the Contracting Parties have also set up arbitral dispute resolution procedures and a review panel as a control mechanism to promote and ensure compliance with and implementation of the AC.[17]

4.2 The Soil Conservation Protocol

The Soil Conservation Protocol was negotiated in 1998 and, since then, has been signed and ratified by all AC Member States with the exception of Switzerland. It came into force on 18.12.2002.[18] In terms of its structure, it is similar to the other AC protocols. It contains sections on general provisions, specific measures, research, training and information as well as on the implementation, monitoring and evaluation of the efficacy of the SoilProt provisions.

4.2.1 General Objectives and Obligations

Art. 1 Para. 2 of the SoilProt firstly lists the general objectives of soil conservation and states "in particular" that the "ecological functions of soil, which are essential elements of the ecological balance, shall be safeguarded and preserved [...]. The restoration of impaired soils shall be promoted". The failure to expressly refer to the climate-protecting function of soil as a carbon pool is conspicuous here. The measures to be taken should aim to ensure soil utilisation in line with the local conditions, the economic use of land resources, the avoidance of erosion and the minimisation of input of substances which are harmful to the soil.[19] The diversity of the soils and characteristic locations are to be preserved; special significance should be attached to the prevention principle in terms of soil conservation.[20] Art. 2 and Art. 3 of the SoilProt contain further general obligations. Specifically, the Contracting Parties shall take the necessary legal and administrative measures to ensure soil protection in the Alpine region. Where the functionality of the soil is faced with serious and sustained damage, "protection shall, as a matter of principle, be given priority over utilisation." Furthermore, the objectives of the Protocol

[16] Art. 9 of the AC.

[17] The Compliance Committee is supported by the Secretariat, see: Schroeder (2006) p. 133.

[18] Protocol "Bodenschutz" dated 16th October 1998, Federal Law Gazette II 2002, p. 1785, 1842: For the status of ratification see: www.alpconv.org/de/convention/ratifications/default.html#Bodenschutz (last visited on 8th January 2015).

[19] Art. 1 Para. 3 of the SoilProt.

[20] Art. 1 Para. 4 and Para. 5 of the SoilProt.

should be taken into consideration in other policies and the possibility of supporting the measures outlined in the SoilProt with fiscal means should be investigated. Art. 4 and Art. 5 regulate participation of local and regional authorities and their transboundary cooperation.

4.2.2 Specific Measures

As per Art. 6, the Contracting Parties undertake to ensure that soils worthy of protection are included in the "designation of protected areas." "Specifically, soil and rock formations which have particularly characteristic features or a particular significance for the documentation of earth's history, shall be preserved".

Art. 7 Para. 1–Para. 4 describe the economical and prudent use of soils in the Alpine region. In this respect, when drawing up and implementing settlement plans and programmes, the Contracting Parties are obliged to take matters regarding soil conservation and especially the economic use of soil and land into account.[21] They shall provide for space-saving and soil-conserving construction and shall seek to keep the development of human settlements within existing boundaries. When evaluating and assessing the spatial and environmental compatibility of large-scale projects, the scarcity of space in the Alpine region is to be factored in. "Disused or impaired soils, especially landfills, slag heaps, infrastructures or ski runs, shall be restored to their original state or shall be recultivated".

Art. 8 Para. 1–Para. 2 refer to the requirement to make economical use of mineral resources and to ensure prudent extraction of the latter. With respect to the use of mineral resources, they shall strive towards ensuring "that preference is given to the utilisation of substitute materials and that recycling options are fully used or their development is encouraged." "When extracting, processing and utilising mineral resources, impairments of other soil functions shall be reduced to a minimum. In those areas which are particularly important for the protection of soil functions and in areas specifically designated as drinking water resources, the extraction of mineral resources shall be foregone."

In Art. 9 Para. 1–3 the Member States undertake to conserve soil in wetlands and moors. High and lowland moors are to be preserved. In the medium term, the Member States are working towards complete discontinuation of the use of peat. Drainage schemes in wetlands and moors are to be limited to the upkeep of existing networks; remedial measures shall be promoted. "On principle, moor soils shall not be utilised or, when used for agricultural purposes, shall be managed so that their characteristic features remain intact."

Art. 10 and 11 deal with the designation and management of endangered areas and areas threatened by erosion. The focus here is on drawing maps of areas facing geological, hydrogeological and hydrological risks (especially mass slides,

[21]Plans and programmes as defined by Art. 9 Para. 3 of the Protocol "Spatial planning and sustainable development" of the AC.

mudslides, landslides), avalanches and floods and to list these in a register.[22] High-risk areas are to be designated as such wherever pertinent. Construction measures in these areas are to use engineering techniques which are as compatible with nature as possible. In addition, local and traditional building materials which suit the local environment are to be used. With respect to measures for minimising water erosion and reducing surface runoff, preference is given to hydraulic, engineering and silvicultural techniques which are close to nature.

Art. 12 and 13 are concerned with agriculture, pasture farming and forestry/ silviculture. The Contracting Parties undertake "to use sound practices in agriculture, pasture farming and forestry which are adapted to suit local conditions" to prevent soil erosion. In terms of the input of substances from fertilisers, herbicides and pesticides, the Contracting Parties shall endeavour to "elaborate shared standards for sound expert practices." Details of the latter are provided. As such, for example, the use of mineral fertilisers and synthetic herbicides and pesticides is to be minimised. Mountain forests are to be preserved where they have a protective function for other assets. They shall be used and maintained in such a way that soil erosion and harmful soil compaction are avoided. Silvicultural measures in line with the local conditions and natural forest rejuvenation are to be promoted.

The provisions on the effects of tourism infrastructures in Art. 14 Para. 1–3 of the SoilProt are relatively specific and binding. According to these, the Contracting Parties shall exercise their influence "in the most appropriate manner" to avoid detrimental effects of tourism activities on Alpine soils. Specifically, soils which have been impaired by tourism are to be stabilised. Moreover, "permits for the construction and levelling of ski runs in forests with a protective function are granted only in exceptional cases and with the proviso that compensatory action is taken, and that such permits are not granted for fragile areas." Chemical and biological additives for grooming ski runs shall only be allowed if they have been proven to be ecologically harmless. When damage to soils and vegetation is identified, the Contracting Parties shall take the measures necessary to remedy this as soon as possible.

Art. 15–17 of the SoilProt are dedicated to soil pollution load. The Contracting Parties undertake to minimise inputs of harmful substances as far as possible through preventive action. To this end, preference is given to measures limiting emissions at their sources. To achieve this, the Contracting Parties shall issue further technical regulations, provide for checks and carry out research programmes and education work. The use of gritting salt is to be minimised. In terms of contaminated soil, the Contracting Parties shall survey and document their environmental liabilities and suspicious landfills (environmental liabilities register). Comparable methods are to be used to analyse the condition of such areas and assess their risk potential. Waste concepts shall continue to be developed and implemented to avoid soil contamination.

[22]This is done in line with comparable criteria for the quantification of erosion agreed between the Contracting Parties, see Art. 11 Para. 1, p. 1 of the SoilProt.

Art. 19–Art. 22 focus essentially on research, monitoring and information. The aim in all areas is to achieve coordinated cooperation. Art. 23–Art. 25 contain provisions for the implementation, monitoring and evaluation of the effectiveness of the SoilProt.

4.2.3 Integration in National Legislation

The SoilProt was negotiated in 1998 and, since then, has been signed and ratified by all Member States with the exception of Switzerland. It came into force on 18.12.2002.[23] Ratification means that it is applicable at national level and, accordingly, is binding for the respective legislators in the Member States, executives and judiciary.[24] The rules of the protocols differ in terms of their regulatory purpose and specificity and thus also in their effect.

Thus one can distinguish between mandates for action for legislators, rules for interpretation as well as directly applicable rules for authorities and courts.[25] For example, in terms of their content, purpose and formulation, Art. 8, Art. 12, Art. 14, Art. 16 and Art. 17 are sufficiently precise for the authorities and courts to decide on specific issues. Where the provisions are less specific, they still guide authorities and courts in their decision-taking.

The effects of the SoilProt on the national legislations and administrative and judicial practices of the Member States are to be assessed individually. On the one hand, it should be stressed that in some Member States it is only thanks to the SoilProt that binding and systematic soil conservation legislation exists at all. On the other hand, implementation of the SoilProt in Austria, for example, caused the competent national and federal state authorities to analyse existing soil conservation legislation extremely closely which, in turn, led to the identification of different calls for action.[26] Moreover, the SoilProt has also developed considerable force in Austrian administrative and judicial practice.[27] For example, the authorities in Tirol denied the construction of a new cable car by invoking, amongst others, Art. 14 Para. 1 of the SoilProt.[28] It is thus crucial here that the legal consequence of Art. 14 Para. 1 Sub-para. 3 is sufficiently clear: "The Contracting Parties shall use their influence in the most appropriate manner to ensure that: permits for the construction and levelling of ski runs in forests with a protective function are

---

[23]Federal Law Gazette II (2002), p. 1785., 1842.

[24]See, e.g., for Germany Art. 59 Para. 2 of the Basic Law.

[25]See: Schroeder (2006), p. 137; see also: Söhnlein (2013), pp. 105–111; Alternative Unterscheidung für das Österreichische Recht, see, e.g., Lebensministerium Österreich (2007); Blecha (2009); also informative: Umweltbundesamt (2007).

[26]For example with respect to the designation and handling of Alpine regions at risk of erosion, e.g., in implementation of Art. 11 of the SoilProt. See: Lebensministerium Österreich (2007).

[27]For Germany see: Söhnlein (2013), p. 105.

[28]See Schroeder (2006), p. 133.

granted only in exceptional cases and with the proviso that compensatory action is taken, and that such permits are not granted for fragile areas." This interpretation was confirmed in court.[29] With respect to Germany, where soil conservation legislation already existed when this protocol came into force, it can be stated prima facie that the overall level of protection afforded to soil has only increased marginally.[30]

5 The SoilProt as a Model for Soil Conservation Agreements in Other Regions?

The adoption of the SDGs by the United Nations in September 2015 has finally and emphatically highlighted just how important the preservation of ecologically intact soil, and in particular mountain soil, is.[31] SDG 15 calls on the international community "to achieve a land degradation neutral world" and "[to] ensure the conservation of mountain ecosystems, including their biodiversity [. . .]."[32] The significance of internationally, institutionally supported and binding agreements in reaching such goals has been stressed repeatedly.[33] The SoilProt of the AC is, to date, the only internationally binding and institutionally supported soil conservation convention. The special status of the latter therefore evokes the question of whether and to what extent the rules it contains can be used as a model for international soil conservation in other regions of the world. Furthermore, it should also be asked whether it can thus contribute to realisation of the SDGs.

5.1 *Political Signal as Regards the Need for International Regulation*

From the perspective of the national legislators there is *a priori* essentially little interest in regulating issues internationally. Indeed the opposite is true: States tend to shy away from the sovereignty costs associated with international regulation as well as the costs incurred in the drafting and implementing of agreements. They also regularly pursue different interests and values and the complexity of the

[29]See Lebensministerium Österreich (2007), pp. 114–115.

[30]Implementation of the mandates to act for the German Federal legislator/Bavarian state legislator is well advanced. See synopsis in: Bayerisches Umweltministerium (2006) available at: http://www.stmuv.bayern.de/ministerium/eu/zusammenarbeit/alpenkonvention/protokolle.htm (last accessed on 18 Feb 2016).

[31]GA Resolution 25th September 2015, UN A/RES/70/1.

[32]Sustainable Development Goals 15.3 and 15.4.

[33]Boer and Hannam (2014), Lee (2006).

problems in question often differs substantially. This, in turn, makes it difficult to find common solutions.[34] In the case of international soil conservation, the multidimensional nature and complexity of the problems and the causes of the loss of ecologically intact soil are extremely important.[35] It is significant here that efforts aimed at the conclusion of an international soil conservation convention affect two core areas of national sovereignty: property ownership (land) of states and food security (food sovereignty).[36] Given these interests, the added value of an international soil conservation convention needs to be thoroughly explained and justified.[37]

This is precisely where the AC with its SoilProt sends an important signal in terms of legal policy. The Contracting States have recognised that a unilateral national approach is only minimally expedient and effective in ensuring Alpine soil conservation. In particular, it was possible to create a greater understanding of how the individual states are affected and involved: Though the loss of ecologically intact soil in neighbouring states may seem remote and far off, it clearly has transboundary impacts! This also becomes evident in the Alps. Uses such as the construction of ski runs and intensive grazing reduce the seepage and storage capacity of mountainous soil. The subsequent soil erosion can result in more frequent flood discharges and mudslides.[38] Moreover, the increase in flood discharges in the Alpine region with its numerous rivers can have significant transboundary effects.[39] It should also be borne in mind that rivers in areas at risk of soil erosion have a higher content of mud, sand, salt, nutrients and harmful substances, which would then also likely be distributed across boundaries.[40]

A factor which is less concrete yet nevertheless relevant is the loss of ecologically intact soil as a material prerequisite for preserving biological diversity. Any drop in ecological quality in the comparably fragile mountain soils has a short- to medium-term, transboundary effect on flora and fauna in the ecologically connected and, for its part, sensitive Alpine environment.[41] Soil erosion further reduces the absorption capacity of soil with respect to atmospheric emissions of CO_2 and hazardous substances in the medium term, which can cause regional as well as global effects.[42]

[34]States represent here complex entities which are naturally influenced on their part by domestic interests. Fundamentally: Putnam (1988), p. 427; Not least they represent the dilemmas of collective *action, see* fundamentally Olson (1965); see most recently Markus (2016), forthcoming.

[35]Linz and Lobos Alva (2015).

[36]Fromherz (2012), pp. 62, 67–68.

[37]This aspect is frequently not dealt with in enough detail in the literature on international soil conservation legislation or results in only very general considerations. Exceptions to this are, in particular Ginzky (2015), p. 199; Fromherz (2012), pp. 107–109.

[38]Geitner (2007), p. 62; quoted in Schumacher and Schumacher (2012), p. 284.

[39]Generally, see Fromherz (2012) p. 69.

[40]Fromherz (2012), p. 69 ff.

[41]See for example Point 5 and Point 8 of the preamble of the SoilProt.

[42]Fromherz (2012), p. 70.

In the case of the Alps, one reason for a common solution to soil problems arises from the interests of the Alpine states in the distribution and mitigation of the economic advantages and ecological disadvantages of transalpine traffic. Spatial and quantitative control of transalpine traffic can only be achieved in an effectual, cost-effective and environmentally friendly manner through joint action by the Alpine states. An additional incentive for a joint approach must also be the fact that the coordinated adoption of soil conservation measures, e.g., emission reduction objectives for local industry, input limits for fertilisers in the agricultural sector and the environmentally friendly design of tourism infrastructures mitigates the above problem of collective action and any competitive disadvantages. Not least, a joint approach could increase the overall capacity of the Contracting Parties to solve the problems at hand: Prior to conclusion of the SoilProt, many of the Contracting Parties did not have their own soil conservation legislation.[43] As a facilitating measure, the SoilProt is of benefit to both categories of states, those which already had soil conservation legislation in place (significantly increased level of protection in the region) and those who had no legislation (increased level of protection in the respective country).

5.2 *Positive and Negative Aspects of the SoilProt*

Against the backdrop of the complex, transboundary, problematic structure of Alpine soil conservation as well as the various challenges in connection with creation of a binding international soil conservation convention, the fact that the SoilProt even entered into force must, in itself, be considered a great success.[44]

Furthermore, the SoilProt provides the legal framework for development of an information and infrastructure platform for the coordinated handling of soil conservation. It contains provisions on cooperation between the Contracting Parties as well as relevant institutions and municipalities. The Contracting Parties "ensure that the national results of the research and systematic observation are integrated in a joint permanent observation and information system."[45] They further create "comparable databases [. . .] to establish possibilities for data exchange."[46] To this end, they shall "reach agreement about soil endangering substances which require priority treatment, and they shall strive for comparable evaluation parameters."[47] In particular they "undertake to map Alpine areas threatened by extensive erosion on the basis of comparable criteria for quantifying soil erosion, and to register those areas [. . .]."[48]

---

[43] Ständiges Sekretariat der Alpenkonvention (2009), p. 60.

[44] Also see Schroeder (2006), p. 138.

[45] See Art. 19 Para. 2 of the SoilProt.

[46] Art. 20 Para. 1 of the SoilProt.

[47] Art. 20 Para. 2 of the SoilProt.

[48] Art. 11 Para. 1 of the SoilProt.

Above and beyond this, the SoilProt pursues a comparatively integrative approach. In addition to the ecological protection aspects with respect to Alpine soil, it also takes the *socio-cultural* and *socio-economic* aspects of soil conservation into account.[49] Socio-cultural objectives are factored in part through a high level of participation by local municipalities as well as by obliging the relevant players to use natural, local and traditional methods and materials which suit the local countryside when implementing construction measures.[50] Socio-economic objectives are pursued through the specific regulation of certain soil-relevant activities such as the extraction of mineral resources, agriculture, pastoral farming and forestry, silvicultural measures and tourism infrastructures.[51] Art. 3 of the SoilProt also requires contracting parties generally to "[...] undertake to take account of the objectives of this Protocol in their other policies [...]".[52]

With regard to the level of protection, one positive element worthy of note is that the regulations of the SoilProt place emphasis on a preventative approach to protection.[53] Additionally, for a convention under international law it contains a comparatively large number of regulations from which categorical or self-executing obligations can be derived. [54] Moreover, in various Member States, the creation/entry into force of the SoilProt as an internationally binding agreement resulted in the adoption of soil convention legislation for the very first time at national level.

It must be stressed that the SoilProt is an important element of the AC. Thanks to its integration in the AC and links with other protocols, soil conservation is now embedded in a more comprehensive ecological and economic regulatory context. This, on the one hand, enables the specific and, in terms of its efficacy from the perspective of international environmental law, ambitious regulation of soil conservation issues. On the other hand, it also allows these questions to be addressed within the framework (of the implementation) of other protocols. Furthermore, implementation of the SoilProt is supported by the institutional framework of the AC.[55] The AC's institutional endowment corresponds to what is usual under international environmental legislation.[56] The Standing Committee in particular has an important role to play in ensuring compliance with the SoilProt.[57] Insofar as the individual Contracting Parties have committed themselves, the SoilProt benefits

[49] For the relevance of the corresponding aspects see in particular: Fromherz (2012), p. 57.

[50] See, e.g., Art. 10 Para. 2; 11 Para. 3; 12 Para. 1, Para. 2; Art. 13 Para. 1 and Para. 2 of the SoilProt.

[51] See, e.g., Art. 8; Art. 12–Art. 14 of the SoilProt.

[52] Clause on taking account of the objectives in other policies in Art. 3 of the SoilProt.

[53] Art. 1 Para. 5 of the SoilProt.

[54] See regulations in Art. 8, Art. 12, Art. 14, Art. 16 and Art. 17 of the SoilProt.

[55] The Member States the SoilProt are always Contracting Parties to the AC, see Art. 26 Para. 2 SoilProt.

[56] Schroeder (2006), p. 134.

[57] See Art. 24 of the SoilProt.

from the dispute resolution system established in the scope of the AC and the extended monitoring mechanism.

The numerous positive aspects of the SoilProt can be set against the few disadvantages. One may be sceptical about whether the institutional structure and design of the AC is currently able to cope with the demanding and dynamic tasks of the AC and its protocols.[58] In addition, with respect to implementation efforts within the scope of the AC, the SoilProt would appear to be of secondary importance compared to the other protocols. Even though the prioritisation of these can be justified on good grounds given the urgency of the issues with which they deal, the problem of soil conservation must not be neglected.[59] Ultimately, a more systematic integration of climate protection objectives in the SoilProt appears expedient.

5.3 *Transferability?*

If one adheres to the functional premise that the purpose of law is to solve social problems, it is in principle conceivable that established legal solutions can be transferred, i.e. be applied to similar social problems arising in other contexts.[60] Since a comparison of the substantive and legal problems of soil conservation in other regions of the world is not possible here, the focus must be upon identifying the particular features of the "international Alpine soil conservation law" context. These features would suggest the limited transferability of the regulations contained in the SoilProt as the legal and factual foundation upon which it is based is characterised by various peculiarities, for which reason this situation differs from the regulatory situations in other parts of the world. Nevertheless, the fundamental structure, individual regulations and, in particular, the experiences gained as part of the national implementation of the SoilProt can, in perspective, still offer significant insights for other possible binding international soil conservation agreements.

In view of the causes behind the loss of ecologically and biologically intact soil in the Alpine region, tourism must undoubtedly be regarded as a significant impact source. The construction of hotels, holiday homes and ski runs are all specific contributing factors here. Equally, the sensitivity of mountain soils to impacts and destruction is, in comparative terms, high.

From a socio-economic and institutional perspective, the Alpine region enjoys affluence compared to many other regions around the world which face the loss of

---

[58]Schroeder (2006), p. 138.

[59]Here it should be noted that the implementation of the other protocols also has a positive effect on the realisation of the regulatory objectives of the SoilProt.

[60]This is the theoretical basic assumption of functional comparative law, see: Zweigert and Kötz (1996), pp. 31–47; Rheinstein and von Borries (1997), pp. 25–28; for critiques see in particular: Frankenberg (1985), pp. 426–429.

ecologically intact soil. Member States' administrative structures are mostly relatively stable and functional. There is a pronounced and institutionalised civic interest in the protection of the Alps which also comprises soil conservation. The development and implementation of the AC were and continue to be shaped and advanced by these institutionalised interests (notably *Commission Internationale pour la Protection des Alpes* [CIPRA], Alpenverein, etc.).[61] The monetary interests of the Alpine tourism industry in preserving the environment are also of relevance here.[62] Moving on, the need for an international agreement on Alpine environmental protection in general and on soil conservation in particular given the transboundary links (traffic, rivers, biodiversity in an ecologically fragile region, see above) is comparatively evident.[63] Not least, the successful conclusion of the AC as a framework convention is another facilitating factor for the creation of an ambitious and relatively specific international agreement on soil conservation. As mentioned above, thanks to its integration in the AC and links with other protocols, soil conservation is now embedded in a more comprehensive ecological and economic as well as regulatory and institutional context.

6 Conclusions

The SoilProt of the AC is, to date, the only internationally binding soil conservation convention; it was developed to protect the soil in the Alpine region. A moderate degree of scepticism should be exercised in terms of the hypothesis that it can act as a model for other regions. The structure, content and value as regards the steering policy of the SoilProt are to be understood first and foremost in the context of the legislative and institutional framework of the AC and, ultimately, as a result of the specific socio-economic and political conditions of the Alpine region. Nevertheless, it does send a positive signal on international soil conservation and thus towards those efforts directed at achieving the SDGs. From an international environmental law perspective, the regulatory design and the ambitious nature of individual regulations can be regarded as exemplary. The SoilProt contains several structural elements and individual regulations which would also appear expedient for other potential international soil conservation agreements.[64] In particular the issue of soil conservation is addressed by taking historical, local, general socio-cultural and socio-economic questions into consideration. In addition, binding measures for sectoral activities such as tourism, mountain farming, silviculture and hazardous substance input are included. Finally, the SoilProt shows that the international

[61] Blecha (2009), p. 1.

[62] See Point 7 and Point 9 of the preamble of the SoilProt.

[63] See Point 11 of the preamble of the SoilProt.

[64] See the contentual suggestions for an international/global soil conservation convention: Boer and Hannam (2014), p. 15 ff. Fromherz (2012), p. 109 ff.; Wyatt (2008), p. 36 ff.

community is capable of recognising the transboundary dimension of soil conservation at least at regional level and that many of the obstacles faced along the path to creating an international soil conservation regime can be overcome.

References

Bayrisches Umweltministerium (2006) Die Bestimmungen der Protokolle der Alpenkonvention und deren Entsprechung im innerstaatlichen Recht der Bundesrepublik Deutschland. Available at https://www.stmuv.bayern.de/ministerium/eu/zusammenarbeit/alpenkonvention/protokolle.htm.Accessed 1 Feb 2016

Blecha S (2009) Die Umsetzung und Anwendung der Alpenkonvention unter besonderer Berücksichtigung touristischer Infrastruktur. Rechtswissenschaftliche Fakultät, Wien

Boer B, Hannam I (2014) Developing a Global Soil Regime. International Journal of Rural Law and Policy, Vol. 2015, No. 1, pp. 1–13, Sydney Law School Research Paper No. 14/85

Ständiges Sekretariat der Alpenkonvention (2009) Die Alpen – Acht Staaten ein Gebiet. Ständiges Sekretariat der Alpenkonvention, Innsbruck

European Environment Agency (2009) Regionaler Klimawandel und Anpassung – die Alpen vor der Herausforderung veränderter Wasserressourcen, Bericht EUA, Nr. 8/2009

Frankenberg G (1985) Critical comparisons: re-thinking comparative law. Harvard Int Law J 26:411–456

Fromherz N (2012) The Case for a Global Treaty on soil conservation, sustainable farming, and the preservation of agrarian culture. Ecol Law Q 39:57–123

Geitner C (2007) Böden in den Alpen – Ausgewählte Aspekte zur Vielfalt und Bedeutung einer wenig beachteten Ressource. In: Borsdorf A, Grabherr G (eds) Internationale Gebietsforschung (IGF Forschungsbericht). Institut für Interdisziplinäre Gebietsforschung, Innsbruck, pp 56–67

Ginzky H (2015) Bodenschutz weltweit – Konzeptionelle Überlegungen für ein internationales Regime. Zeitschrift für Umweltrecht 4(15):199–209

Körner C (1999) Alpine plant life: functional plant ecology of high mountain ecosystems. Springer, Berlin/Heidelberg

Lebensministerium Österreich (2007) the Alpine Convention: Handbuch für ihre Umsetzung. Bundesamt für Land- und Forstwirtschaft, Umwelt und Wasserwirtschaft, Wien

Lee YH (2006) Nachhaltiger Bodenschutz – international, europäisch und national. Universitätsverlag der TU Berlin, Berlin

Linz F, Lobos Alva I (2015) Boden und Land in der internationalen Nachhaltigkeitspolitik – von der globalen Agenda zur lokalen Umsetzung. Zeitschrift für Umweltrecht 4(15):195–199

Markus T (2016) Die Problemwirksamkeit internationaler Übereinkommen am Beispiel des Klimaschutzes. ZaöRV

Meusburger K, Alewell C (2014) Soil erosion in the Alps: experience gained from case studies (2006-2013). Federal Office fort h Environment, Bern

Norer R (2002) Die Alpenkonvention: Völkerrechtliches Vertragswerk für den Alpenraum. Institut für Wirtschaft, Politik und Recht – Universität für Bodenkultur, Wien

Olson M (1965) The logic of collective action – public goods and the theory of groups. Harvard University Press, Cambridge

Permanent Secretariat of the Alpine (2010) People and pressures in the mountains, the facts at a glance. Ständiges Sekretariat der Alpenkonvention, Innsbruck

Poulenard J, Podwojewski P (2006) Alpine Soil. In: Lal R (ed) Encyclopedia of soil science, vol 1, 2nd edn. Taylor & Francis Group, London/New York, pp 75–79

Putnam R (1988) Diplomacy and domestic politics: the logic of two-level games. Int Organ 42(3):427–460

Quillacq P, Onida M (eds) (2011) Environmental protection and mountains – is environmental law adapted to the challenges faced by mountain areas? Permanent Secretariat of the Alpine Convention, Innsbruck

Rheinstein M, von Borries R (1997) Einführung in die Rechtsvergleichung. Beck C. H, München

Schroeder W (2006) Die Alpenkonvention – Inhalt und Konsequenzen für das nationale Umweltrecht. Natur und Recht 28(3):133–138

Schumacher J, Schumacher A (2012) Die Alpen und der Klimawandel – Herausforderungen für die Alpenkonvention. In: Bosecke T et al (eds) Meeresnaturschutz, Erhaltung der Biodiversität und andere Herausforderungen im "Kaskadensystem" des Rechts. Springer, Berlin/Heidelberg, pp 261–285

Söhnlein B (2013) 20 Jahre Alpenkonvetion – Bilanz und Perspektiven aus juristischer Sicht. BayVBl:105–111

Steiniger K, Weck-Hannemann H (2002) Global environmental change in alpine regions: recognition, impact, adaption and mitigation. Edward Elgar Publishing Ltd, Camberley Surrey

Umweltbundesamt (2007) Anregungen zur regionalen Umsetzung der Alpenkonvention. Umweltbundesamt, Dessau-Roßlau

Veit H (2002) Die Alpen – Geoökologie und Landschaftsentwicklung. Ulmer, Stuttgart

Villeneuve A, Castelein A, Mekouar MA (2002) Mountains and the law – emerging trends. FAO, Rome

Wyatt A (2008) The dirt on international environmental law regarding soils: is the existing regime adequate? Duke Environ Law Policy Forum 19:165–208

Zweigert K, Kötz H (1996) Einführung in die Rechtsvergleichung: Auf dem Gebiete des Privatrechts. Mohr Siebeck, Tübingen

Are Soils Taken into Consideration by the IPBES Assessment on Land Degradation and Restoration?

Maylis Desrousseaux, Janne S. Kotiaho, and Florent Kohler

1 Introduction

The Intergovernmental Platform on Biodiversity and Ecosystem Services (IPBES) was established in 2012. It is an international science policy platform the goal of which is to bring independent scientific assessments of biodiversity and ecosystem services into the Convention on Biological Diversity (CBD) process.[1] The interesting short history of the IPBES consultation process reveals that the idea of IPBES has been brewing at least since early 1990s in United Nations Environment Program (UNEP), but that its establishment repeatedly failed.[2] This failing has been attributed to fear of loss of national sovereignty in controlling how biological diversity should be conserved.[3] Today, however, the first assessments are in progress, and the very first one on pollinators, pollination and food production is ready and was approved by the IPBES plenary i.e. governments in February 2016.[4]

[1]Larigauderie and Mooney (2010).

[2]Ibid.

[3]Mooney (2009) and Larigauderie and Mooney (2010).

[4]Press Release: Pollinators Vital to Our Food Supply Under Threat, IPBES, February 26th of 2016.

M. Desrousseaux (✉)
Aix-Marseille University, Marseille, France
e-mail: maylis.desrousseaux@univ-amu.fr

J.S. Kotiaho
University of Jyväskylän, Jyväskylä, Finland
e-mail: janne.kotiaho@jyu.fi

F. Kohler
University of Tours, Tours, France
e-mail: florent.kohler@gmail.com

H. Ginzky et al. (eds.), *International Yearbook of Soil Law and Policy 2016*,
International Yearbook of Soil Law and Policy,
DOI 10.1007/978-3-319-42508-5_13

The first IPBES work programme for the period 2014–2018 is designed to put the Platform on the right path, firmly establishing its working modalities, deliverables, credibility, relevance, legitimacy and reputation.[5] In the work programme it is mentioned that the intention is to pave the way for the incremental strengthening of the science-policy interface for biodiversity and ecosystem services across scales, sectors and knowledge systems. Analytical work and the several assessments initiated under the work programme will be guided by the Platform's conceptual framework.[6]

In essence the idea is that IPBES would have a similar synthesizing role to the Intergovernmental Panel on Climate Change (IPCC).[7] The IPBES is frequently compared to the IPCC, but these two platforms diverge on many points, the most important being the fundamental knowledge gaps "surrounding biodiversity, ecosystem services, and the relationships between them".[8] Plus, the methodology is different and the IPBES process pays a particular attention to Human sciences, including the law. Moreover, the experts are supposed to use an array of sources including the so called "grey literature". This exception in the scope of international expertise is interesting for legal research as it allows indigenous local norms and regulation to be taken into consideration. Indeed, majority of knowledge in this field is not published in peer-reviewed law journals.

The IPBES assessment that may be most relevant for soil is on Land Degradation and Restoration. This thematic assessment is realized at a global scale and it is supplemented by regional approaches divided in four reports, corresponding to four regions: Africa, Americas, Asia and the Pacific, and Europe and Central Asia. Such a task has never been accomplished and no doubt that its reading will bring new elements for decision makers and will hopefully raise the awareness on the necessity to improve soil protection. This specific aspect will be developed in this article.

Land and soil issues are highly connected but still different. Their differences are frequently overlooked by international, regional and national law regimes and this fact increases the lack of a proper soil status and therefore also protection. Some institutions are aware of this situation and in order to avoid confusion, the European commission has expressively differentiated soils from land: the European thematic strategy for soil protection states that "While soil is the physical upper layer of what is usually referred to as "land", the concept of "land" is much wider and includes territorial and spatial dimensions. It is difficult to separate soil from its land context. However, this communication focuses on the need to protect the soil layer as such, due to its unique variety of functions vital to life".[9] This thematic strategy aimed at characterizing the specificity of soil issues and we will follow this distinction along this article. In the IPBES Land Degradation and Restoration assessment, the

[5]http://www.ipbes.net/work-programme.

[6]Diaz et al. (2015).

[7]Larigauderie and Mooney (2010) and Lundquist et al. (2015).

[8]Brooks et al. (2014).

[9]COM (006).

element targeted is the land but in many situations the focus of the law isn't that specific and law regimes use equally "land" or "soil". Furthermore, for the coherence of the Land Degradation and Restoration assessment among the different chapters, definitions of land degradation and restoration are given. Even though one might find these definitions limiting, they present the undeniable advantage to allow relevant comparison points in a domain where we observe important discordances between different sciences. Indeed, concepts and perceptions of land degradation and restoration diverge among social sciences, agronomists, ecologists, etc.[10]

Thus, according to the IPBES scoping document for the Land Degradation and Restoration assessment, "degraded land" is defined as the state of land which results from the persistent decline or loss in biodiversity and ecosystem functions and services that cannot fully recover unaided within decadal time scales. "Land degradation", in turn, refers to the many processes that drive the decline or loss in biodiversity, ecosystem functions or services, and includes the degradation of all terrestrial ecosystems.[11] From this assumption, this article will discuss (2) the potential of the integration of soil issues in the IPBES Land Degradation and Restoration (LDR) assessment from a scientific point of view, and will demonstrate (3) in what ways this process will open new perspectives to improve the legal protection of soils at an international level.

2 The Integration of Soil Issues in the Land Degradation and Restoration Assessment

In spite of the fact that the IPBES assessment is focused on land, the attention given to ecosystem services reveals that soils are included, to the extent that soil services contribute directly or indirectly to the ecosystem services of land. As a result, (Sect. 2.1) soil is expressively targeted in the scoping document of the IPBES. (Sect. 2.2) It appreciates the importance to improve soil knowledge, (Sect. 2.3) including indigenous and local knowledge.

2.1 The Mention of Soils in the Scoping Document of the IPBES

The chapter four of the LDR assessment is about *Status and trends of land degradation and restoration and associated changes in biodiversity and ecosystem*

[10]Chapter 2 "Concepts and perceptions", LDR assessment, in process.

[11]Background Document, 2015, IPBES Thematic Assessment on Land Degradation and Restoration (LDR), Appendix 3.

functions. As soil degradation is connected to land degradation, it couldn't ignore the different threats that soils are facing. The chapter details 11 forms of degradation: soil erosion, contamination, compaction, sealing, and sedimentation, loss of organic matter, soil and water salinization, degradation of freshwater systems, invasion of alien species, changes in natural fire regimes and pollution as a land degradation process.[12] These threats are more or less equivalent to the list established in 2006 in the Proposal of a soil framework directive of the European Union that was unfortunately abandoned in 2014.[13] Moreover, soils serve as an illustration on the Chapter three on *direct and indirect drivers of LDR:* "Direct drivers of degradation (e.g., unsustainable levels of biomass extraction and extractive industries) can result directly in degraded land, including reduction in the productivity of land, or in processes such as soil erosion due to unsustainable land management techniques, and natural drivers, such as floods, wind and drought, that result in land degradation".[14]

Consequently, soils are an element of the IPBES process, which contributes to the global recognition that soil knowledge must be part of the assessment and we can expect that under the cover of land, soil issues will be indirectly addressed, especially by the Chapter 6 that deals with the identification of responses to avoid land degradation and restore degraded land and by the Chapter two about "Concepts and perceptions of LDR".

2.2 *The Need for Soil Knowledge*

At a regional level, the past 10 years have been rich of initiatives, especially thanks to the European Commission. A large part of soil issues has been studied, such as soil erosion[15] or soil sealing, and policy makers can now rely on strong documents[16] due to not having a legally binding basis. At the international level, the United Nations, *via* the Food and Agriculture Organization (FAO) or the United Nation Convention to combat desertification (UNCCD) secretariat, have also produced a lot of literature. As soils are not directly taken in consideration by any UN convention, it is generally observed that they are indirect beneficiaries of diverse Treaties. The Kyoto protocol for instance mentioned agricultural soils and their role

---

[12]Ibid., p. 13.

[13]Withdrawal of obsolete Commission proposals, 2014/C 153/03, *OJEU*, 21st may 2014.

[14]Background Document, 2015, IPBES Thematic Assessment on Land Degradation and Restoration (LDR), Appendix 3, p. 13.

[15]Grimm et al. (2002), p. 44.

[16]Jeffery et al. (2010), p. 136; EC (2012) Guidelines on best practice to limit, mitigate or compensate soil sealing.

in carbon sinks in order to limit greenhouse gas emissions by limiting land use change and encouraging agroecological practices.[17]

This IPBES assessment is a great opportunity to continue to fill the knowledge gaps on soil biodiversity, by following the example of the European Atlas of soil biodiversity and the Millennium Ecosystem Assessment (MEA). In 2005, the chapter on biodiversity of the MEA understood the importance of the species richness in soils for many ecosystem processes and admitted in the same time that this habitat has been relatively poorly studied compared with "aboveground systems".[18] On the contrary, drylands have been the topic of a whole chapter in the MEA and soil formation and soil conservation have been identified as "key supporting services of dryland ecosystems, the failure of which is one of the major drivers of desertification".[19]

A report on the status of the world soil resource has been released by the Intergovernmental Technical Panel on Soil (ITPS) and published in the beginning of 2016 and one of its deliverable is to identify soil-related knowledge gaps that constrain the achievement of sustainable development.[20] The ITPS purpose is similar to the IPCC but for soils and is composed of 27 soil experts from across the seven FAO regions. Among the 600 pages of this major report, regulation systems are slightly mentioned as part of national and regional policies, but the content is quite interesting to the extent that it identifies different systems (regulation, incentives etc.) to point out the benefits and disadvantages of each of them at regional levels. This report gives a steady basis for further research and confirms between the lines that there is a need for legislation, even though in many countries, as China for example, the main problem comes from the gap between the Environmental protection laws and their implementation.[21]

2.3 *A Process Integrating Indigenous and Local Knowledge*

The IPBES assessment pays a particular attention to Indigenous and local knowledge (ILK). In the context of land degradation and restoration, ignoring ILK would have created a gap in the assessment report, considering the large and various array of agricultural practices, harvest, and representations of land among a community of about 370 millions of indigenous people living in more than 70 countries around the world.[22]

[17]Kyoto protocol, 1997, art. 3.

[18]Fitter (2005).

[19]Safriel and Adeel (2005), p. 628.

[20]Intergovernmental Technical Panel on Soil (ITPS) (2016), p. 604.

[21]Mu et al. (2014).

[22]Baptiste (2015) oral presentation, Bonn, Fellow's training session on ILK.

Since 1992 and the adoption of the Convention of Biological Diversity, ILK have known a growing recognition. Article 17 states that the exchanges of information include exchange of results of technical, scientific and socio-economic research, as well as information on training and surveying programmes, specialized knowledge, indigenous and traditional knowledge as such and in combination with the technologies referred to in article 16 paragraph 1. Therefore, they have been progressively included in research projects until the Aichi Target 18: it plans that by 2020 the traditional knowledge, innovations and practices of indigenous and local communities relevant for the conservation and sustainable use of biodiversity, and their customary use of biological resources, are respected, subject to national legislation and relevant international obligations, and fully integrated and reflected in the implementation of the Convention with the full and effective participation of indigenous and local communities, at all relevant levels. The IPBES assessment, then, is a way to achieve this target, which will certainly contribute to the definition of new indicators for sustainable land management and furthermore, open new perspectives for the protection of soils. The methodology to include efficiently these types of knowledge isn't an easy task and the participatory approach has been so far the preferred way to make sure that Indigenous people are part of the assessment and that diverse worldviews are respected.[23] However, as legal research is constrained by legal sources, the access to indigenous data is an issue especially because it is not always clear to determine whether or not a traditional practice has become a custom defined as a usage, which have acquired the force of law, and should be assessed as an expression of a legal perception.

Many elements will come out of the LDR assessment and we can already imagine that soil and indigenous knowledge will be connected, regarding the value of soil given by indigenous people. For instance, an interesting aspect that should be explored is the cosmological dimension of soils as an expression of cultural diversity, a dimension which is heavily impacted by extractive activities such as precious ore mining (Yanomami, Bushmen or San) or changes in land use, especially wetlands (Maori). A law protecting biocultural diversity should integrate the threat on cosmological representation of soils, necessary to the social reproduction of the populations affected.

For the Yanomami (Brazil and Venezuela), goldmining affects the balance between sky and earth, unleashing pathogenic substances that were buried after a cataclysm. This cataclysm—the fall of the sky—allowed a new world to emerge, thanks to Omäme, Yanomami's demiurge. The violence accompanying gold extraction—pollution, explosions, robberies, murders—is also perceived as a cosmological disorder, as Yanomami's society is founded on sharing and non-accumulation.[24]

The case of the Bushmen in Namibia and Botswana is also dramatic. Bushmen live in harsh ecological conditions and are one of the rare human groups that still

[23]Berkes (2004).

[24]Albert (1993).

survive on hunting and gathering. In the Kalahari Desert, most of the resources, including water—are situated underground. Their social reproduction—and their survival—is based on the extension of desert, where food—particularly roots and tubers—is scarce and scattered.[25] Extractive projects (diamond mining) threaten these subsistence activities by destroying resources and fragile ecosystems, luring the game away, then freezing huge territories, complicating the circulation of these hunters and gatherers. According to Survival International—an NGO struggling for Indigenous Rights—the creation of Nature sanctuaries in Botswana is actually the result of vested interests, those of diamond mining companies.[26] As a result, San's way of life has virtually disappeared, while the protected areas will possibly be subject to mining concessions.

For Maori's populations, soil is the expression of balanced ecosystem and cosmic order.[27] The concept of "Mauri" expresses the transcendence of a spiritual/physical principle according to which degraded soils are spiritually damaged, while the connectedness between humans and nature is broken. They read environmental degradation as a deep, underground process threatening social cohesion and the survival of Maoris themselves. A slight modification in land cover or species distribution is integrated into a holistic approach linking ecosystem health and culturally significant places. This knowledge led the Government of New Zealand to build an integrated scientific/indigenous approach to restore wetlands based on Maori memory and classification system.[28]

These three cases illustrate that, beyond ecological threats, the damages inflicted to soils have deep social and representational implications. The figure of "rootedness" is literally adequate to express how soil degradation might affect social cohesion and reproduction. Global negotiations about soils should take into consideration not only the ecological potential of soils, but also the relational values established between human populations and their survival as social-ecological entities.[29] To take the Brazilian example, the recently published Law about mining in Indigenous territories,[30] threatens their protection status and contradicts the policy in favour of cultural diversity, as this diversity is produced and reproduced by special relations established with an above ground and underground territory, which integrity should be considered as a priority. At the opposite, New Zealand Government relied on Maori ancestral knowledge to restore wetlands according to their ecological/spiritual articulation. However, Indigenous and Local population should not be considered only through their knowledge and practices—which is the main focus of IPBES assessments. They are, among human population, those who managed to create and maintain an intimate relationship with their environment,

[25]Tanaka (1980).

[26]http://www.survivalinternational.org/tribes/bushmen.

[27]Harmsworth and Roskruge (2014).

[28]Harmsworth (2002).

[29]Chan et al. (2016).

[30]PL n° 1610/96 modified.

suggesting that relational values associated with the diversity of cultures is to be considered and explored.

As for local knowledge, IPBES assessment is bound to take into account several anterior assessments such as IAASTD[31] and OHCHR[32] report on the right of food, both of which insisting on rescuing local knowledge on seed and breeding varieties adapted to local ecological conditions.[33] The balance food security paradigm is challenged by a new transnational paradigm—food sovereignty[34]—aiming at empowering local populations and rehabilitating peasant and indigenous knowledge,[35] by restoring soil carrying capacities through natural inputs, mixed crops, local seed varieties and agroforestry.[36] These new approaches, among which agroecology, is an original way of dealing with urgent ecological issues while reconnecting human necessities to ecosystemic potential.[37]

3 New Perspectives for the Protection of Soils

The IPBES assessment is part of the process of a rising awareness on soils, (Sect. 3.1) which empowers institutional collaborations at the international level. (Sect. 3.2) Proposals about how soil regulation should be reshaped stem from this collaborations, (Sect. 3.3) and open new perspectives for a complementary and integrated approach.

3.1 Empowering Institutional Collaborations on Soils

Although the IPBES assessment is focused on land, soils are integrated in the scoping document. New partnerships, more or less formal, with existing organizations will be developed. Collaboration or assistance could be expected at various stages, in particular with the United Nations Convention to Combat Desertification. Its science-policy interface and its Committee on Science and Technology are going to be a key user of and a key contributor to the assessment on land

[31]UNEP (2008) International Assessment of Agricultural Knowledge, Science and Technology for Development (IAASTD) Summary for Decision Makers; UNEP (2008) Agriculture at a crossroad. Summary for decision makers.

[32]OHCHR (2011) "Agroecology and the Right to Food", O. DE SCHUTTER, 16th Session of the UN Human Rights Council [A/HRC/16/49].

[33]Scoones (2009).

[34]Altieri (2009).

[35]Rosset (2013).

[36]Deléage (2011).

[37]Rabhi (2006).

degradation and restoration. Collaboration should also be developed with the Global Soil Partnership and its International Technical Panel on Soils (ITPS).[38]

However, partnerships are proven to be insufficient if they are not accompanied with legal initiatives and "In the globalized world of the 21st century, soil sustainability not only depends on management choices by farmers, foresters and land planners but also on political decisions on rules and regulations, marketing and subsidies, while public perceptions are perhaps the most important issue".[39] Now is "time for legislation" is the statement of Luca Montanarella, reminding that if no improvement is made, soils will become soon a matter of security.[40]

3.2 *Rethinking Legal Instruments for Soil Biodiversity*

Most of the time, law is perceived as a tool and not as a scientific domain in itself, which creates a gap between what can be expected from policy makers and how the law can effectively translate new concepts or new goals. The increasing number of international or local assessments interrogates the purpose of the law, especially in environmental law. Eve Truilhé-Marengo observed that if it is clear that scientific discourses have been a powerful force for environmental law and for the protection of biodiversity in particular, the difficulties of implementing this discourses are emphasized at the international level.[41] The sovereignty principle is one of them and remains a strong obstacle against the empowerment of soil protection. Indeed, unlike water, soil isn't perceived as a moving element, which explains partially why there is no international convention on soil. Moreover, protecting soil biodiversity would not even be a powerful lever to minimize the sovereignty principle, regarding that the CBD recalls in its preamble that States have sovereign rights over their own biological resources.

One of the deliverable of the IPBES assessment is to provide policy support tools to decision makers in order to conduct to legitimate and relevant decisions. Policy support tools are expected to be based on science and other knowledge systems that can inform, assist and enhance relevant decisions, policymaking and implementation at local, national, regional and international levels to protect nature, so promoting nature's benefits to people and a good quality of life. But this deliverable raises the question of how the law will take into consideration this new knowledge?

---

[38]Background Document, 2015, IPBES Thematic Assessment on Land Degradation and Restoration (LDR), Appendix 3.

[39]Keesstra et al. (2015), p. 3.

[40]Montanarella (2015), p. 33; concept of security drafted by Günter Brauch and Oswald Spring (2009) *Securitizing the ground, grounding security, Desertification land degradation and drought, UNCCD issue paper* n°2, p. 52.

[41]Truilhé-Marengo (2012).

To properly translate what could be called the "scientific reality" in the law, relevant indicators are needed, to avoid the risk of a counterproductive legal framework which could become a driver of land degradation! In this context, international knowledge complemented by experience based knowledge are highly necessary and the IPBES assessment will contribute to the definition of these indicators. It has a complementary approach of the ITPS soil report, as the latter doesn't take Human sciences into account. Nevertheless, on a scientific aspect, the soil report underlines the difficulties to define indicators of sustainable land management in a domain where there is no scientific consensus. The need for good and clear indicators in protection laws is clear, acknowledging that "the policy challenge depends heavily on organized systems for monitoring soil condition and understanding the relationship with land management".[42]

Despite the fact that the Convention on Biodiversity expressively includes terrestrial ecosystems in the definition of biodiversity, soil biodiversity isn't protected. One of the reasons is that the listing systems of endangered species that is currently use at international and national levels to identify and protect species, isn't suitable for most of the micro-fauna that lives into soils.[43] Indeed, soils would benefit from the development of a proper environmental governance, meaning a governance respectful of ecological processes and managed at coherent territorial scales. Above all, the difficulty to embrace the numerous functions and services provided by soils, require environmental law to be based on an ecosystem approach as described by Likens: "The ecosystem approach, embracing complexity and inclusiveness rather than avoiding them, provides an important ecological tool for identifying, describing, and addressing multifaceted environmental problems".[44] As it is not the case, we observe a mismatch between the ecological reality given by the IPBES assessment and the many realities given by the law at various levels.

On this aspect, the Land Degradation and Restoration assessment will bring an analysis on how the law perceives land degradation and restoration. The lack of perception of soils as a natural media is one of the reason why land degradation can't be efficiently addressed and why land restoration thresholds are most of the time beneath the restoration of functioning ecosystems. It will then represent a starting point of further research and will perhaps contribute to renew existing but obsolete legal concepts.

3.3 Developing a Complementary Approach

The legal recognition of soils would serve the implementation of other regimes of protection such as the Water Framework Directive at the European level.

[42]Intergovernmental Technical Panel on Soil (ITPS), technical summary, 2016, p. 84.

[43]Chaussod (1996), p. 261.

[44]Likens (2014), p. 103.

Reunifying ecological and legal perceptions of soils would simplify the law. As highlighted by Giuseppe Sgorbati and Nicoletta Dotti, The European union network for implementation and enforcement of environmental law (IMPEL) carried out a project to establish the reasons for incomplete or lack of implementation of European environmental law across Europe (2015). In this study, among the causes of the insufficiency of the Water Framework Directive in achieving the "good ecological status" for water, we find the lack of soil protection. Indeed, they qualify the creation of a specific regulation as a "key implementation challenge".[45] Soils can no longer be considered separately and a renewed governance is needed.

There are high expectations towards the post-2015 Development Agenda and the Sustainable Development Goals (SDGs). Many of them take soils into account and the 15th SDG is directly related to soil: "By 2020, ensure the conservation, restoration and sustainable use of terrestrial and inland freshwater ecosystems and their services, in particular forests, wetlands, mountains and drylands". Furthermore, a study about the implementation of the land degradation neutral world target has underlined that "a number of goals expressly mention the relevance of land". The authors have enlightened connections between land and soil in Goal 2 (End hunger, achieve food security and improved nutrition and promote sustainable agriculture), Goal 3 (Ensure healthy lives and promote well-being for all at all ages), Goal 7 (Ensure access to affordable, reliable, sustainable and modern energy for all), Goal 11 (Make cities and human settlements inclusive, safe, resilient and sustainable), and 12 (Ensure sustainable consumption and production patterns) etc..[46] If it is certain that Ecosystem services contribute to "nearly all land-related SDGs",[47] the scientific community still needs to figure out how soil science could have a concrete contribution. Thus, "A clear framework linking SDGs, ecosystem services and soil functions will also pave the way towards a more relevant contribution of the soil science community to on-going major global and regional ecosystems assessments related to land and soils".[48] In other words, the IPBES assessment on Land degradation and restoration will provide it.

4 Conclusion

Soil is a "dark realm" to remember the words of Rachel Carson.[49] Already in the 1960s, she strived to catch people's attention toward what we would name nowadays "soil biodiversity": "Life not only formed the soil, but other living things of incredible abundance and diversity now exist within it". Forty years later, the

[45]Sgorbati and Dotti (2015), p. 10.

[46]Dooley et al. (2015), p. 5.

[47]Dominati et al. (2014).

[48]Keesstra et al. (2015).

[49]Carson (1962), p. 53.

scientific community definitely knows more about soils. Knowledge is a lever to improve soil protection, not only to counter the argument that it is impossible to properly legally frame an object that the scientific can't seize, but also because a better knowledge on soil will increase the awareness that a "Global soil regime" is absolutely necessary.[50] With the IPBES assessment, the role of soils to preserve and restore land should be enlightened. Most of the chapters of this assessment are asked to do so, directly and indirectly. Thus, knowledge gaps on soils in social sciences as much as life sciences will be identified, giving new opportunities to policy makers in the construction of a national and international soil regulation.

References

Albert B (1993) L'Or cannibale et la chute du ciel: Une critique chamanique de l'économie politique de la nature (Yanomami, Brésil). L'homme, 349–378
Altieri MA (2009) Agroecology, small farms and food sovereignty. Mon Rev 61(3)
Berkes F (2004) Rethinking community-based conservation. Conserv Biol 18(3):621–630
Boer B, Hannam I (2015) Developing a global soil regime. Int J Rural Law Policy Soil Gov, Special edition 1/2015, 1–13
Brooks T, Lamoreux JF, Soberon J (2014) IPBES ≠ IPCC. Trends Ecol Evol 29(10)
Carson R (1962) Silent spring, First Mariner Book editions, p 381
Chan KM, Balvanera P, Benessaiah K, Chapman M, Díaz S, Gómez-Baggethun E et al (2016) Opinion: why protect nature? Rethinking values and the environment. Proc Natl Acad Sci 113 (6):1462–1465
Chaussod R (1996) La qualité biologique des sols, évaluation et implications. Etude et Gestion des Sols 3(4):261–278
Deléage E (2011) Les mouvements agricoles alternatifs. Informations Sociales 2(164):44–50
Diaz S et al (2015) The IPBES conceptual framework—connecting nature and people. Curr Opin Environ Sustain 14:1–16
Dominati E, Mackay A, Green S, Patterson M (2014) A soil-change based methodology for the quantification and valuation of ecosystem services from agro-ecosystems: a case study of pastural agriculture in New Zealand. Ecol Econ 100:119–129
Dooley E, Roberts E, Wunder S (2015) Land degradation neutrality under the SDG's: national and international implementation of the land degradation neutral world target. ELNI Rev 1+2:2–9
Fitter AH (2005) Darkness visible: reflections on underground ecology. J Ecol 93:231–243
Grimm M, Jones R, Montanarella L (2002) (Revised) Soil Erosion Risk in Europe, European Soil Bureau, Institute for Environment & Sustainability, JRC Ispra, p 44
Harmsworth G (2002) Coordinated monitoring of New Zealand wetlands, Phase 2, Goal 2: Maori environmental performance indicators for wetland condition and trend. A Ministry for the Environment SMF project–5105. Landcare Research Contract Report LC, 102(099), 65
Harmsworth G, Roskruge N (2014) Indigenous Maori values, perspectives, and knowledge of soils in Aotearoa-New Zealand. In: The soil underfoot: infinite possibilities for a finite resource, p 111
IPBES (2015) Background document, IPBES thematic assessment on land degradation and restoration (LDR), p 43
ITPS (2016) Report on the status of the world soil resource, p 650

[50]Boer and Hannam (2015).

Jeffery S, Gardi C, et al (eds) (2010) European Atlas of soil biodiversity. European Commission, Bureau des publications de l'Union européenne, Luxembourg, p 136

Keesstra SD et al (2015) FORUM paper: the significance of soils and soil science towards realization of the UN sustainable development goals (SDGs), journal SOIL-88

Larigauderie A, Mooney HA (2010) The intergovernmental science-policy platform on biodiversity and ecosystem services: moving a step closer to an IPCC-like mechanism for biodiversity. Curr Opin Environ Sustain 2:9–14

Likens GE (2014) The ecosystem approach for understanding and resolving environmental problems. In Watson R (ed) Environment and development challenges: the imperative to act. University of Tokyo Press, p 103

Lundquist C et al (2015) Engaging the conservation community in the IPBES process. Conserv Biol 29(6):1493–1495

Montanarella L (2015) Govern our soils. Nature 528:30–33

Mooney HA (2009) The ecosystem-service chain and the biological diversity crisis. Trans Roy Soc B 365:31–39

Mu Z, Bu S, Xue B (2014) Environmental legislation in China: achievements, challenges and trends. Sustainability 6(12):8967–8979

Rabhi P (2006) Conscience et environnement, Gordes, éd. du Relié

Rosset P (2013) Re-thinking agrarian reform, land and territory in La Via Campesina. J Peasant Stud 40(4):721–775. doi:10.1080/03066150.2013.826654

Safriel U, Adeel Z (CLA) (2005) Ecosystems and human well-being: current state and trends, dryland systems, ch. 22, pp 623–662

Scoones I (2009) The politics of global assessments: the case of the International Assessment of Agricultural Knowledge, Science and Technology for Development (IAASTD). J Peasant Stud 36(3):547–571

Sgorbati G, Dotti N (2015) Perspectives and actions to improve water quality in European Union Member States. ELNI Rev. 1+2:10

Tanaka J (1980) The San, hunter-gatherers of the Kalahari: a study in ecological anthropology. University of Tokyo Press, Tokyo

Truilhé-Marengo E (2012) Gouvernance de la biodiversité, droit et expertise scientifique, Research project, CERIC. Aix-Marseille University

Part III
Cross-Cutting Topics

Soil Degradation and Migrations in the Age of the Global Environmental Crisis: A Policy-Making Perspective

Grammenos Mastrojeni

1 A Quest for Certainty and the Need to Act Now

Ideally, all policies should be rooted in an accurate description of the reality they intend to act on. And when such reality is complex and subject to interpretation, it is all the more essential to evaluate it according to rigorous quantitative findings.

Yet, when it comes to the nexus between the conditions of soils and migration patterns, available data is scarce and interpretations appear uncertain, if the question is looked upon with a technical, statistical, or anyhow purely scientific approach. Science is growing and increasingly accurate in evaluating the global state of soils: a recent survey by FAO and the Intergovernmental Technical Panel on Soils sums up conclusions,[1] depicting an accelerating worldwide trend of land degradation. But its further societal consequences, including migrations, are not investigated with comparable means.

On the other hand, "grey literature"—mainly produced by strategy think tanks, and international or non-governmental organizations—is assertive in claiming that indeed there is a cause-to-effect link between the degree of vitality of soils and movements of populations, and that it has to be urgently coped with. Is it a mere advocacy bias? (Fig. 1)

Correlations between soils and migrations represent, at a first glance, but a portion of a still much debated and much broader issue: the impacts of environmental modifications on development, social cohesion, human rights, security and peace. Within this yet largely unexplored overarching theme, comes another broader question: is there a link between environmental degradation and

[1] Status of the World Soils Resources, Rome (FAO and ITPS), 2015.

G. Mastrojeni (✉)
Italian Development Cooperation, Rome, Italy
e-mail: grammenos.mastrojeni@esteri.it

H. Ginzky et al. (eds.), *International Yearbook of Soil Law and Policy 2016*,
International Yearbook of Soil Law and Policy,
DOI 10.1007/978-3-319-42508-5_14

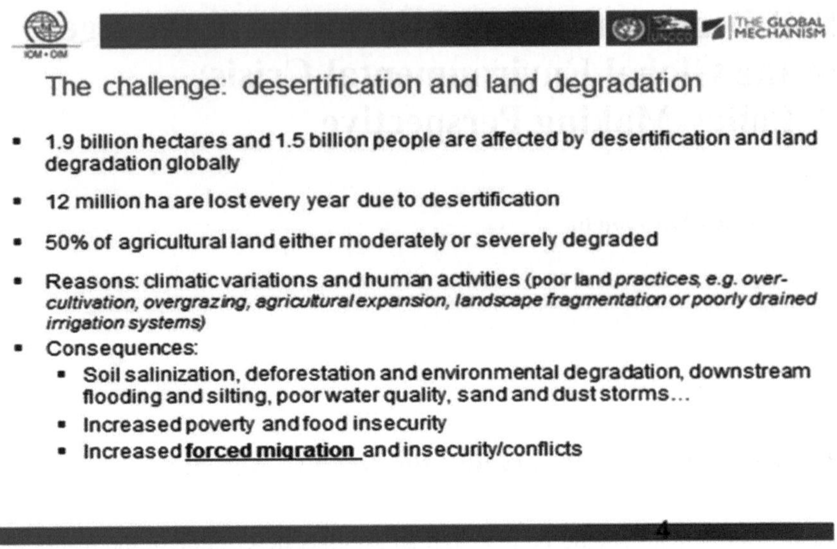

Fig. 1 IOM and UNCCD asserting the nexus

migrations? The reason is obvious: if environmental modifications grow into a migration push factor it is mainly because they erode local livelihoods in all their dimensions. In turn, the soil degradation—displacements nexus looks but a segment of the environment—migrations problem (Fig. 2).

Most peer reviewed articles about the broader environment-migrations issue open up cautioning that there is no consensus, and that data is fragmented. Based on such uncertainty, in general, literature tends to fall in two categories: "maximalists", who claim that the environment directly causes people to be forced to leave their homes; and "minimalists" who suggest that the environment is only a contextual factor in migration decisions.[2]

Under the umbrella of such a fundamental divergence about environment and migrations, comes the problem of collecting globally significant and reliable data: a daunting task, since it has to be carried out at the crossroads of complex interactions among societal, economic, cultural, and political factors, each isolated and verified, and yet correlated, within diverse ecosystems, after assessing the latters' evolution and rate of change. On top of this complexity, findings have to be ascribed to categories of actors—the "environmental migrants" and "environmental refugees"—which are themselves contended or, at least, not unanimously defined and, in any case, sub-specified in further groups based on heterogeneous criteria: voluntary or forced migrants, rural to urban, rural to rural, intra-national, international, seasonal or permanent, and so on. To say nothing of another fundamental divide in perceptions, charged with moral and political implications, likely to inject

[2]Among others, Morrisey (2012).

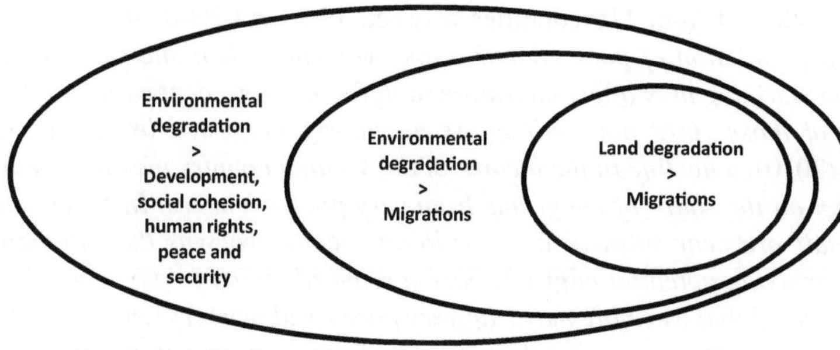

Fig. 2 Part of a broader problem

bias even in scientific approaches: are migrations a problem or a solution[3]? Are population movements a form of adaptation?

As François Gemenne clearly puts it, *four themes permeate the literature on the [environment—migration] nexus: that research is impeded by a lack of empirical studies; driven by a climate change dominated agenda; abundantly supplemented by 'grey' literature; and marked by disciplinary divides.*[4]

When it comes to further isolating a specific role of soils, within the overall issue of environmentally induced displacements, things get worse; but they also get better. Worse because in a "climate change dominated agenda" peer reviewed literature is not abundant and is inherently based on local findings that lack the critical mass to be proven of global relevance. Better because at least one pattern seems to emerge out of the different approaches, areas, methods and indexes, many of the latter being only indirect indicators of the state of soils: droughts, crop yields, and vegetation cover dynamics, among others.

The one finding that seems to be recurrent is described in a literature overview by Clak L. Gray, pointing out that *studies suggest that the most common relationship will be for out-migration to be higher where soil quality is poor. Nonetheless, both migration theory and a subset of previous studies indicate that the opposite effect is also possible, i.e. that out-migration might be higher where soil quality is high. The key theoretical issue is that the demands of migration, such as access to migrant networks and the cost of housing, can be high relative to household resources and thus create barriers for entry that exclude poor households from particular migration streams (White and Lindstrom 2005). In extreme cases, a poverty trap might prevent agricultural households from both purchasing agricultural inputs and sending out-migrants. Consistent with the view are studies showing that international migrants often originate from land-rich households.*[5]

These findings about land mirror dynamics also recurrent in the climate change-migration nexus, where—as related by Cristina Cattaneo and Giovanni Peri, based

[3]Atlas des migrations environnementales, 2016, Paris (IOM-OIM), p. 10.

[4]Gemenne (2011), p. 1.

[5]Gray (2011), p. 3.

on data collected from 116 countries between 1960 and 2000—*by impoverishing the rural population of poor countries and worsening their income perspectives, longterm warming may affect migration in different ways, depending on the initial income of those rural populations. As previously suggested by studies such as Mayda (*2010*), a decline in the income of the sending country may have a depressing effect on the share of emigrants from very poor countries. In these countries, individuals are near subsistence, so a lower income worsens their liquidity constraint, implying potential migrants have a reduced ability to pay migration costs. In this case, global warming may trap very poor rural workers who become unable to leave agriculture, worsening their poverty. To the contrary, in countries in which individuals are not extremely poor, a decline in agricultural income may provide incentives to migrate to cities or abroad. Decreasing agricultural productivity may encourage a mechanism that ultimately leads to economic success for migrants, benefiting their country of origin and shifting people out of agriculture into urban environments. The inverted U shape of migration rates as a function of income per person in the countries of origin is usually rationalized in this type of framework.*[6]

Awareness of the inverted U curve as a recurrent pattern is of course one first certainty to build on. But is it enough to guide urgently needed political choices? There is a gap between large scale dynamics perceived as occurring in the real world that need a pragmatic response, on one hand, and their rigorous quantitative assessment: in other words, threatening cycles that link land degradation to instability are being repeatedly observed as taking place at an ever faster pace, and policy makers need to act now; but from a scientific perspective the furthest one can venture to say is that they are no more than "commonly" considered real.

Two gaps of knowledge look most problematic:

- literature has so far mainly examined the link between soils conditions and human displacements only with reference to agricultural productivity, neglecting the chain reactions that could result in displacements due to the degradation of lands that do not host agriculture; like forests, wetlands and other more or less pristine ecosystems on one end of the spectrum, or lands supporting urbanization, infrastructure, industry, tourism, or mining, on the opposite end;
- current analysis have adopted a business as usual overlook on the soils-migrations nexus, trying to identify what generally happens; yet, sadly enough, the situation policy makers have to manage has evolved beyond business as usual, and the role of soils has to be understood in the context of a potential chain collapse of environmental balances, in which every ecosystem variable reacts to the modification of any other variable.

At this stage, policy makers possibly need to harness and manage a global environmental emergency, with all its socio-economic implications. In this perspective, the following pages aim at describing, in a systemic set, the network of

[6]Cattaneo and Peri (2015), p. 2.

interactions around soils that are perceived as relevant for stability by the political machinery; in the hope that this would encourage the scientific community to move in swiftly, and transform into verified facts what is nowadays a critical, reasonable, but still no more than untested perception.

2 Compelling Policy Choices and a Special Status for Lands

The grey science of political analysts and geo-strategists has been slower than Academia, and has only recently caught up reaching a basic consensus: environmental balance matters for livelihoods, peace and stability.[7]

The general cause-to-effect link that has been identified is that, in the first place, some environmental modifications directly impact human beings: for instance, heat waves, water pollution, particulate, or the hole in the ozone layer directly put at risk human health. Together with these, other environmental modifications trigger cumulative cycles of instability when they affect human socio-economic organization, even if they do not directly concern human physiological well-being. By suppressing or displacing ecosystem services upon which societies have organized their livelihoods, environmental modifications cause insecurity and uncertainty in planning for the future, well beyond the dimension of agricultural yields; insecurity and uncertainty, in turn, pave the way to poverty and disputes around vanishing services, fragility, possibly violent conflicts and, all together, this mix integrates a push factor for both voluntarily chosen or forced migrations.

This link is not a new product of the recent, fast pace, human induced, environmental degradation; it has always been at work. Our global ecosystem is subject to spontaneous fluctuations that, historically, have impacted the stability of numerous human communities, have led many to disperse or migrate or, according to various researches, have even proven the tipping point towards the demise of entire civilizations.[8]

Quantitative verifications have been run to identify spontaneous fluctuations and natural cycles that created socio-economic stress and therefore instability: mankind,

---

[7]While academic literature and warnings from environmental international organizations started to proliferate 30 years ago, it wasn't before 2009 that some timid official recognition of the peace and stability dimensions of environmental stress surfaced as a shared concern in the official world of policy makers. The turning point can be pinned down in the 2009 UN Secretary General Report to the General Assembly *Climate Change and its Possible Security Implications*. Currently, reflections on these links are supported within G7 that, i.a., commissioned the 2015 Report *A New Climate for Peace, Taking Action on Climate and Fragility Risks*. The exception is to be found in military organizations: notably, NATO started an early evaluation for security purposes. See *Conflict and the Environment: Proceedings of the NATO Advanced Research Workshop*, 1997.

[8]Among various, Hodell et al. (1995), p. 391; Peterson et al. (2003); Zhang et al. (2006), p. 459; Gee (2004), p. 411; Huntington (1971), p. 173.

for instance, until now could do nothing to prevent the socio-economic impacts of El Nino Southern Oscillation (ENSO) on the southern Pacific and beyond.

ENSO is a recurring 5 years cycle—due to a coupled and resonant interaction between oceanic and atmospheric temperatures—that causes alternate phases of drought or heavier rainfall: during the El Nino phase, when the ocean warms up, a drier climate affects soils fertility, while during La Nina part of the cycle, a more humid climate restores soil's productivity. And it did not come as a surprise the statistical proof that in affected areas El Nino phase has historically brought with it a roughly double probability of conflict compared with La Nina periods and that ENSO may have had a role in 21 % of all civil conflicts since 1950.[9] Nor could our ancestors understand and manage the consequences of periods of colder weather in Asia, when—according to a research that examined almost a millennium since the year 1000—the probability of conflict increased up to 2.24 times due to diminishing agricultural yields.[10]

This recently understood relationship is presently being exacerbated by human induced systemic modification of the environment that, if left unchecked, could inject pervasive and global instability. The pivotal factor in this nowadays crucial dynamic is not the kind of environmental alteration—climate change, acid rain, biodiversity loss or else; instead, the sensitive variables and triggering fulcrums of instability are the ecosystem services that all disparate environmental changes affect: for instance, even a significant evolution in climate will not grow into an instability engine in a given location, if it leaves ecosystem services vital and unattained.

Therefore, while it is imperative to restore environmental health as such—and solve the roots of the problem—in the medium term and tactically it is essential to identify and protect useful ecosystem services: the two goals are largely overlapping but, operationally, do not always coincide.

Preserving ecosystem services that provide human stability—as a clear goal on its own, besides preserving environmental balance as such—is essential; at least because it is the way to defuse the most dangerous, and too often neglected, positive feedback loop that could jeopardize our chances to secure a functional planetary environment. The notion that human induced modifications in the environment could trigger amplifying feedback loops in the eco-sphere is becoming familiar to the public, mainly thanks to loops and tipping points that have been clearly identified by science in the dynamics of climate change: permafrost thaw, snow and ice albedo, forest fires-carbon release cycles, and so on. Chain collapses in biodiversity are also coming in the spotlight, like the one feared as a consequence of the sudden disappearance of phytoplankton—the base of marine food chain—due to ocean acidification, caused in turn by the absorption in the seas of increased levels of CO_2 in the atmosphere.

---

[9]Hsiang et al. (2011), p. 438; Lee et al. (2013), p. 1. A projection of severe near-surface permafrost degradation during the 21st century; Lee (2009).

[10]Zhang et al. (2007), p. 19214.

These concerns are construed conceiving mankind as the initiating actor of a pattern of change in nature, but in some cases are analyzed and projected in future scenarios as if they only developed within the "natural world", with humanity as a non-reactive spectator. This approach overshadows the fact that the greatest unknown variable for the future refers to human behavior in the context of a growingly dysfunctional ecosystem.

While the ecosystem keeps challenging human organization with its ever swifter modifications, the fundamental question is whether humanity will be able to stay united in order to restore the balance of the ecosystem itself in a rational way, or rather turn to irrationally competitive behaviors that could continue feeding its disruption. The commonly feared and condemned scenario is a "business as usual" protracted neglect of natural balance; but the truth is that a worse turn is looming and could take the shape of a catastrophic feed-back loop far worse than business as usual: if the impairment of ecosystem services is pushed beyond a given threshold, it triggers a certain degree of systemic insecurity, societal and institutional fragility, instability and conflict which, in turn, will paralyze the international community's capacity to unite and manage rationally the ecosystem itself; or, worse, lead to those blind behaviors and deep scars to nature typical of exacerbated competition, warfare, or instability situations. This, in turn, could worsen environmental degradation creating even greater instability and conflict in a dangerous globally growing cycle.[11]

Defusing this loop is imperative; and it necessarily passes through granting a sufficient and fairly distributed access to those ecosystem services essential for an orderly economy and social cohesion.

This policy oriented perspective—protect the services to make sure we will remain able to rescue the environment—paves the way to a further gap between science and decision making by giving lands a special status, and one that would make little sense in a purely scientific approach. Science would spontaneously investigate, without setting up artificial priorities or rankings, the effects of all environmental modifications on nature and mankind, within the interdependencies that connect everything to everything in our planetary balance: there is no point, for science, to conclude that a biodiversity-migrations nexus is less important than a soils-migrations one, or to prioritize services against root causes.

On the contrary, from a policy perspective and if we agree that preserving ecosystem services that provide human stability is essential, land emerges with a special status. Soils indeed—in parallel with oceans and seas—are the sponge, the aggregators, and the concrete containers of ecosystem services. Therefore, the "vitality" of soils—and seas—is the end aggregate variable on which all sorts of environmental modifications concentrate those effects that are likely to destabilize society: the main connection central between environmental and human balance. And, at the same time, it is a practical—although really not technical—variable to handle, simply because soil is material and localized.

[11]Austin and Bruch (2000).

No matter what the originating causes of environmental modification are—if they primarily concern climate, water, glaciers, biodiversity or else—the way they reflect on the capacity of soils to provide services represents the fundamental mechanism of involvement of mankind in a potential cycle of instability, societal collapse, conflict and displacements, that will in turn lead to even greater ecosystem failures.

The good side of this central and fundamental link between humanity and land is that it could work also in the opposite direction: chances are that protecting soils could trigger a broad peace, stability and ecosystem recovery cycle, extending far beyond soils themselves.

3 Gray Complexity: Protect Our Lands!

At this stage of the analysis, having identified soils as a "practical object" of policies, some systemic order needs to be made to define the social, economic and political functions that lands play in the long chain that connects environmental degradation to instability, especially to its present most visible symptom: tragic waves of displacements.

To do so, nonetheless, we have to venture even further in an area characterized by a visible shift between science and policy approaches, that implies relying on blurry and unchecked indicators. Especially one: the "vitality" of soils, a non-technical, never clearly named and, paradoxically, hardly quantifiable measure of the aptitude of soils to continue providing the services they have been traditionally harboring. But that generic "vitality" is what, substantially, the policy community is looking at.

From a policy making point of view, it is enough to "make sure that soils continue doing their jobs". This description, simplistic as it is, could fit the purpose and it is the only practical one available to strike delicate and urgent choices, but with a great risk: since it is the product of mere policy thinking—driven by so imperfect and yet so powerful tools as experience, historical similarities, and common sense—it may just reflect an only apparent reality. And at this junction lies, once more, the main meaning of these pages: sending out an SOS to science, called to check on the following non rigorous—but not irrational—policy assessment of the complex interactions that concur to form the nexus between soils and the looming exodus. Checking the connection knots identified by policy makers would not come spontaneous to the hard science community, because these knots and interactions reflect a different way to categorize reality; they are assembled through the observation lens of a governance problem, not on the basis of an unbiased distant observation of mankind and nature. But they serve a purpose, and science confirmation is necessary.

In a nutshell, the policy approach to the soils-stability connection is being more and more oriented by two reasonably sound observations, although rigorously they would barely qualify as little more than assumptions:

(a) taken as core aggregators of ecosystem services, soil proper management emerges as a crucial challenge for an orderly progress of mankind as a whole. In its obvious and first implication, this means that direct organization of soil use has to be improved. But the stakes are now so high for human stability, that it furthermore implies that soils need to be protected also from the impacts of spontaneous ecosystem fluctuations, and from other forms of man-induced environmental modifications different from direct land mismanagement. And, beyond preserving the healthy soils that are left, it is necessary—and potentially a remarkably cost-effective investment—to recover the vitality of lands that have lost it;
(b) soil degradation tends to display the same chain of societal consequences in every ecosystem, and their magnitude does not only depend on the bio-physical features of a given territory; it is rather a direct function of local fragility in the human context. All forms of environmental degradation, generally affecting the state of soils, act according to a definition that the U.S. Department of Defense and NATO have focused referring to climate change: "crisis and conflict accelerators" or "threats multipliers". The idea that environmental stress will prove first an "accelerator" rather than a stand-alone cause of conflicts, instability, and migrations as an end result, reflects the notion that ecosystem services depletion can be absorbed and countered in richer societies, especially if they provide emergency safety nets or social and productive assistance to concerned families, and if they have means to access the global market to compensate local depletion. On the contrary, stress on ecosystem services overburdens the cohesion and security structure of socially fragile or poorer communities; there it initiates or amplifies latent tensions and conflicts, that have nonetheless the potential to spread globally, so that it is clearly a common interest of mankind to give priority to the protection and recovery of poorer and fragile communities lands. In this scenario, no nation can consider itself safe and isolated, even if it is solid enough to face environment degradation on its own territory, or if it is temporarily benefitting from environmental modifications.

This policy vision results in defining a set of relevant interactions focused by decision makers, better described as a flux diagram (see Fig. 3).

In this mindset, soil propelled population displacements represent the final result of a combinatory game; a complex chain of possible interactions—which mix is actually at work basically depends on local contexts—triggered once the disappearance of ecosystem services harbored by soil concurs to making a certain area hard or even impossible to live in. In the continuum between "harder to stay than it once was" and "impossible to remain", the inverted U curve applied against income becomes a useful tool to forecast local outcomes and compare different situations.

Beyond this grid of analysis that can be applied locally, another layer of complexity has to be added due to the transmission chain provided by global markets. It is a fact that not only economic growth, but also the impoverishment of soil in richer countries, fuels large acquisitions of land in poorer regions, that at

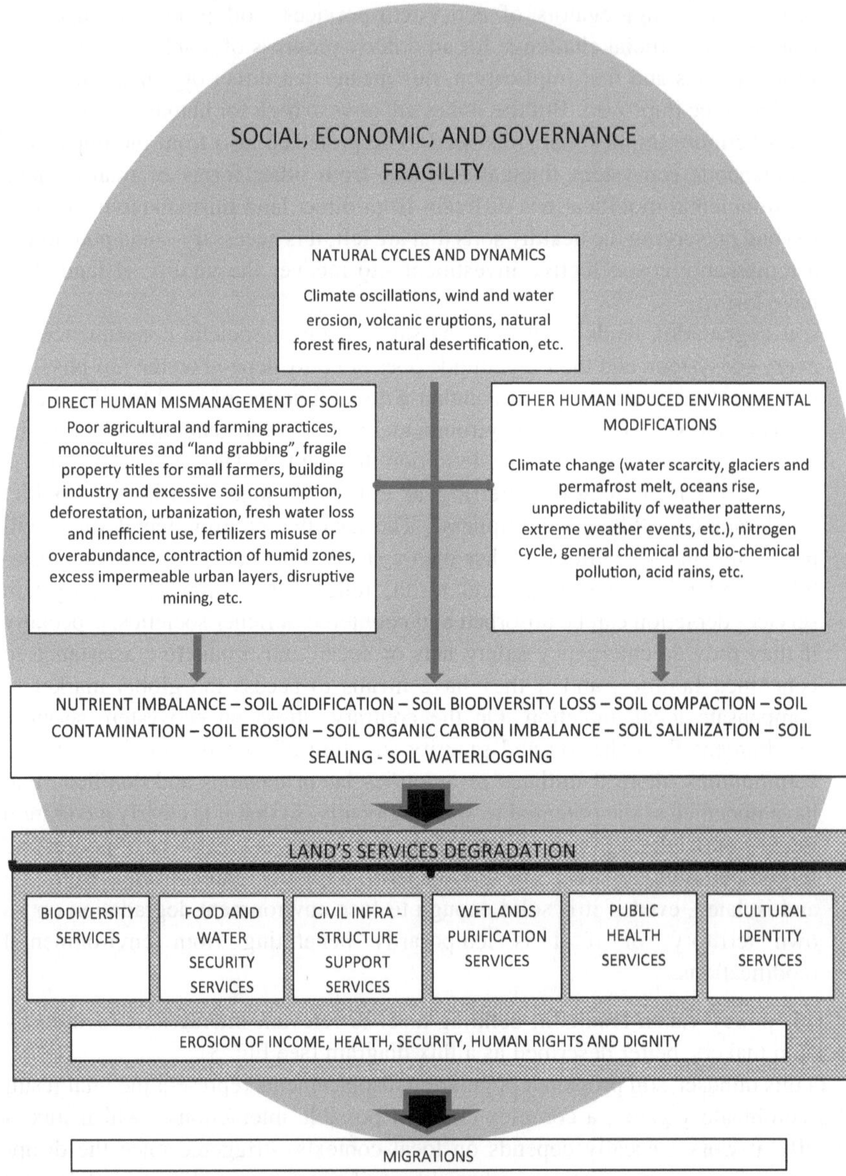

Fig. 3 The scope of a combinatory game of possible causes

times take the obscure face of "land-grabbing". Often these surfaces, to the detriment of more soil protective family-scale uses, are exploited as industrial farming, monoculture, chemical fertilizers intensive productions. This way, a land degradation problem in a richer portion of the planet transforms in a threat to socio-economic empowerment and also in soil degradation dynamics in another, weaker, part of the world.

Out of all this complexity, the important aspect is that, through the above described prism of observation, it looks like chain reactions involving land degradation—and that push livelihoods beyond the "stay threshold"—are frequent.

4 The World According to Grey

Of course, the policy-making community is not monolithic in its perceptions of the problem. The diagram above represents an attempt to put together in an orderly way the core and most commonly shared assessment; but it is likely to be met by each think tank and organization with disappointment, because it does not highlight enough one or another factor valorized in their analysis: for instance, demography, energy, food market trends, are also concurring and relevant dimensions in the soils-migrations nexus; but, comparing the various narratives, the central and purposely basic thinking can be reduced to this interactions network.

Here, 50 shades of grey characterize conclusions and even more their sources, all reliable but of heterogeneous nature. Besides the available science, the policy community has run an assessment of the soil—stability issue based on results of past interventions, seemingly similar and frequently heard individual or communal stories, historical precedents, strategic projections and so on.

Out of these, and through the above described observation prism, a picture of the worldwide problem has emerged, mainly catalyzed in positions expressed by FAO, IOM and UNCCD.[12] Basically it says that it is hard to quantify current volumes of specifically land-degradation driven displacements: too many factors interact. Seemingly, it is unwise to venture in selective projections about the future numbers of a soil caused exodus. But it also firmly asserts that the potential is disturbingly vast. And this—the potential before the outcome—is not guessed by policy practitioners; it is drawn by science.

According to this vision, this threatening potential is condensed in a few figures: worldwide some 1.9 billion hectares and 1.5 billion people are affected by land degradation. Approximately 50 % of agricultural land is either moderately or severely degraded. And every year, the productivity of 12 million hectares of land is lost entirely. This is the amplitude of the field where the destabilization and migration engine can be possibly put in motion. It is being made worse because we are struggling to combat land degradation while also battling climate change. Land degradation and climate change form a 'feedback loop' whereby land use and agricultural practices increase emissions. At the same time, the loss of soil and vegetation significantly reduces potential carbon sinks. The result is more

[12]Most explicitly on this issue, *Migration and Desertification*, UNCCD Factsheet Series n. 3, Bonn, 2014.

greenhouse gases in the atmosphere feeding a dynamic cycle. Land degradation, biodiversity loss and climate change feed off each other.

Natural resources such as drinking water are set to become ever more limited. Many crops are unlikely to survive in certain locations. Food security which is already a significant concern will become even more challenging.[13] The IPCC predicts a decline in production of 2 % per decade for the rest of the century. This is all happening at a time when a growing population—expected to hit more than 9.5 billion by 2050—will demand more food.

This shrinking natural services base is indeed directly undermining the livelihoods and food security of many millions of people. In these kind of contexts, already present in 168 countries, the field is open for the above described combinatory game of interactions to start and grow into a chain reaction, resulting in massive population displacements.

If this vision lacks enough quantitative investigation, on the other hand it displays one of the typical strengths of sound scientific theories: predictive power. Some observed situations stand out as a foretelling warning. Just one example, in Latin America: as a result of increasing drought and falling productivity, more than 60 million people in the Sertão region of northeast Brazil migrated from rural to urban areas between 1970 and 2005.[14] This is a worrying trend for Brazil, but also for the world at large because it is being repeated again and again—in communities all over the planet.

As pressure on the natural resource base mounts and supply of food and water dwindles, competition will inevitably rise. Populations in at-risk areas are simply running out of coping mechanisms locally. Desperate people are increasing looking for a "way out". And migration is one way out.

In recent years, internal migration in poorer countries grew to record levels, and the projected scale of forced migration in the coming decades will exceed anything experienced previously. Grey—but still trustworthy—estimates suggest that by 2020, 60 million people could have moved from degraded parts of sub-Saharan Africa towards Europe and North Africa[15] and that 200 million people may be permanently displaced environmental migrants.[16] Other estimates suggest that 2.4 billion people could be living in areas subject to intense water scarcity by 2025.[17] If migration is not an option, extremism and conflict is another way out or, simply, the two could coincide and foster reciprocally.

One example is in the Lake Chad region. In just 5 years, the Lake Chad region has become a hot-bed of Boko Haram terrorist activities. The roots of this

---

[13]UNHCR, http://www.unhcr.org/pages/49e4a5096.html.

[14]UNCCD, policy brief on migration, http://www.unccd.int/Lists/SiteDocumentLibrary/Publications/Migration%20policy%20brief%20Final%20draft.pdf.

[15]http://www.unccd.int/Lists/SiteDocumentLibrary/secretariat/2014/ES%20Statements/Vienna%20Speech%20210314.pdf.

[16]Migration, Environment and Climate Change – Assessing the Evidence, Geneva (IOM), 2009.

[17]*Land Based Adaptation and Resilience*, Bonn (UNCCD), 2014.

extremism can in part be traced back to the combined effects of land degradation and drought. Lake Chad is a critical wetland area. In times of drought, it serves as a seasonal migration area for people from Cameroon, Chad, Niger and Nigeria. But the population has grown rapidly—from 22 million in 1991 to 38 million by 2012. It is forecast to reach 50 m in 2020. Two out of three people in the area are Nigerian. Under pressure, the lake has receded into Chad. It shrank from about 25,000 sq km in 1963 to less than 2000 sq km by 2001. Agriculture, the lifeline of the local economy, declined just as demand for food rose. The population became poorer and more marginalized.

Nigeria used to be a food exporter. As agriculture collapsed, Nigeria now needs to import food. 90 % of all the livestock in Nigeria was concentrated in the 11 northern states that are now severely affected by desertification. Internal migration accelerated, and this led to clashes among farmers; between farmers and pastoralists; as well as between land-users and fisher folk. Northern Nigeria has suffered more than 100 violent conflicts, of varying intensities, since 2003.

It is this outcome—the combined loss of livelihoods, competition within and among communities and the strategic nature of Lake Chad for the future and survival of entire populations—that Boko Haram is exploiting. It is mobilizing fighters from among the disillusioned youth and unemployed. And, today, there are more than 3.5 million refugees from Nigeria alone in the region.[18]

Is it a situation completely defined by peculiar regional balances? If we look at the tensions mounting around the Sea of Aral, similarities are easy to identify.

The Sea of Aral, the Amu Darya and Syr Darya rivers are drying up,[19] already announcing one of the gravest global environmental disasters of modern times,

---

[18]http://documenti.camera.it/leg17/resoconti/commissioni/bollettini/pdf/2015/03/19/leg.17.bol0409.data20150319.com03.pdf.

[19]The Aral Sea region crisis directly concerns Turkmenistan, Kazakhstan and Uzbekistan, and affects Tajikistan and Kyrgyzstan indirectly.

As stated in a Letter dated 12 September 2013 from the Permanent Representative of Uzbekistan to the United Nations, addressed to the Secretary-General, "The Aral Sea catastrophe stands as convincing evidence of the interplay" between the environment and strategic security. For this reason, the countries in the region affected by the catastrophe are increasingly drawing the attention of the international community to the fact that the destruction of the Aral Sea will have damaging effects not just on the immediate area, but on the entire world [...]. Until 1960, the Aral Sea was one of the largest closed bodies of water in the world. It was 426 km long and 284 km wide, with an area of 68,900 km^2, a volume of water of 1083 km^3, and a maximum depth of 68 m. The Aral Sea region had a large variety of flora and fauna; its waters contained 38 species of fish and a range of rare animals; it was the habitat of 1 million saiga antelopes; and its flora included 638 species of higher plants. The Aral Sea played a vital role in the development of the regional economy, its industries, sources of employment and sustainable social infrastructure. In the past, the Aral Sea was among the richest fisheries in the world: 30,000–35,000 tonnes of fish were caught annually in the waters of the Aral Sea region. More than 80 % of those living along the Aral Sea shore were employed in catching, processing and transporting fish and fish products. The fertile lands of the Amu Darya and Syr Darya deltas and the rich grazing lands provided employment for more than 100,000 people in livestock rearing, poultry breeding and raising agricultural crops.

affecting the countries of Central Asia and their population of some 60 million. Its socio-economic and humanitarian consequences make it a direct threat to sustainable development in the region, and to the health, livelihoods and future of the people living there, as much as the dramatic shrinking of lake Chad is jeopardizing security in Africa. This environmental disaster has had serious economic, social and health consequences for the ethnic Karakalpak population, which is native to the region immediately around the sea. A 2011 report by the United Nations Environment Programme (UNEP) on the Amu Darya River shed further light on the serious social, economic and health impacts of the Aral Sea crisis on the Karakalpak population. They have lost their traditional livelihoods and are being forced to move away from the sea to find work and healthier environmental conditions.[20]

Disputes are currently kept under control through governance and diplomacy, but the situation has a potential to worsen, given that the forecasted melting of Himalayan glaciers could disrupt the entire water cycle on which agriculture, industry and infrastructure rely in both central and South Asia.

These stories, one might object, are about water not land. But the affected variables—the certainties that are swept away, pushing people to move or fight—are mainly embodied in the natural services that soils once provided and cannot offer any more in the changed situation.

In the Sahel, Horn of Africa and across much of the Middle East, throughout fragile mountain or insular ecosystems, the story is much the same. Our failure to

---

The Aral Sea also served to regulate the climate and mitigated the sharp fluctuations in the weather throughout the region, exerting a positive influence on living conditions, agriculture and the environment. In winter, arriving air masses heated up over the waters of the Aral Sea. In summer, they cooled down over the same waters.

The problems of the Aral Sea arose and expanded into a threat in the 1960s, as a result of the feckless regulation of the major cross-border rivers in the region—the Syr Darya and Amu Darya, which had previously provided some 56 km^3 of water to the Aral Sea each year. A jump in the population in the area, urbanization, intensive land development and the construction of major hydrotechnical and irrigation facilities on the water courses of the Aral Sea basin carried out in previous years without regard for environmental consequences led to the desiccation of one of the most beautiful bodies of water on the planet. Within a single generation, an entire sea was virtually destroyed. The process of environmental degradation continues, and the Aral Sea region is becoming a lifeless wasteland.

Over the past 50 years, the total outflow from rivers into the Aral Sea has fallen almost 4.5 times, to an average of 12.7 cubic kilometres. The area of the sea's surface is eight times smaller than it was, and the water volume has decreased by more than a factor of 13. The water level, which until 1960 had reached a maximum of 53.4 m, has fallen by 29 m. Salinity has increased by more than 13–25 times and is now 7–11 times higher than the average mineralization of the world's oceans.

The sand-salt Aralkum desert, with a surface area of more than 5.5 million hectares, is inexorably taking over the Aral region and now covers the dried-up portion of the sea that was once home to a wealth of flora and fauna and served as the natural climatic regulator of the adjacent areas. Constant environmental risk, with its negative impact on the quality of life, health and, most importantly, the population's gene pool, now affects not only the areas around the Aral Sea, but the whole region of Central Asia.

[20]Environment and Security in the Amu Darya Basin, Nairobi (UNEP), 2011.

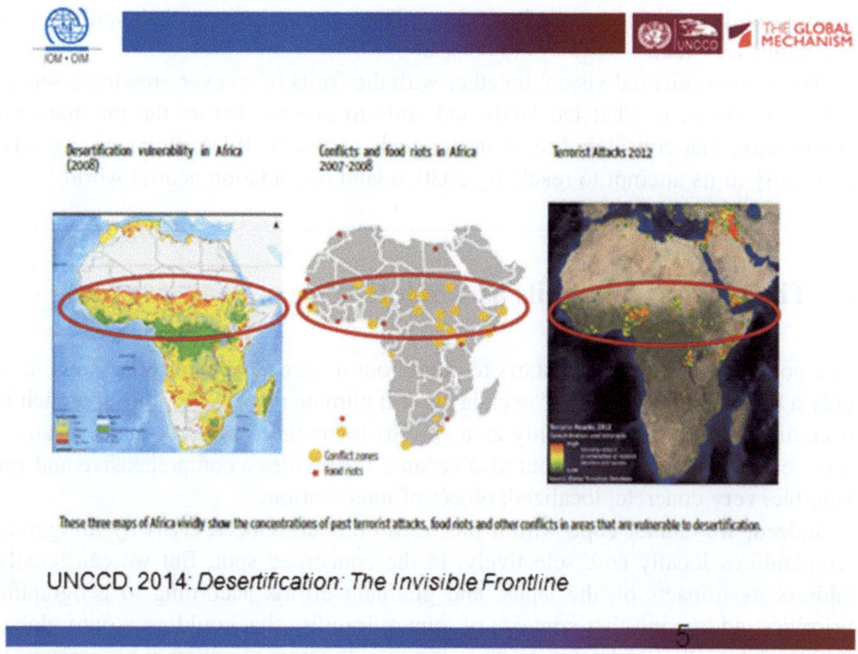

Fig. 4 From an IOM—UNCCD presentation

act early on land degradation, also in the face of water scarcity and climate change, is leaving a trail of destruction in its wake. Food insecurity, internal displacement, end-of-chain international migration, extremism and political instability pile up, one on top of each other.

In general, research has not reached the result of proving quantitatively a cause-to-effect link; but it has shown at least a recurrent association: that 40 % of all interstate conflicts over the last 60 years occurred in areas marked by soil vitality fragilization[21]; and these, in turn, acted as a driver of displacements. This association is for instance recurrently considered in strategic analysis about the Sahel region[22] (Fig. 4).

It has now become a common place, but a true one: the Syrian crisis is doubtless acting as a propeller of massive forced population displacements and Syria—between 2006 and 2010—experienced the worst drought in its history. It forced 1.5 million Syrians off the land and into the cities where they clashed with refugees from other regional conflicts. Once again, the initiating force was external to soils: drought. But the direct cause of instability has been that drought prevented soils

[21]http://www.italy.iom.int/conf/panel3/Session2/GMspeech.pdf.

[22]http://www.unccd.int/Lists/SiteDocumentLibrary/Publications/Final_Security_second_issue_7_march_14%20low%20res.pdf.

from providing their expected services. The potential for global spill-over is becoming very real.

This policy oriented vision, together with the fruits of an ever growing science-policy interface, is what led lands and soils to emerge before the international community, that considered them in the goals of the 2030 Development Agenda, especially in its attempt to reach, by 2030, a land degradation neutral world.[23]

5 The Good News: Soils as a Solution

The policy perspective thus starts to single out a special relevance for soils: it is only a beginning because the "agenda" is still climate driven. Yet, this approach is becoming more visible not only as a way to focus the weak link in the chain—services harbored by lands—but also because it provides a comprehensive and yet tangible, very concrete, localized, object of intervention.

Indeed, we cannot cope with a localized climate-induced crisis by mitigating temperatures locally and, selectively, in the concerned spot. But we can locally address its impacts on the lands, and graduate efforts according to geographic priorities, addressing first contexts of human fragility that could grow into global hotspots of conflict and population movements. In terms of policy, if not science, it makes a huge difference.

In many societies land is a delicate political issue. Property rights, fair distribution, competition between urban and rural uses, between grazing and planting, between wild nature conservation and agricultural expansion, between tourism and industrial occupation, are just a few of the dilemmas that many communities, especially the less organized ones, still need to solve in a balanced way.

Therefore, ascribing lands a social and environmental function that goes beyond its traditional and already problematic uses could prove complicated in numerous contexts. Nonetheless, for many other aspects social and environmental investments on land are bound to prove less conflictive, more cost effective, more accessible to fragile societies, and swifter to profit, than the majority of innovations called upon in order to face the global environmental crisis.

The deep reason lies in the basic and essential nature of the link between human communities and their territories: the first and very concrete nursing grounds of their livelihoods, values and identity. Given such an essential identification, sound management of land creates cumulative and synergic cycles of environmental and societal sustainability in which both actors, mankind and nature, reciprocally protect each other; conversely, mistreated lands tend to degrade themselves in the first place, while disrupting societal balance and peaceful progress. In this sense, any choice about land has the potential to be a choice between stability or disorder.

[23]Target 15.3 of the 2030 Development Agenda.

Land is so directly relevant to mankind that it is only normal for it to be frequently a delicate political and economic issue; it represents the basic interface between nature and production, as it serves as the most immediate and concrete value extracting context. At the same time and for the same reason, human relationship with soils is so primary, deeply rooted and spontaneous that:

- practical protection and recovery interventions are generally within the reach of less organized, technologically endowed, or rich communities;
- results are generally short term and visible, readily understandable by socially and educationally less sophisticated groups, making it easier for them to subscribe to individual, family scale, or larger collective initiatives;
- as a result, in many contexts effective interventions do not need to be overly structured and endlessly supported by top-down, complex and costly organization;
- almost every community has a portion of degraded lands that have become marginal on the market, and thus accessible for socially and environmentally oriented interventions without a political or economic competition too unrealistic for non-dominant subjects; but, globally summing up these surfaces, the protection or restoration of their vitality could make a crucial difference for planetary balances.

In other words, land presents itself as an accessible mean of synergic empowerment for both nature and human communities, both concurring in long term sustainability. Beyond these dimensions, land care appears particularly promising because generally it is not perceived as that kind of investment "for the sake of the environment" that implies a "sacrifice" of human interests: a suffered trade-off that emerges in other sectors and that feeds the rather common impression that we have to "give up something" in the name of nature.

Proper land care ordinarily initiates simple, concrete, understandable and short impact cycles of human and environmental co-recovery and co-growth. It achieves it on a multidimensional scope that seems to spontaneously match those human deep needs that, once satisfied, build the foundations of a peacefully productive society but, once denied, directly create conditions of abuse, tension, poverty, conflict, and the urge to flee.

Vital soils do not only provide a production opportunity. They tend to provide fairer income distribution, solicit community cohesion and cooperation, foster gender equality, encourage education, support cultural identification, generation to generation solidarity and transmission of knowledge, and favor projects for the future rooted in local communities. In this perspective land, as the first and foremost renewable asset, once well managed could provide the true booster of a balanced and not conflictive development: empowerment and trust in the future. And, in the same perspective, it embodies in a simple and practical way all the complexity of the interconnected goals of the 2030 sustainable development agenda.

Given this favorable potential—although every situation is different and largely defined by local balances—in terms of policy it is reasonable to expect in most contexts a rebalancing feed-back loop out of proper care, protection, and recovery of soils. One that acts locally, but also globally mitigates the environmental instability that, one way or the other, favors population displacements.

Energy has been the core issue in climate negotiations. Replacing fossil fuels with renewables, boosting efficiency in energy production and consumption, giving up subsidies to traditional sources will be crucial aspects to solve. Yet, is it reasonable to expect the energy sector alone to face the huge challenge of climate change?

It is neither practical nor wise to concentrate on one single business the responsibility of such a fundamental adjustment of our economies. Instead, the sum of growing but moderate emission savings achievable by each of the sectors that produce greenhouse gases would take us on a safer path, and one that would not impose shocks or a traumatically rapid transformation to one industry singled out as responsible. Among the various parallel actions that can be undertaken to reduce emissions, a better management of lands, together with the recovery of already degraded areas, looks extremely promising: globally addressing the root causes of instability, including climate change and biodiversity loss, while locally protecting fragile communities.

A land based strategy for climate change, stability, and migrations, comes in two faces: on one hand, it is about stopping the accelerating trends of land degradation, that turn increasingly vast areas from natural carbon wells into sources of emissions, while causing instability. On the other hand, we can start at once rehabilitating the vast expanses of agonizing soils and redraft them in the fight against global warming.

The costs of rehabilitating an acre of land differ enormously, ranging from less than a 100 U$ in the most favorable contexts, to the tens of thousands necessary to restore complex coastal biomes, for instance. Nevertheless, most lands in those regions that are turning into instability hotspots can be recovered with costs not exceeding 250 U$ per ha, and these extents are abundant in the areas where growing migration waves originate, especially in the Sahel region. Their recovery, in proper conditions, can turn each acre in a carbon well so effective that it can outcompete the cost-effectiveness of the most advanced investments in renewable energies. This alone could justify the endeavor, but the most relevant aspect is that it triggers a cascade of significant side benefits. Especially if recovered lands are entrusted to small scale farmers, it has the potential to generate:

- effective carbon absorption;
- biodiversity protection;
- food security;
- retention of humidity and consolidation of water sources and hydrological balance;
- community cohesion;
- the creation of agricultural surplus to reinvest in commerce and crafts;
- local, family, and gender empowerment;
- a trench against land grabbing, by local populations that have recovered a sense of value of their individual or collective properties;

- dignified livelihoods that defuse temptations to turn towards illegal economy, fanaticism and violence;
- the rehabilitation of origins and identity, transmission of precious knowledge from one generation to the next, a path of modernization of traditional productive approaches and therefore a non-conflictive integration in broader economic growth, and
- the last advantage of the list, decreasing migration pressures, that would come as the final product.

These gains are not only "collateral" to the goal of lowering emissions, an unexpected gift in unrelated dimensions. They are part of a coherent cycle of holistic balance, also providing precious basic conditions for an effective fight against climate change.

6 Conclusions: Soil and the 2030 Development Agenda

In 2016 a new era has started for development. A new Agenda, focused on 17 goals, will set the course of international development efforts until 2030, building on the results achieved in pursuing the previous international framework, known as the Millennium Development Goals. The latter, a list of eight simple and easy to understand objectives, paved the way to a much more articulated architecture: the goals have been multiplied in the 2030 Agenda, specified in 169 sub-targets, and subjected to a hopefully rigorous monitoring through a set of quantitative indicators. Yet, its more complex articulation is only the surface of a deeper revolution in perspective. The true novelty in the 2030 Development Agenda is not that it focuses more goals in a more detailed way, but that, finally and assumingly, it fully reflects a new intuition about the world we live in: a global balance encompassing the idea that the goals themselves are interconnected and synergic.

Another way to describe this novelty of the 2030 Agenda is to say that it also takes care of the environment. Besides the fact that 4 out of 17 goals directly refer to the health of the ecosystem, the inclusion of the environment further implies all the advancements in the new Agenda. Indeed, introducing the environment is different from considering another supplementary set of goals for mankind to reach, rather meaning that classical development goals have to be redefined and managed within a reactive system that surrounds us. It portrays a radical shift in perspective, as we start to look at the future of mankind not as an absolute, but within the grid of relationships and balances that shape the functioning of a broader system we belong to: one that is common and therefore has to be managed together; and one that, like a home a family shares, has to be kept in balance in all its elements, both human and physical.

The infinitely complex interactions network harnessed within Earth's balance can be described in the terms of a mathematical instrument often used to account for dynamic balance systems: a matrix. Matrixes portray—in compound systems like planet Earth—how the whole situation evolves as a result of the variation of its elements. To describe system Earth, and monitor the preservation of its stability, we need a matrix that interconnects in an overarching final balance, through algorithms, a set of sub-balances. It implies that a variation in one sub-balance can affect other related sub-balances and that these, in turn, can reverberate on further elements, so that if the cycle gains enough strength, it can modify the global state of the system.

And indeed, the two last development Agendas hint in this direction: probably as an intuitive consequence of a perceived but not rationalized truth, once environment came into the picture of development, the idea of a global balance matrix started unintentionally to emerge. The images chosen to communicate both the Millennium Goals and the 2030 Agenda do look like a matrix, at least in the way they were marketed to the public, ascribing each goal to a box.

Especially if we look at the more evolved 2030 Agenda in this perspective, we recognize functions that connect, for instance, "life on land" with "quality education" that, in turn, reflect on "no poverty" and "zero hunger" which, again, are factors in "peace, justice, and strong institutions", the end result of which could, again in turn, reshape "life on land" and "quality education". In other terms, we are recognizing that we are coping with trans-sector local, regional, or even global feedback loops. What the Agenda says, therefore, is that we cannot cope with isolated problems, nor can we isolate the environment from mankind. We rather need to revert dangerous feed-back loops that involve both, and identify structural interference knots in a resonant cycle of growing unbalance between the human and natural dimensions: these feed-back loops are the engines and root causes of displacements.

In this global cycle perspective, "everything is connected with everything"; true. But some dimensions and goals seem characterized by more general impact multipliers, both in the cycle of instability we are facing, and in the recovery cycle we need to revert to. Land is one (Fig. 5).

Indeed, revitalized lands give rural communities—and entire nations—a concrete solution for their livelihoods and therefore subtract them from the dynamics of poverty, insecurity and conflict: the same that would otherwise trap them in a cycle of structural incapacity to take care of their environment or, in the worst cases, that would push them to actively destroy it.

The same all-round rebalancing cycle could be triggered by protecting, for instance, biodiversity or other sensitive environmental variables. It also has to be done, but it is more difficult to turn these interventions into general policies: the difference is that land is local, concrete, and universally understood as a value by people; the same people who, significantly, once compelled to migrate go through a sense of loss of their land. We are children of our lands and our link to the soils we walk on is much more than productive, rather looking like an overall mutual belonging relationship. It is not surprising that correcting our approach to lands entails such a broad spectrum of both natural and social healing solutions. Can we afford to neglect it?

Fig. 5 Soil in the SDGs

References

A projection of severe near-surface permafrost degradation during the 21st century

Austin J, Bruch CE (eds) (2000) The environmental consequences of war. Cambridge University Press, Cambridge

Cattaneo C, Peri G (2015) The migration response to increasing temperatures. National Bureau of Economic Research, Cambridge

Gee H (2004) Treeless at Easter. Nature CDXXXI:411

Gemenne F (2011) How they became the human face of climate change – the emergence of 'climate refugees' in the public debate, and the policy responses it triggered. In Piguet E et al (eds) Migration and climate change. Cambridge

Gray C (2011) Soil quality and human migrations in Kenya and Uganda. Glob Environ Chg 21 (2):421–430

Hodell DA, Curtis JH, Brenner M (1995) Possible role of climate in the collapse of classical Maya Civilization. Nature 375:391

Hsiang SM et al (2011) Civil conflicts are associated with the global climate. Nature CDLXXVII:438

Huntington E (1971) Climatic change and agricultural exhaustion as elements in the Fall of Rome. Q J Econ XXXI:173

Lee JR (2009) A brief history of climate change and conflict. Bull At Sci (online), August 14 2009

Lee HG, Zhang D, Brecke P, Fei J (2013) Positive correlation between the North Atlantic oscillation and violent conflicts in Europe. Clim Res LVI:1

Mayda AM (2010) International migrations: a panel data analysis of the determinants of bilateral flows. J Populat Econ 23(4):1249–1274

Morrisey J (2012) Rethinking the debate on environmental refugees: from 'maximalists and minimalists' to 'proponents and critics'. J Polit Ecol 19:37

Peterson LC, Haug GH et al (2003) Climate and the collapse of Maya Civilization. Science CCXCIX:1731

White MJ, Lindstrom DP (2005) Internal migrations. Berlin

Zhang D et al (2006) Climatic change, wars and dynastic cycles in China over the last millennium. Clim Chang LXXVI:459

Zhang D et al (2007) Global climate change, war, and population decline in recent human history. Proc Natl Acad Sci USA CIV:19214

FAO: Voluntary Guidelines on Responsible Governance of Tenure of Land, Forests and Fisheries—Relevance, Reception and First Experiences in Implementation

Michael Windfuhr

1 The Voluntary Guidelines on Responsible Governance of Tenure of Land, Forests and Fisheries in the Context of National Food Security

The "Voluntary Guidelines on Responsible Governance of Tenure of Land, Forest and Fisheries in the Context of National Food Security" (VGGT) were developed during a 2 years process of intense negotiations in an open-ended working group of the United Nations Committee on World Food Security (CFS), a format that allows also non-state actors to participate in the negotiations. In the realm of the CFS, the civil society has developed an own civil society mechanism (CSM) as the private sector has developed its own private sector mechanism (PSM).[1] Both mechanisms are organizing the respective inputs from their constituencies. The CFS got a renewed mandate during and after the World Food Crises 2007/08. It was created to overcome the clear lack of guidance and governance at the global level and within the UN-Systems during the World Food Crises. The first substantive issue taken up by the CFS was the development of a new international legal instrument in relation to land tenure governance. The issue was seen as particularly important because the investments in land and other forms or land transfers increased fast after 2008 in many parts of the world as a consequence of the changing price trend during the World Food Crisis. Already in 2009, the CFS decided to develop the VGGT. The negotiations took place from 2010 to 2012. Finally, the document was

[1]More information can be found on the website of the CFS: The United Nations Committee on World Food Security (CFS 2016b) has an own website (2016), as the Civil Society Mechanism (www.csm4cfs.org) and the Private Sector Mechanism (www.agrifood.net) have.

M. Windfuhr (✉)
German Institute for Human Rights, Berlin, Germany
e-mail: windfuhr@institut-fuer-menschenrechte.de

H. Ginzky et al. (eds.), *International Yearbook of Soil Law and Policy 2016*,
International Yearbook of Soil Law and Policy,
DOI 10.1007/978-3-319-42508-5_15

adopted unanimously, which was a surprising and strong signal for negotiations on such a complex issue as land governance is.[2]

The political momentum for such an ambitious negotiation process was created through the trends that culminated in the World Food Crisis 2007 and 2008. As one of the results of the crisis, which can also or even better be described as a food price crisis, there was a substantial increase of agricultural investments. The increasing prices during and after that crises initiated a wave of land investments. Different actors are involved in that new wave of investments in land: (1) Some of these investments are made by other states or state enterprises. Being afraid that world markets could close again as it happened at the peak of the World Food Crises in 2007, state actors bought land in order to produce food for the supply of their own population. (2) Other investments in land were fuelled by speculative reasons: Since the financial crises in 2009 as well as the low interest policy of most central banks, land is seen as a safe haven. (3) A third group of investors particular private companies, became interested to invest in agriculture. After years of low and depressed world market prices which hindered investments in agriculture for decades, the sudden price trend change triggered many investments in land. One of the strongest trends that supported and still supports private investments is the increasing worldwide demand for biofuels.

Several of these investments caused problems with persons using or living on the land, for example issues of forced evictions, inadequate compensation, non-communication with and neglected participation of affected individual or communities accumulated. The term "land grabbing" was introduced by research publications from civil society organizations (CSO) or books that were describing the trend. The term is used to indicate that several of those investments are made also with land already used by persons living and working on the land. They often belong to weaker segments of the society without documented user or property rights. The problem started to gain prominence at national and international forums.[3]

The relevance of the land governance problems can also be seen in the political response to it. Several intergovernmental organisations started already in 2009 to

[2]This adoption is of high importance to the new CFS because it is the first international standards setting instrument adopted by the CFS. It was possible to reach a consensus for such a complicated issue in the course of 2 years. Therefore, it can be seen as a first good signal about the capacity of the CFS to regulate also complex global issues with respect to food security challenges.

[3]The International Land Coalition developed a definition of Land Grabs during its 2011 assembly in Tirana, Albania: *"Land grabs are acquisitions or concessions that are one or more of the following: (i) In violation of human rights, particularly the equal rights of women; (ii) not based on free, prior and informed consent of the affected land-users; (iii) not based on a thorough assessment, or are in disregarded of social, economic and environmental impacts, including the way they are gendered; (iv) not based on transparent contrasts that specify clear and binding commitments about activities, employment and benefits sharing; and (v) not based on effective democratic planning, independent oversight and meaningful participation"* (Art. 4 of the Tirana Declaration). For more information check Grain's (2016) website, Anseeuw et al. (2012, p. 29), the World Bank's (2011) or the International Land Coalition's (2011) article.

adopt seven principles for responsible agricultural investments in order to provide some guidance to investors.[4] The idea of describing standards for responsible agricultural investments was also taken up by the CFS that decided to develop firstly the VGGT and later the Principles for Responsible Agricultural Investments (RAI) that were negotiated between 2012 and 2014 in an open-ended working group of the CFS. These principles were negotiated in a similar process like the VGGT and received broad support from states but criticism from CSOs.[5] Core objective of the RAI is to define state obligations and other actors responsibilities toward all other issues related to agricultural investments, except the governance of land tenure.

1.1 Strong Support for the VGGT After Adoption and During the Follow-Up Process

The VGGT have gained tremendous support during the decision making process, but also in the follow-up process afterwards. Several reasons can be identified that contributed to the support and recognition:

1.1.1 Good and Comprehensive Negotiations Process

The drafting process was generally considered to be adequate as proved by the result of a very comprehensive document. This positive recognition of the VGGT is, for one, the result of the transparent and inclusive 3 years multi-stakeholder process which provided the basis for drafting. The VGGT have also received recognition for being a detailed and well elaborated framework, which is concrete enough to provide guidance to governments and governmental institutions. The document is based on the agreed language from other international negotiations and must be seen as an international legal instrument that collects, combines and further develops many relevant standards, which is available for governance of tenure of land, forests and fisheries. The G8 summit in 2013 also endorsed an initiative to support the implementation of the VGGT.[6]

[4]World Bank/FAO/UNCTAD/IFAD (2009).

[5]FAO (2014): The Committee on World Food Security (CFS) endorsed the Principles for Responsible Investment in Agriculture and Food Systems on October 15th, 2014. The civil society mechanism of the CFS decided, after the active engagement of CSO in the negotiation process to not endorse the principles, because they are not fully oriented towards support small holder investment but describe also rules and standard that can be applied to larger scale investments.

[6]See paragraph 44 of the G8 declaration of Lough Erne.

1.1.2 The VGGT Address All Relevant Stakeholder

Adequate for the problem of investments in land, the VGGT address all actors involved bearing obligations and responsibilities in that context. The annex "in the context of national food security" was chosen in analogy to the *Voluntary Guidelines to Support the Progressive Realization of the Right to Adequate Food in the context of national food security* (Right to Food Guidelines) which were adopted in 2004 unanimously by the FAO Council. As with the Right to Food Guidelines, the purpose of the annex in reference to national food security is to highlight the importance of national action and national obligations for the implementation of these guidelines. The intention of the drafters was to make clear that it is first and foremost the state which is the duty bearer and needs to guarantee a fair and proper land administration and management of land transfers. The state has to set the rules for investors and has to make sure that the rule of law is applied in all cases of land investments. Therefore, it is not astonishing that the majority of the paragraphs of the VGGT are directed toward the state.

The primary objective of the framework is to "improve governance of tenure of land, fisheries and forests. They [the Voluntary Guidelines] seek to do so for the benefit of all, with an emphasis on vulnerable and marginalized people, with the goals of food security and progressive realization of the right to adequate food, poverty eradication, sustainable livelihoods, social stability, housing security, rural development, environmental protection and sustainable social and economic development" (Guideline 1.1 VGGT).

At the same time, the document of the VGGT is far from being restricted to responsibilities of host states. It discusses also the role of those states which invest directly in other countries abroad, either directly or through state owned companies or sovereign wealth funds. As the cases of large scale land investments during the last years show, several of the largest land deals were done by other states. Therefore, the VGGT take into consideration the role of states which act abroad, both as own actors but also as home states of business enterprises who invest abroad. For the first time ever, obligations of home states for business actions abroad are explicitly recognized in an inter-governmentally negotiated document: "When States invest or promote investments abroad, they should ensure that their conduct is consistent with the protection of legitimate tenure rights, the promotion of food security and their existing obligations under national and international law, and with due regard to voluntary commitments under applicable regional and international instruments (Guideline 12.15, VGGT)." This formulation, which was proposed by the African Group during the negotiations process, goes beyond the agreed language and recognizes the obligation of home states with regard to the support for economic activities of businesses abroad.

In Guideline 3.2 of the VGGT, the document addresses directly the responsibilities of non-state actors. The wording of this paragraph has essentially been taken from the *United Nations Guiding Principles on Business and Human Rights*

(UN Guiding Principles, UNGP).[7] It highlights that "business enterprises should act with due diligence to avoid infringing on the human rights and legitimate tenure rights of others".[8] The VGGT require from business enterprises that "they should include appropriate risk management systems to prevent and address adverse impacts on human rights and legitimate tenure rights". Guideline 3.2 of the VGGT follows the UN Guiding Principles in describing what is expected from business enterprises. Moreover, further provisions in the VGGT also set out what is expected by business enterprise (see below). All in all the number of paragraphs dealing with the direct responsibilities of business enterprises are far below the number of paragraphs addressing the obligations of states. The document is consistently differentiating between state obligations and responsibilities of private actors and thus complies with the core differentiation of the UN Guiding Principles on Business and Human Rights. For this reason, the importance and co-responsibility of private actors is recognized.

1.1.3 Clear Human Rights Based Understanding and Accountability

The content of the VG Land is based on a human rights understanding. It connects land governance issues to existing human rights obligations form binding treaties and describes the accountability of all relevant actors in that field. While it is a voluntary document, it describes what state parties are bound to by binding human rights treaties. Therefore, it can be seen as best practice application of a human rights based approach for one policy area. It requires from governments an adequate regulation of land tenure through governmental acts and activities at all levels. The accountability of states acting abroad is taken up as well as the responsibilities of other actors. The framework is therefore well developed on the roles of different stakeholders in guaranteeing human rights standards related to land tenure. Moreover, it covers all areas of government, (from local, over district and/or regional to national) and offers a clear description of human rights principles, inter alia how to set up processes from meaningful participation up to recourse mechanism in section 3b under the title: "Principles for Implementation". These "principles of implementation" relate to all processes of land registration, land transfer and also to the daily work of land administration. The ten principles for implementation are general principles for the implementation of all human rights. They describe how states shall organize processes which allow all affected person to have a fair role in these processes: They should not suffer from discrimination, they should not be overseen and if problems occur they should have access to justice. The Principles are: Human dignity, non-discrimination, equity and justice, gender equality, holis-

[7] United Nations (2011).

[8] Quote from VGGT (2012: 3/2).

tic and sustainable approach, consultation and participation, rule of law, transparency, accountability, and continuous improvements.[9] When these principles are applied, it is likely that no infringement of human rights takes place in land related processes.

Additionally, the VGGT focus on the concerns of particular vulnerable groups, inter alia it takes up vulnerable groups in rural areas, such as small holder farmers, pastoralists, fishermen and hunter as well as indigenous peoples. The meaning of governance is filled with a government responsible to its human rights obligations and to the respect to a rule of law. Governance goes hand in hand with sensitivity to people and their participation at all levels of processes related to land as well as with reversibility. Affected persons or communities can file complaints when needed and when they do not feel treated adequately. Finally, it is a strong and relevant document that takes existing forms of gender discrimination into consideration with reference to land registration, land transfers and land administration.

1.2 Legal Interpretation and Relevance of the VGGT

The Voluntary Guidelines have a focus on national implementation, which is also shown with the annex to the title "…in the context of national food security". The annex shall highlight the importance of national action and national obligations for the implementation of these guidelines. The national focus was also added by donor states in order to clarify that the main burden of implementation also related to finances is a national one and that implementation is more than a call for development aid.

States at home are the primary duty bearer under international law. For all human rights states have three different levels of obligations. First, they have to respect that they do not violate human rights through own actions. They have to protect all persons against human rights infringements of third parties, including business. Moreover, they have to prove that the utilise the maximum of its resources to fulfil the human rights approach. This understanding of state obligations has become the standard to characterize state obligations in general according to a human rights based approach. For the issues of land governance the first two levels are of particular importance. Respect means that the state itself shall not violate the rights of persons through own action. States are not allowed to forcefully relocate or evict persons without adequate compensation and due process. The obligation to protect means that the state is the principle guarantor of human rights, when other actors, such as other states, their foreign wealth funds, or business enterprises of other private actors invest in land. They have to guarantee that the rights of all persons living in the country are protected and secured. At the core of most

[9]The principles are also referred to in the RAI-negotiations document see footnote 5 above FAO (2014).

problems related to large scale land acquisition is the host state that do not recognize the user or tenure rights of people and/or that do not protect them adequately, in conflicts concerning that land with other state or non-state actors.

States acting abroad is also an important component in the context of the large scale land acquisition, because a substantial proportion of the large scale land acquisition documented by the land matrix are done by states abroad. Those investments are made by actors whose actions can be attributed to a state acting abroad, such as by sovereign wealth funds. The state's obligations are the same as for a state acting abroad: respect, protect and fulfil. The VGGT require that all investors including states investing abroad "should do no harm, safeguard against dispossession of legitimate tenure right holders and environmental damage, and should **respect** human rights" (VG Land 12.4). The VGGT require from states when they act on their own "that their conduct is consistent with the protection of legitimate tenure rights, the promotion of food security and their existing obligations under national and international law" (VG Land 12.15). While generally it is debated in international law how far extraterritorial obligations of human rights for states go, the VG Land are at least clear that states have to respect human rights in other states when acting abroad. Moreover, states are also requested to protect human rights standards with respect to companies or other private actors domiciled in the country and investing abroad.

States do also have obligations when actions in their role as development policy actors through donor support to host states or through policy advice in the areas of land policy, investment policy or overall economic policy are implemented. This role is increasingly important because many host states receive a substantial package of policy advice. Due to the fact that large scale land acquisition are particular happening in states with a weaker governance system, the importance of policy advice given to these states or a group of states is an important component of aid policies and at the same time an huge chance for implementation of the VG Land.

The responsibilities of private actors in the VG Land are taken up in the part "General principles" of the VGGT by a direct reference to the United Nations Guidelines on Business and Human Rights: "Non-state actors including business enterprises have a responsibility to respect human rights and legitimate tenure rights. Business enterprises should act with due diligence to avoid infringing on the human rights and legitimate tenure rights of others. They should include appropriate risk management systems to prevent and address adverse impacts on human rights and legitimate tenure rights. Business enterprises should provide for and cooperate in non-judicial mechanisms to provide remedy, including effective operational-level grievance mechanisms, where appropriate, where they have caused or contributed to adverse impacts on human rights and legitimate tenure rights. Business enterprises should identify and assess any actual or potential impacts on human rights and legitimate tenure rights in which they may be involved (VG Land, 3.2)."

The description of obligations and responsibilities in the VGGT follow those in the UN Guiding Principles, currently the most authoritative description of the required human rights conduct of business. Business enterprises do not have human rights obligations as states do, but they should respect human rights (Principles 14) wherever they operate and they should do this independent from the situation in the given country and from the role of the government. According to UNGP "The responsibility to respect human rights is a global standard of expected conduct for all business enterprises wherever they operate".[10] It exists independently of States' abilities and/or willingness to fulfil their own human rights obligations, and does not diminish those obligations. And it exists over and above compliance with national laws and regulations protecting human rights." Private Actors will implement their responsibility to respect human rights by checking their potential impacts already in advance and by carrying out risk and impact assessments in regard of the implementation of due diligence. UNGP 17 formulates this responsibility as follows: "In order to identify, prevent, mitigate and account for how they address their adverse human rights impacts, business enterprises should carry out human rights due diligence." Business enterprises shall assess the risks involved with their operations if possible ex ante, during and ex post of operations and investments. The UNGP require also that they cover risks related to all business activities i.e. along the whole value chain. They are responsible for their direct labour relations (Principles 12) but also for the products and services they buy through their business relationships (Principle 13). To sum up: business enterprises need to check carefully their potential impacts. Moreover, private actors "shall not undermine States' abilities to meet their own human rights obligations, including by actions that might weaken the integrity of judicial processes (Commentary to Principles 11)."

1.3 Structure of the VGGT

The main content dimension of governance of tenure of land which structure the text of the VGGT are: land registration, all forms of land transfer and land administration. The structure is clear and adequate.[11] In all three main parts of the VGGT, the document is precise in describing its core concerns:

(1) **Land registration** is the activity where most problems occur in terms of recognizing legitimate tenures rights or where they historically happened. Many land transfer processes suffer from a weak baseline situation. In the historic recognition of land users often groups were overseen, e.g. in traditional land use settings were individual access rights are often unknown, or in socialist settings, where formally all land belongs to the state and the factual land use

[10]United Nations (2011), (Commentary to Principles 11).

[11]The part on land registration is covered by chapter 3, the part on land transfers in chapter 4 and the part on land administration of chapter 5 of the Voluntary Guidelines (2012).

patterns are often not adequately registered. Land registration is therefore the starting point for any meaningful recognition of those persons that use land.

(2) **Land transfer** can be managed in a better way, when land registration is done in a good, sensitive and comprehensive way. If the current status of land registration is incomplete or if several land users are not formally recognized or identified, land transfer processes bear a large risk of infringement of human rights. Land transfers processes, in situations of incomplete land registration and in which third parties are involved, need to research the current land use situation in detail in order to understand the harm to potentially affected persons. Furthermore, it helps to check, if mitigation measures can be taken in order to compensate affected persons' losses. Where mitigation is impossible or conflicts are emerging about a particular project and its impacts land transfers might become problematic or should not pursued.

(3) **Land administration**. As with the two other main issues of the VGGT, the land administration part is sensitive to potential governance problems such as corruption, access to grievance mechanisms, etc. The content of the VGGT describes in detail, what should be expected from a human rights perspective.

The understanding of land policies in the VGGT are based on the term legitimate tenure rights. The term was introduced and strongly promoted by Brazil during the negotiation. Brazil's central idea behind the term is the conviction that the text needs to pay utmost attention to those individuals or groups who have no recognised user relations to land or who make their living by using forests products collected from land. Such sensitivity to all those individuals and groups depending on land without any formal recognition is a key feature of the VGGT and come from their human rights orientation. It is of importance due to the fact that these groups are often overseen or neglected within the processes of land registration, admissions or transfers. The term 'legitimate' was promoted by Brazil at the same time in order to limit potential claims in the registration of land use or income from land to those who have a stable relationship to their access to land. It should help governments when registering land, or investors when planning to buy or rent land to recognise carefully all those who have a history of usage of that land and not necessarily those who have moved into an area under registration shortly before.

1.4 Limitations of the Document

The VGGT have been written as a tool to improve the governance of tenure of land, forests and fisheries. It has therefore a focus on the role of governments. Such an orientation is particularly important because governance in land issues is at stake in many countries and need urgent improvements. At the same time states are not always functioning well. They might be unable or unwilling to establish and to budget the necessary governance institutions properly. States can be complicit and being biased in supporting elites in their access to land, while discriminating weaker segments of the societies as well as persons or groups living in vulnerable situations. So while the improvement of governance capacities of states is needed, they might be weak allies in doing so. Is it realistic to expect from governments as much as the VGGT does or would it be more advisable to focus on the role of other actors than states?

The VGGT must be seen as a normative text describing that without a good governance system for tenure of land forests and fisheries, it is very difficult to adequately deal with conflicts over tenure, ownership and use of land. Therefore, the VGGT describe in detail how governance should be set up ideally. While private actors also have responsibilities to avoid harm and to infringe on human rights, the governance setting needs to be existing and set up by the state. A document that describes in detail what can and should be expected from states/ governments, has also the potential in helping both civil society groups and private actors to understand what they can and should expect form governments. Furthermore, the VGGT can also be used to monitor the state performance in land tenure governance. The text does not discuss what to do, when governance is absent or relevant institutions does not function properly. This must not be seen as a weakness of the document, because it is written as a normative endeavour to describe what type of governance would be needed in order to guide states how to do it. The task of developing adequate land policies in situations of weak governance remains to be one of the largest challenges for land policy.

Another limitation of the VGGT is their focus on land tenure. It is not comparably strong on the issues of governance of forests and fisheries. While many governance issues are similar in terms of access to land, forest and fisheries, the VGGT are detailed and rich of experiences referring to governance of tenure of land, but incomplete when it comes to governance of access to or tenure for forests and fisheries. While all general principles (participation, no-corruption) remain the same and can be applied for all areas, the tenure of land, forests and fisheries, the text is not very detailed with respect to tenure rights related to forests and fisheries.

2 Implementation and Follow-Up

The implementation and follow-up activities of the VGGT are quite rich and undertaken by the different actors addressed in the Guidelines. Compared to the implementation processes of many other international legal instruments, the first 4 years of implementing the VGGT can be seen as a success story. In the following second part of this article the core activities of different actors are briefly described and important guidance documents are introduced. The text will end with the summary of challenges and open issues that need to be discussed in order to support and speed up the implementation process. Nevertheless, it must be noted that many activities have just been started, but it is way too early to evaluate the state of the progress and its achievements so far.

One can identify an enormous amount of different implementation activities and initiatives for the VGGT. They have been referenced by the G7 governments and pilot country collaboration has been started in that context. They are used by a number of multilateral institutions and inter-governmental agencies to guide their land policy related work. Several civil society initiatives have been started to promote the use of the VGGT as a tool to challenge inactive governments or as a monitoring frame to document and measure government activities related to governance of tenure land. Also the private sector has started initiatives to implement the guidelines. Some companies have made direct commitments to take up the VGGT as guidelines for their own policies.[12] Moreover, they have also formed a private sector corporate responsibility initiative, the Interlaken group. The Interlaken group is a multi-stakeholder network that includes companies and investors involved with land and agricultural investments. The group has produced an own guide "Respecting land and forest rights: a guide for companies"[13] in order to help companies to align their operations with the VGGT.

A key actor for the implementation process has been the Food and Agriculture Organization of the United Nations (FAO), which has developed a wide range of resource and tool books for a variety of actors and issues related to the implementation of the VGGT. The FAO has published a serious of technical guides to increase the capacity of implementing the VGGT. The guides cover the following issues: "Governing land for woman and men", "Improving governance of forest tenure", "Respecting FPIC, the free, prior and informed consent", "Implementing improved tenure governance and fisheries"; "Safeguarding land tenure rights in the context of agricultural investment", "Responsible governance of tenure and law".[14] The FAO has also started to incorporate the VGGT into its regular work. The FAO organized more than 30 awareness-raising workshops in all continents.[15] Other donor agencies, multilateral institutions and financial institutions have developed a range of products to guide the application of the VGGT for various actors.[16] Several other multilateral actors supported the voluntary guidelines such as the OECD and FAO guidelines for private sector investment and agricultural value chains are applying the VGGT.[17] The World Bank and International Finance Corporation (IFC) have formally endorsed the VGGT and are trying to integrate its provisions into their safeguard policy review.

---

[12]Several companies have made public statements in support and announced to align their activities to the VGGT, inter alia Coca Cola and Nestlé.

[13]The guide can be found on the website of the Interlaken Group (2015).

[14]Additional city guides are planned inter alia on pastoralism, tenure of the Commons, technologies for land administration, the private sector, public resources.

[15]A very good overview about all follow-up activities is given in a recent publication sponsored by the UK-Government: Hall et al. (2016).

[16]The FAO implementation guides and tools can be found on the VGGT website of the United Nations Committee on World Food Security (2016a).

[17]The OECD-FAO (2016).

In addition to the technical guides developed by the FAO a wide range of other toolkits, frameworks, handbooks have been developed by development actors. They have been written with the aim to assist governments or companies or others to develop agricultural investments in compliance with the standards described in the VGGT. Several donors have prepared toolkits and guides both for states and for private actors in order to foster agricultural investments in developing countries. The Agence Fraincaise de Devéloppement (AFD) developed a guide to due diligence of agribusiness projects that affect land and property rights. The United State Agency for International Development (USAID) developed a guide for private sector investment in land. Some of the donors met in 2015 and produced in cooperation with the African Union's Land Policy Initiative and the New Alliance an "Analytical Framework for Land-based Investments in African Agriculture".[18]

Some of the toolkits focus on a broader set of issues that are relevant to make agricultural investments more responsible. This is a point of overlap to the content of the Principles for Responsible Investments in Agriculture and the Food System (CFS-RAI 2014), which has been adopted in the CFS in 2014. The CFS RAI cover much more issues related to agricultural investments than land tenure problems, for example on labour regulations, water resources and water management etc. Their implementation process is just about to start.

Together with all the other grits and toolkits one can describe the developments of such a large number of toolkits also critically as a proliferation of instruments to support responsible agricultural investments. It reflects the urgent need of donors to have proper guidance on the issue how to do agricultural investments in a correct format—given the substantive criticism on large-scale land investments, which were collected in cabinets between 2009 and 2013 by civil society organisations by journalists[19] and by the International Land Coalition. The proliferation of tools and guidance books carry the risk that they all pick and choose certain elements of the VGGT, but hardly contribute to a consistent understanding of the standards incorporated in the voluntary guidelines.

The donor community established in 2013 a coordinating working group, the Global Donor Working Group on Land (GDWGL) comprises 23 bilateral and multilateral donor agencies as members. They decided to help operationalizing the VGGT. The WG has actively promoted a land indicator in the Sustainable Development Goals. A starting point for the GDWGL has been the G7 countries commitment to align their development assistance to the VGGT. Following that decision several bilateral partnerships between national governments and the G7 has been developed to implement the VGGT. Bilateral and regional partnerships have been prioritised in Africa. Since 2009 the African Union had developed the

---

[18]USAID (2015); The Interlaken Group and the Rights and Resources Initiative (RRI 2015); The French Ministry of Foreign Affairs and the French Development Agency (AFD 2014) have also produced a Guide to due diligence of agribusiness projects that affect land and property rights.

[19]On the issue of land grabbing, several books and articles has been published inter alia: Bommert (2012), Pearce (2012).

AU Framework and Guidelines on Land Policy in Africa,[20] which takes up many issues that later were reflected in the VGGT. Since June 2014, FAO and the Land Policy Initiative (LPI) of the African Union are collaborating to integrate the implementation of the two tools. Both frameworks are promoted at high-level intergovernmental meetings and in the New Partnership for Africa's Development (NEPAD and the Comprehensive Africa Agricultural Development Programme (CAADP). The EU is supporting programmes for the implementation of the VGGT and the Framework and Guidelines on Land Policy in Africa currently in 10 countries.[21] After the adoption of the VGGT, the debate in the African Union intensified on how to respond to the "land rush". This lead to the development of "Guiding Principles on Large-scale Land-based Investments (AU GP)".[22] The text gives guidance to states and could help to implement the VGGT at national levels in Africa. It is taking up many formulations from the VGGTs. The donor communities is currently also support the conversion of the Land Policy Initiative of the African Union to an African Centre for Land Policy (Hall et al. 2016, p. 36). Several donors have developped with the Land Policy Initiative an analytical framework on how to apply the VGGT as private investor (New Alliance 2015).

The VGGT have been strongly supported by several Latin American Countries, particularly Brazil, where the National Security Council on Food and Nutrition has put the guidelines at the centre of its national food security priority list. The VGGT has been taken up in national policy formulation also in Bolivia and Colombia. In South-east Asia the FAO has established partnerships with Cambodia, Laos, Myanmar and Vietnam. In all these countries the implementation is just at the start and an evaluation of the effectiveness of the VGGT in guiding country policy is still to be done, when more activities and time has been used for implementation. The FAO reports and the overview on implementation progress presented by Hall et al, come to the conclusion that it is visible already that the VGGT have the potential to shift the national policy discourses, particularly if several actors demand the implementation, such as governmental and civil society actors, but also donor agencies.

Very active in the promotion of the VGGT have been civil society organisations. Some larger international non-governmental organisations (NGOs) have started their own programs for the implementation of the guidelines. ActionAid, for example, has embarked on a Land For campaign, with programs in plays in four countries from Guatemala, Haiti to Senegal and Zambia. In 2013 Oxfam International has started a campaign called "behind the brands" which targets the 10 largest multinational companies in the food and beverage sector to implement the VGGT. The Food First Information and Action Network (FIAN), the International Young

[20]African Union (2009) AU Framework and Guidelines on Land Policy in Africa. http://www.uneca.org/sites/default/files/PublicationFiles/fg_on_land_policy_eng.pdf (access 12 May 2016).

[21]The first ten countries are: Angola, Burundi, Cote d'Ivoire, Ethiopia, Kenya, Malawi, Niger, Somalia, South Sudan and Swaziland. In a new phase five additional country programmes will follow in Cameroon, Ghana, Guinea Bissau, Sudan and Uganda, see Hall et al. (2016, p. 35).

[22]African Union et al. (2014).

rights organisation for the right to food, has established an action research program in several African countries and works together with a lot of local NGOs and social movements to spread the information about the voluntary guidelines. The International Land Coalition, a membership let network of more than 200 civil society and intergovernmental organisations, has also started to pursue a strategy to coordinate its members in national platforms on land governance as promoted in the VGGT. A broad coalition of more than 350 organisations worldwide started "The Land Rights Now" campaign in 2016. The network includes members and affiliates of the International Land Coalition (ILC), ActionAid, Oxfam and many national and local civil society organisations. The campaign is aiming to build the momentum for the implementation of the VGGT, particularly with a focus on indigenous and community land rights. The number of publications prepared by civil society organisations for the implementation of the VGGT is quite large and increasing.

3 Conclusion

Key to the implementation of the voluntary guidelines will be in the long run the prove how much such a soft law instrument as the VGGT is able to change or guide governance behaviour at the national level. While the promotion tools, the support of multi-stakeholder actors and inter-governmental organisations are all well developed in the 4 years of implementation of the guidelines, the national and local reception, adoption and policy changes are still at the beginning. The current overview of the lessons learnt from the implementation of the voluntary guidelines prepared for UK aid come to the conclusion that the VGGT are arguable at a critical juncture: "For years on, enormous commitment connectivity is evident yet progress in some areas has been slow" (Hall et al. 2016).

Progress is slow in concrete policy changes at the national level. In many countries political elites or governments are not particularly eager to implement the VGGT. What is needed, seems to be political pressure on states and governments. An important gap that needs to be overcome has been the limited pressure on states from poor rural populations, which are by nature of politically weak. Governments need to see the economic gains from an increased carefully managed influx of agricultural investments which contributes to sustainable rural development, compared to a fast income from unregulated selling or leasing of land to national or foreign investors.

The VGGT have become the normative reference point for standards related to land tenure. The broad recognition and acceptance of this standard is a huge success in such a complicated policy field. The importance of that standard is reflected in the inclusion of a land indicator in the Sustainable Development Goals (SDGs). The current interest in agricultural investment is at the same time large and the proliferation of toolkits and guides whole to do investments in the correct manner is reflecting the needs of investors for regulation which helps to minimise risks related to land-based investments. Civil society organisations have already criticised the efforts by states donor agencies and also some NGOs in helping the corporate sector to implement the VGGT. They are afraid that this leads to a further concentration of land and other resources in an agribusiness sector while the excess smallholder farmers and rural communities to productive resources might become even more problematic. In 2015 a joint statement was developed by La Via Campesina and more than 40 social movements and grassroots organisations which calls on states, UN agencies, research institutions "to withdraw and refrain from all initiatives that aim at abetting the corporate sector and private investors to use the tenure guidelines for the pursuit of business interests, thus supporting the corporate capture of resources, public policy spaces and human rights".[23]

It will be very interesting in the years to come to follow-up the further implementation work of the VGGT. It is a very dynamic process based on a broadly agreed authoritative standard, the VGGT. The final success of the VGGT will depend on the positive policy changes they might initiate at national levels. When governance of land tenure is becoming more human rights based and sensitive to particular weak segments of rural populations, the VGGT can be considered to be a successful tool. The transfer of an international "soft law" instrument into hard policy and legal changes at the national level needs to happen.

References

African Union (2009) AU framework and guidelines on land policy in Africa. http://www.uneca.org/sites/default/files/PublicationFiles/fg_on_land_policy_eng.pdf. Accessed 12 May 2016

African Union, African Development Bank, UN-Economic Commission for Africa (2014) Guiding principles on large scale land based Investments in Africa. http://www.uneca.org/sites/default/files/PublicationFiles/guiding_principles_eng_rev_era_size.pdf. Accessed 12 May 2016

Anseeuw W, et al (2012) Transnational land deals for agriculture in the global south – analytical report based on the land matrix database. Bern/Montpellier/Hamburg, p 29

Bommert W (2012) Bodenrauch: Die globale Jagd nach den Äckern der Welt, Bastei-Lübbe

CFS-RAI (2014) CFS (Committee on World Food Security) principles for responsible agricultural investments. Adopted in 2014. www.fao.org/cfs/docs1314

Food and Agriculture Organization of the United Nations (FAO) (2012) Voluntary guidelines on the responsible governance of tenure of land, fisheries and forest in the context of National Food Security, Rome. http://www.fao.org/docrep/016/i2801e/i2801e.pdf. Accessed 15 May 2016

Food and Agriculture Organization of the United Nations (FAO) (2014) Principles for responsible investment in agriculture and food systems. http://www.fao.org/fileadmin/templates/cfs/Docs1314/rai/Endorsement/CFS_RAI_Principles_For_Endorsement_Ver_11_Aug_EN.pdf. Accessed 15 May 2016

French Ministry of Foreign Affairs and the French Development Agency (2014) Guide to due diligence of agribusiness projects that affect land and property rights. Operational Guide, http://www.foncier-developpement.fr/publication/guide-to-due-diligence-of-agribusiness-

[23]See the International Statement of the Guidelines on the Responsible Governance of Tenure at a Crossroads (2015).

projects-that-affect-land-and-property-rights. Accessed 12 May 2016. G8 declaration of Lough Erne (Paragraph 44), https://www.gov.uk/government/uploads/system/uploads/attachment_data/file/207771/Lough_Erne_2013_G8_Leaders_Communique.pdf. Accessed 9 May 2016

Grain (2016) Resources on landgrabbing, www.grain.org. Accessed 11 May 2016

Hall R, Scoones I, Henley G (2016): Strengthening land governance: lessons from implementing the voluntary guidelines. Legend State of the Debate Report 2016. https://landportal.info/sites/landportal.info/files/Strengthening%20Land%20Governance.pdf. Accessed 17 May 2016

Interlaken Group and the Rights and Resources Initiative (2015) Respecting land and forest rights. risks, opportunities, and a guide for companies. Support for aligning your operations with the Voluntary Guidelines on the Responsible Governance of Tenure. http://www.interlakengroup.org/downloads/Brochure.pdf?vsn=660A50E. Accessed 11 May 2017

International Land Coalition (2011) Securing land access for the poor in times of intensified natural resources competition. Report of the ILO international conference and assembly of members, Tirana, Albania, 24–27 May 2011

New Alliance for Food Security and Nutrition in Africa (2015) Analytical framework for investors under the new alliance: due diligence and risk management for land-based investments in agriculture. https://www.growafrica.com/sites/default/files/Analytical-framework-for-investors-under-the%20new-alliance%20%287%29_0.pdf. Accessed 12 May 2016

OECD-FAO (2016) Guidance for responsible agricultural supply chains. http://mneguidelines.oecd.org/OECD-FAO-Guidance.pdf. Accessed 11 May 2016

Pearce F (2012) Land Grabbing: Der globale Kampf um Grund und Boden (The Land Grabber: The new fight over who owns the earth). Kunstmann, Berlin

The Guidelines on the Responsible Governance of Tenure at a Crossroads. International Statement (2015) http://viacampesina.org/en/images/stories/pdf/2015_TG_Statement_final_EN.pdf. Accessed 14 Apr 2016

United Nations (2011) UN-guiding principles for business and human rights. Report of the Special Representative of the Secretary-General on the issue of human rights and transnational corporations and other business enterprises, John Ruggie: Guiding principles on business and human rights: implementing the United Nations “Protect, Respect and Remedy” Framework, U.N. Doc. A/HRC/17/31 issued 21 March 2011. http://www.ohchr.org/Documents/Issues/Business/A-HRC-17-31_AEV.pdf. Accessed 15 May 2016

United Nations Agency for International Development (2015) Operational guidelines for responsible land-based investment. http://usaidlandtenure.net/documents/operational-guidelines-responsible-land-based-investment. Accessed 17 May 2016

United Nations Committee on World Food Security (2016a) http://www.fao.org/cfs/cfs-home/activities/vggt/en/. Accessed 17 May 2016

United Nations Committee on World Food Security (2016b) http://www.fao.org/cfs/en/. Accessed 17 May 2016

World Bank/FAO/UNCTAD/IFAD (2009) Principles for responsible agricultural investment (PRAI) that respects rights, livelihoods and resources, knowledge exchange platform for responsible agro-investment (RRAI). Accessed 1 Mar 2016

World Bank (2011) Raising global interest in farmland. Can it yield sustainable and equitable benefits? The World Bank, Washington D.C. http://siteresources.worldbank.org/DEC/Resources/Rising-Global-Interest-in-Farmland.pdf. Accessed 17 May 2016

Evaluating the Role of Private Land Tenure Rights in Sustainable Land Management for Agriculture in Kenya

Robert Kibugi

1 Introduction

Kenya continues to face a challenge of declining quality of land, in a situation where agriculture is a major contributor to the national Gross Domestic Product (GDP), and to rural livelihoods. The role of law and policy tools, in providing solutions that govern prevention and reversal of land degradation is important. Importantly, the purpose of land tenure rights is critical, as they provide a direct entitlement to land owners to make choices on how they utilize land. As such these can be enhanced with sustainability obligations, and can complement the role played by land use legal provisions, such as agriculture laws. This paper comprises four sections. Section 2 is the background, highlighting the economic significance of agriculture to Kenya, and the negative impact of land degradation to this contribution. Section 3 examines the international policy priorities on land degradation and their implications to Kenya. Section 4 explores the contribution of tenure rights to exercise of stewardship choices by land owners. Section 5 review the role of land use law in protecting the quality of land health, and thereby stopping and reversing degradation, by examining the role of agriculture law in Kenya. Section 6 is the conclusion.

The author has utilized, with modifications, arguments on tenure rights, and stewardship, earlier developed for this Doctoral Thesis in Law at the University of Ottawa. The thesis can be retrieved at: https://www.ruor.uottawa.ca/handle/10393/20268.

R. Kibugi (✉)
University of Nairobi, Nairobi, Kenya
e-mail: rmkibugi@gmail.com

H. Ginzky et al. (eds.), *International Yearbook of Soil Law and Policy 2016*,
International Yearbook of Soil Law and Policy,
DOI 10.1007/978-3-319-42508-5_16

2 Background to Agriculture and Land Degradation in Kenya

Land is a complex resource composed primarily of soil, water and biodiversity.[1] Thus the management of such a complex resource sustainably is a challenge for a country like Kenya. This is because Kenya relies extensively on agriculture for contribution to its GDP, but is experiencing productivity decline attributable to degradation in quality of land.

2.1 The Economic Contribution of Agriculture

According to Kenya's government, in the 2015 Economic Review of Agriculture,[2] the agriculture sector is the backbone of Kenya's economy and the source of livelihood for majority of the rural population.[3] This implies that an increase in challenges facing agriculture in Kenya, such as land degradation, would negatively affect the economy and many livelihoods. The scope of what is formally considered agriculture is rather wide, and this is evidenced by the definition of agriculture activities, in the law, to mean "cultivation of land, and the use of land for any purpose of husbandry, aquaculture, and food production," which includes crops cultivation, fishing, pasture, and agroforestry.[4] Agriculture contributes about 26 % of the country's GDP and employs about 75 % of the population.[5] In 2014, the sector recorded a lower growth rate of 3.5 % compared to the 5.2 % growth in 2013.[6] Although agriculture did not achieve high growth it was by far the leading sector in contribution to GDP, followed by manufacturing sector with a share of 10.0 %.[7] In this regard crops growing, animal production and fishing and aquaculture contributed 19.7 %, 4.9 % and 0.8 % respectively (25.4 %); while the balance came from forestry and other support activities.[8]

[1]UNCCD (2014), p. 4.

[2]Kenya (2015), p. xi.

[3]Kenya (2015), p. xi.

[4]Section 2, Agriculture, Fisheries and Food Authority Act, No. 13 of 2012.

[5]Kenya (2015), p. xi.

[6]Kenya (2015), p. xi.

[7]Kenya (2015), p. xi.

[8]Kenya (2015), p. 1.

2.2 Reviewing the Status of Agriculture Land Degradation

Since 2004, the Kenya government has admitted that national agricultural productivity, in totality, had been on a steady decline that is attributable to many reasons, but that high depletion of soil fertility and land degradation are most prevalent.[9] Through the 2004 *Strategy for Revitalizing Agriculture* (SRA), it was reported that about 80 % of the population lives in rural areas, relying on agriculture as the principal economic activity, and most of these people live in poverty and food insecurity.[10] The Strategy highlighted that the country was experiencing a significant decline in soil fertility, increasing land degradation, and that extension services were ineffective, leaving farmers with limited sources of new skills, knowledge or technology on how to overcome the challenges.[11]

This, according to Kenya's Draft Policy on Land Reclamation,[12] threatens food security, and employment in agriculture based industries.[13] This is worsened by the significant and serious forms of soil erosion occurring in many parts of the country, as the removal of the top fertile soils eventually leads to formation of hard pans, rills and gullies.[14] With soil surface degradation, degraded lands lack adequate plant cover to help reduce surface retention and infiltration, resulting in higher surface runoff.[15] This results not only in reduced soil moisture content in the soil profile, but also increased rates of soil erosion, with a negative effect on plant growth, and a reduction of biomass production.[16] A new agriculture policy was adopted in 2010, the 2010–2020 *Agriculture Sector Development Strategy* (ASDS), which noted that Kenya 'has lost some of her well known biodiversity resources mainly due to population increase, habitat destruction, desertification, over-exploitation of species and conversion through deforestation and drainage of wetlands for agriculture and settlement.'[17]

It is important to note that in the foregoing analysis, land degradation in Kenya is identified as resulting in declining quality of land, and subsequent lower levels of total national agriculture productivity. This is consistent with the definition of land degradation, which according to the United Nations Convention to Combat Desertification (UNCCD) Secretariat, arises when there is any loss or reduction in the biological or economic productive capacity of the land resource base.[18] This land degradation is generally caused by human activities, exacerbated by natural

[9]Kenya (2004), p. 17.

[10]Kenya (2004), pp. 1–2.

[11]Kenya (2004), pp. 15–17.

[12]Kenya (2011).

[13]Kenya (2011), p. 5.

[14]Kenya (2011), p. 5.

[15]Kenya (2004), p. 10.

[16]Kenya (2004), p. 10.

[17]Kenya (2010a), p. 65.

[18]UNCCD (2014), p. 4.

processes, and often magnified by and closely intertwined with climate change and biodiversity loss.[19] Such activities, and natural process include: soil erosion caused by wind and/or water; deterioration of the physical, chemical and biological or economic properties of soil; and long-term loss of natural vegetation.[20]

The unsustainable exploitation of land productivity through anthropocentric activities it appears, contributes significantly to land degradation. This is exacerbated by natural processes, such as climate change, through extreme weather events, such as drought. Thus land degradation poses a complex problem, which in Kenya, law and policy have attempted to address, in order to enhance agriculture productivity, while promoting stewardship through sustainable land management. It is important to note, as discussed below, that international law and policy priorities addressing land degradation, such as the Sustainable Development Goals, intersect with, and support Kenyan development prioritization.

3 International Policy Priorities on Land Degradation, and Impact on Kenya

International law and policy is useful in providing principles on land stewardship, which can be adapted nationally, through law, policy or development plans. These international approaches include the Sustainable Development Goals (SDGs), adopted in 2015, and the Rio + 20 Outcome document. The SDGs highlight the challenges of land degradation, and the implication this has on food security, and environmental quality. In SDG No. 2, which focuses on food security, there is a goal to enhance sustainable food production systems and implement resilient agricultural practices that increase productivity and production, that help maintain ecosystems, and among other things progressively improve land and soil quality. More directly, SDG No. 15 aims to protect, restore and promote sustainable use of terrestrial ecosystems, sustainably manage forests, combat desertification, and halt and reverse land degradation and halt biodiversity loss. The global commitment is to achieve this through combating desertification, and restoring degraded land and soil, including land affected by desertification, drought and floods, and strive to achieve a land degradation-neutral world.[21] Perhaps existing initiatives, such as continental/regional and national policy and legal prioritization of sustainable interventions to protect quality and health of land, are indicative of how this global commitment can be implemented.

In this regard, Kenya is part of the African Union's Comprehensive Africa Agriculture Development Programme (CAADP).[22] The CAADP is organized into

[19]UNCCD (2014), p. 4.

[20]UNCCD (2012), p. 8.

[21]UNGA (2015), para 15.3.

[22]African Union, NEPAD, undated.

four pillars, with the first one being "to extend the area under sustainable land management and reliable water control systems." The objective of this pillar is to reverse resource degradation and ensure broad and rapid adoption of sustainable land and forestry management practices by the small-holder and commercial sectors. The anticipated activities include community-based reforestation and conservation; interventions in soil health, restoring soil fertility with mineral fertilizers or agroforestry. Kenya signed a compact with the African Union in 2010, committing to implement the CAADP through the 2010–2020 ASDS.[23]

The SDGs, have a target achievement goal of 2030, similar to Kenya Vision 2030,[24] the current national development plan that aims to attain national economic middle income status by the year 2030. As part of the strategy for economic empowerment, the Vision recognizes the need for sustainable land use, noting that current land use practices in the country are incongruent with the ecological structure of the country.[25] Thus the SDGs, such as No. 2, which focuses on food security, through enhanced sustainable food production systems and resilient agricultural practices, is very relevant to Kenyan national development objectives.

4 Exploring the Structure and Function of Private Tenure Rights Under Kenyan Law

The role of property rights in management of land, and making choices in land is critical. This paper argues that property rights in land provide the most direct legal entitlement and decision making authority to a land owner, to make choices of utilization of the land. The paper further examines the conceptual structure of tenure rights, and reviews whether tenure rights alone are sufficient to guide land owners in adoption of stewardship, when making land use choices.

4.1 The Constitutional Foundations of Tenure Rights and Sustainable Management

Indeed, the Constitution of Kenya, in place since 2010, specifically supports this argument, when in article 40, it provides for the right to own property, including land, and then links this right to land use in article 60, through the provision that land shall "be held, used and managed in a manner that is equitable, efficient,

[23]Kenya (2010b).

[24]Kenya (2012).

[25]Kenya (2012), p. 23.

productive and sustainable."[26] This applies to all categories of land in Kenya: public, private and community land.[27]

To reinforce the linkage between the right to own land, and the obligation to sustainably utilize it, the Constitution sets out principles to guide this sustainable approach to land management, including security of tenure rights; sustainable and productive management of land resources; and sound conservation and protection of ecologically sensitive areas.[28] The required application of sustainable development paradigms in land management, including for agriculture is not optional as the Constitution, in article 60, applies the mandatory term "shall." In addition, the Constitution contains principles and values of national governance,[29] whose application in development, and implementation of any statutes or public policy is mandatory. These principles and values of national governance include the principles of sustainable development, which has now obtained constitutional legal status in Kenya, as explained below.

In article 10, the Constitution sets out "principles and values of national governance", that will be applied whenever there is constitutional application, making and implementation of new law, and making and implementation of public policy. This clause is mandatory in application, and is written "the national values and principles of governance in this Article bind all State organs, State of officers, public officers and all persons ..."[30] These principles, include that of sustainable development, whose application, as noted above, is required by the Constitution to be mandatory. There has not been, by the time of writing, any authoritative judicial or statutory interpretation of the full legal import of these mandatory principles. Nonetheless, this paper argues that, for instance, application of property rights in land, should be interpreted to include the irreducible duty to exercise these rights sustainably. Article 60, highlighted above, explicitly supports this view through the requirement for land to be held, used and managed in a manner that is equitable, efficient, productive and sustainable. In terms of land use, under police powers of the state, set out in article 66—the requirement in article 60 for land to be used sustainably extends this mandatory application of sustainable development in governance of land resources in Kenya.

In sustainable land management, the constitutionally mandated principles that combine security of tenure, sustainable and productive management, together with conservation the demonstrate the legal complexity of land. This is because sustainable land management creates a linkage between secure tenure rights on the land itself, and the consequential ability, knowledge or willingness to manage the soil, water and biodiversity sustainably. Thus the management of this complex resource, requires innovation on legal tools, and a recognition of the role that tenure rights

---

[26]Constitution of Kenya (2010), Article 60(1).

[27]Constitution of Kenya (2010), Article 61.

[28]Constitution of Kenya (2010), Article 60(1) [b, c & e].

[29]Constitution of Kenya (2010), Article 10.

[30]Constitution of Kenya, Article 10.

play, in providing the right holder the primary entitlement to the land, and therefore an ability to make decisions on the management of the soil, water and biodiversity. It is this linkage that the paper reviews, in order to evaluate whether tenure rights provide an optimal set of legal tools to impact land use decision making by right holders in a manner that mandates stewardship. In addition, in the next section, the paper examines the s sufficiency of land use regulatory regimes and how they can be improved to support sustainable land management, in order to promote quality of land health, and address degradation. This analysis is timely because, as evident from the Rio + 20 outcome document, *The Future We Want*,[31] the economic and social significance of good land management, including soil, is recognized due to its contribution to economic growth, biodiversity, sustainable agriculture and food security, and improving water availability.[32]

4.2 The Conceptual Structure of Land Tenure Rights

Property rights to resources such as land and water—which contribute to agricultural productivity—play a fundamental role in influencing the patterns of natural resource management especially by determining *who* can do *what* with a particular resource, such as a parcel of land, and sometimes also *when* and *how* they can do it.[33] In the case of ownership and use of land, it is these property rights that are referred to as tenure rights. These tenure rights are fundamental to making decisions on the socio-economic utilization of land, and the measures or steps necessary to sustain its environmental quality. Tenure rights in land, when vested in people either individually or collectively, represent the most direct legal entitlement for land owners or occupiers to have control and make decisions over use or management of land, or how to exclude other people. This is because an operative tenure system normally defines methods by which individuals or groups acquire, hold, or transfer interests in land. It has to do with how rights to land and other natural resources are assigned within societies, and determines *who* holds *what* interests in *what* land.[34] Land tenure sets apart access to land and control of land. Control is the command an individual has over a particular piece of land and the derived benefits[35] and it is based on a recognized right, whether indigenous or formal. Access denotes ability to utilize land, but not necessarily ownership, and may include some decision making power over aspects of productivity. In either case, land tenure implies that the holder of the property rights in land would be involved in regular decision making over which land use choices would achieve the desired economic

[31]UNGA (2012).

[32]UNGA (2012), para 205.

[33]Meinzen-Dick et al. (1997), p. 1303.

[34]Kameri-Mbote (2005), p. 6.

[35]Lastarria-Cornhiel (1997), p. 1318.

goals, such as agricultural productivity from, and stewardship choices over, the land. The question, in dealing with land degradation that affects agriculture productivity, is whether the breadth of tenure rights in land offer any legal tools that guide right holders toward any stewardship responsibilities to integrate the environmental quality (soil, water, biodiversity) of the land, with the socio-economic choices they make on how to utilize the land. This is evaluated in the discussion below, on the structure of tenure rights in Kenya.

4.3 Evaluating the Structure and Application of Statute Based Land Tenure in Kenya

The process of tenure conversion from indigenous to individualized land tenure has been implemented on a massive scale in Kenya starting with British Colonial authorities and carried on by post-independence governments. The conversion of African indigenous land tenure was mooted by the 1955 Report of the East African Royal Commission.[36] This report did not focus on land stewardship but rather promoted individualized tenure as possessing great advantages such as giving the individual a sense of security in possession of land, and in enabling the purchase and sale of land.[37] It was mainly intended to drive economic transformation through individual (rather than communal) ownership of land, and enhanced agriculture. A specific process of formalizing individual registration to land was launched by the 1955 Swynnerton Report,[38] which was commissioned by the Colonial government of Kenya.

The Swynnerton report focused on legal and policy methodologies that could be pursued in order to intensify African agricultural productivity. It suggested that realizing sound agricultural development depended upon a system of land tenure which would avail to the African farmer a unit of land and a viable system of production.[39] The report also proposed registration of individual property rights in land such that the farmer would also be provided with security of tenure through an indefeasible title, to encourage him to invest his labour and profits to develop the farm and apply it as collateral. Thus in addition to seeking to confer tenure security, the Swynnerton plan also assumed this would impact agricultural productivity positively. Security of tenure is one of the constitutional principles set out in article 60, as being central to sustainable management of land in Kenya. Thus, understanding the legal, and normative content of tenure security is imperative. John Bruce, and Shem Migot-Adholla have argued that security of tenure exists when an individual perceives that he or she has rights to a piece of land on a

[36]Secretary of State for the Colonies (1955), at p. 323, para 77.

[37]Secretary of State for the Colonies (1955), at p. 323, para 77.

[38]Colony and Protectorate of Kenya (1955).

[39]Colony and Protectorate of Kenya (1955), p. 9.

continuous basis free from imposition or interference from outside sources.[40] This security also includes the ability to reap benefits of labour and capital invested in that land.[41] The Food and Agriculture Organization (FAO) suggests that where there is no security of tenure, or it is weak, households are significantly impaired in their ability to secure sufficient food and to enjoy sustainable rural livelihoods.[42] Such weakness, according to resource economist Tony Scott, could be linked with the duration of, and the assurance of the rights over the land in question.[43]

This correlation between tenure rights (as entitlement to make land use choices for productive uses like agriculture) and ability of the rightholder to reap benefits of labour, implies that tenure rights could be applied to maintain land in good environmental quality to support productivity.

In terms of the breadth of the rights comprising secure tenure, the FAO suggests a structure for the breadth of tenure rights in land to include[44]:

i). *Use rights:* rights to use the land for grazing, growing subsistence crops, gathering minor forestry products, etc.
ii). *Control rights:* rights to make decisions how the land should be used including deciding what crops should be planted, and to benefit financially from the sale of crops, etc.

The user and control rights, signify the legal right and ability of a person, holding the tenure rights, to make decisions on how they can use their land to achieve their desired objectives. In agriculture land use, these objectives revolve around enhancing agricultural productivity, and presumably an explicit duty of stewardship to support the quality of the land that underpins productive agriculture. Typically, secure tenure would be conferred when a definite (including perpetual) duration is defined, and there is an assurance over the defined (user and control rights) over the land in question. Statute law confers security of tenure either through freehold tenure, absolute tenure, or leaseholds. In considering the role of tenure security in sustainable land management, the challenge is whether the breadth of rights include a stewardship role to conserve and protect the quality of the land (soil, water, biodiversity) health.

[40]Bruce and Migot-Adholla (1993), p. 3.

[41]Bruce and Migot-Adholla (1993), p. 3.

[42]Food and Agriculture Organization (FAO) (2002), p. 18.

[43]Scott (2008), p. 5.

[44]Bruce and Migot-Adholla (1993), pp. 9–10. There are also Transfer rights: right to sell or mortgage the land, to convey the land to others through intra-community reallocations, to transmit the land to heirs through inheritance, and to reallocate use and control rights.

4.4 Do Statutory Tenure Rights in Kenya Facilitate Stewardship in Agriculture Land Use Choices?

The content of individual or private land rights are mainly manifested through the 2013 Land Registration Act, which replaced two earlier (now repealed) statutes: the Registration of Titles Act,[45] and the Registered Lands Act.[46] Under the 2013 Land Registration statute, the registered owner of a parcel of land is granted an absolute quantum of rights. Thus section 24(a) provides that "the registration of a person as the proprietor of land shall vest in that person the absolute ownership of that land together with all rights and privileges belonging or appurtenant thereto." The absolute estate is indefeasible except on grounds specified by the Land Registration Act, and can only be limited by interests or claims such as the leases, charges and other encumbrances and other conditions and restrictions that maybe shown in the register. Limitations also occur through overriding interests, specified in section 28 of the land registration statute, which provide the Kenyan State and other persons with superior rights over the rightholder in certain circumstances such as compulsory acquisition, wayleaves for public utilities, spousal rights over matrimonial property, etc.

These statutory protections demonstrate assurance on the existence of the right, while the right to transfer the land demonstrates that the duration of tenure security is, legally, subject to the decision of the right holder to undertake a disposition. In practice, these rights show that the registered land owner enjoys the legal right to control and use the land. The quantum of tenure rights conferred with this absolute estate impacts sustainable agricultural land use as they give the registered owner (or assignees) the legal rights to make land use decisions. Based on the above analysis of constitutional provisions linking tenure rights and sustainable development, the quantum of statutory tenure rights should, ideally, include a duty to protect the quality of the land when making such land use decisions.

The Kenya government, despite decades of land reform towards formal tenure, acknowledges persisting overall poor agricultural productivity in the country,[47] with consistently declining soil fertility and higher land degradation identified as the major contributor to the falling productivity in agriculture.[48] In this regard, David Attwood, contends that the conventional view of land titling (through formal registration of right holders) automatically enhancing the adoption of soil conservation measures or new technology for higher agricultural productivity has been inaccurate.[49] This validates reasoning that contrary to previous arguments,

[45]Cap 281 Laws of Kenya, (now repealed).

[46]Cap 300 Laws of Kenya, (now repealed).

[47]Kenya (2009), p. 18.

[48]Kenya (2004), pp. 15–17.

[49]Attwood (1990), pp. 668–669.

conferment of statutory tenure, without more (such as stewardship obligations), does not increase quality of land husbandry.

The 2012 FAO voluntary guidelines on responsible governance of tenure rights[50] recognize this challenge. They argue that no tenure right should be absolute, including private ownership, but that in fact such rights should be balanced by duties, perhaps such as those duties that create environmental safeguards,[51] contingent on the right holder, to prevent or minimize land degradation and loss of biodiversity. It is the placement of these duties in the statutory structure on tenure rights that is challenging, as it is necessary to attain the most optimal outcomes on land stewardship, during land use decision making by tenure right holders.

In Kenya, it is the agriculture legislative framework that set out the conditions of land use that those practicing agriculture should observe. This is applied as legal entry point to limit the rights of a land owner, through the constitutional power of the State to regulate land use in the country.[52] Kenya also has in place the framework Environmental Management and Coordination Act[53] that governs environmental management practices, which include agriculture land use. In addition, in May 2016, Kenya enacted a national Climate Change Act,[54] which upon implementation will prescribe land use measures necessary to implement adaptation and mitigation interventions. Nonetheless, the role of agriculture law is superior since is it the primary land use statute, and it also manifests the police power of the Kenyan State with respect to agriculture. In the next section therefore, the paper examines the approach taken by agriculture legislation, and evaluates whether there is a linkage between the land use approach to regulating stewardship, and the tenure rights of a land owner.

5 The Role Played by Agricultural Land Use Law in Safeguarding Sustainable Land Management

The regulatory power of the state over land use in Kenya, as noted above, is established by the Constitution. Article 66 of the Constitution empowers the State to "regulate the use of any land, or any interest in or right over any land, in the interest of defence, public safety, public order, public morality, public health, or land use planning." A sustainability basis for the exercise of this power can be derived from the mandatory constitutional requirement, in article 10, to apply sustainable development principles when making, or implementing any law or public policy. The application of the police power of the State is usually undertaken

[50]FAO (2012), p. 6.

[51]FAO (2012), p. 24.

[52]Constitution of Kenya (2010), Article 66.

[53]Environmental Management and Coordination Act, Act No. 8 of 1999, as amended in 2015.

[54]The Climate Change Act 2016 was assented into law by the President of Kenya on 6 May 2016.

through legislation, including on physical or spatial planning, and development control. In Kenya, in addition to the spatial planning rules under the physical planning legislation,[55] major land uses, such as forestry, or agriculture are governed by sector specific laws, which function as land use planning laws, as well as regulatory laws. For instance, the Forestry Act 2005, in Part 1, provides for administration and State regulation in management of forests, while at section 35 provides tools for forest land use planning, through mandatory management plans. Thus, in the case of agriculture, in 2012 new legislation was enacted to give effect to provisions of the 2010 Constitution. The 2012 Agriculture, Fisheries and Food Authority Act (AFFA) was enacted to give effect to the land use powers of the Kenyan State, and to overhaul the 1955 Agriculture Act[56] that had been operation since before Kenya was an independent country. In this section, the paper reviews two possible legal tools that can supplement tenure rights, by placing a direct responsibility on land owners to explicitly be duty-bound to the application of stewardship in making land use decisions. It is imperative to test whether the new agriculture law attempts to give effect to the mandatory sustainable development principles set out in article 10 of the Constitution.

5.1 *Application of Land Husbandry Requirements to Tenure Holders Through Land Preservation Rules*

Under the new agriculture legislation, attention has been paid to governance of land use practices, in a manner that preserves the quality of land health. Section 22 of the Agriculture, Fisheries and Food Authority Act (AFFA) legislation, provides for development of rules for the "preservation, utilization and development of agricultural land . . ." Although these rules are yet to be developed, at the time of writing, their normative content if focused on sustainability will strengthen the focus on land quality. According to AFFA, these, rules will be designed to prescribe the manner in which land owners (whether or not in actual occupation of the land) manage the land in accordance with rules of good estate management. The approach of linking the stewardship obligation to the land owner is crucial to note, because the obligation continues to attach on the primary tenure holder, even where there are other occupiers, for instance on a periodic lease to use the land in question. The rules in question may also prescribe how the land shall be farmed, in accordance with rules of good husbandry. The AFFA law sets out the threshold for fulfilment of the legal obligation here, by defining the meaning of the "rules of good estate management," and "rule of good husbandry." Good estate management is deemed to have been attained where, having regard to the character and situation of the land, a land owner skilled in husbandry can maintain efficient production in quality and quantity

[55] Physical Planning Act, Cap 286 Laws of Kenya.

[56] Cap 318 Laws of Kenya, (now repealed).

of the produce. Good husbandry skills will ensure the efficient production in quality and quantity of produce while "while keeping the land in a condition to enable such a standard to be maintained in the future." It is the requirement of good husbandry that draws the link between the obligations of the land owner/occupier, the agriculture productivity needs, and stewardship role of the land owner when practicing good estate management.

Good land husbandry plays an instrumental role in stopping and reversing land degradation. Francis Shaxon defines good land husbandry as the 'active process of implementing and managing preferred systems of land use in such ways that there will be an increase - or, at worse, no loss - of productivity, stability, and usefulness for the chosen purpose.'[57] The FAO defines land husbandry to mean a system of 'water, soil fertility and biomass management'[58] that is essential to sustainable land use. FAO further suggests that sound land husbandry requires that farmers 'concentrate on the sound management of good, productive land before it becomes degraded, since there is a faster and larger increase in yields on deep soil than on exhausted stony soil.'[59] The concept of land husbandry in relation to sustainable agriculture therefore requires a constant and dynamic process of land stewardship by land owners or occupiers to safeguard the environmental quality of the land. Supporting this view, Shaxson, Tiffen, Wood & Turton, have argued that farmers (as land owners or occupiers) as the main 'husbanders' of the land are central to successful implementation as it is these farmers who 'consider' then decide to 'reject' or 'integrate' husbandry options into land use decision making.[60] In this sense, the application of the user and control rights aspect of tenure, described earlier, is demonstrably important. This is because only the land owner that enjoys sufficient tenure security, and particularly user and control rights that empower decision making will be in a position to apply the husbandry requirements under agriculture law.

5.2 The Utility of Agroforestry in Enhancing the Duties of Land Owners Towards Stewardship

The Agriculture, Fisheries and Food Authority Act (AFFA) defines agriculture to include agroforestry activities, where those agroforestry activities are ancillary (and beneficial) to the use of land for agricultural purposes.[61] Also known as farm forestry, it serves a multiplicity of functions including sustainable land management. These forests are usually planted and cultivated by land owners, who often

[57]Shaxson (1993), p. 105.

[58]Roose (1996), p. 24.

[59]Roose (1996), p. 24.

[60]Shaxson et al. (1997), p. 6.

[61]Agriculture, Fisheries and Food Authority Act (AFFA), s 2(e).

make the decisions over the desired output, thus the existency of sufficient tenure security encompassing user and control rights is crucial. The trees may be indigenous or exotic species that have been domesticated to serve particular local purposes.[62] These purposes are broad and wide, and may vary from community to community. The trees and accompanying biodiversity may be a major source of food for rural communities, enhancing food security, but also promoting dietary balance, diversity and good health.[63] They also provide fodder; enhance security of land rights by serving as demarcation points marking boundaries; and may also serve as water catchments, or to prevent soil degradation[64] that has significantly outstripped the ability of land in Kenya to regenerate soil without assistance. When implemented properly farm forestry provides legal tools with a constructive role in stabilizing soil, reducing soil erosion and enhancing soil fertility through agroforestry.[65]

Most of the land under which farm forestry is undertaken, is agricultural land, thereby bringing these land use activities under the direct authority of the Agriculture, Fisheries and Food Authority Act. However, since the new agriculture law was enacted, the pace of making new regulations has been slow, such that earlier subsidiary legislation under the now repealed Agriculture Act, remain in operation.

This including regulations made under section 48 of the repealed legislation, which empowered the Minister to make rules imposing mandatory obligations on land owners to take land preservation measures such as afforestation and reforestation, intended to prevent or reverse soil erosion. It is under this legal authority that the *Agriculture (Farm Forestry) Rules, 2009* were enacted. The farm forestry rules define farm forestry, in similar terms as the *Forest Act,*[66] as 'the practice of managing trees on farms whether singly or in rows, lines, boundaries or in woodlots or private forests.'[67] The farm forestry rules, *prima facie,* set a clear legal responsibility on land owners or occupiers by creating a mandatory requirement for every land owner or occupier to 'establish and maintain a minimum of 10 % of the land under farm forestry...'[68] The objectives of the rules, in addition to maintaining the 10 % tree cover are

- Conserving water, soil and biodiversity
- Protecting riverbanks, shorelines, riparian and wetland areas
- Sustainable production of wood, charcoal and non-wood products
- Providing fruits and fodder, and
- Carbon sequestration and other environmental services

---

[62]FAO (2002), First Part, p. 20.

[63]FAO (2002), First Part, p. 22.

[64]FAO (2002), First Part, p. 22.

[65]FAO (2002), First Part, p. 22.

[66]Section 3.

[67]Rule 3.

[68]Rule 5(1).

These objectives reflect some legal concern with sustainability of land use practices. The objectives also suggest that a variety of possibly complex land use practices would be necessary to ensure the conservation, and environmental protection aims are integrated with socio-economic objectives of farming. Unlike the *Forest Act* in which case a landowner voluntarily applies for their farm forest to be registered, these rules make farm forestry compulsory.

The Farm forestry rules provide for inspection and enforcement of compliance by agricultural officers acting as inspectors. The inspector is authorized to enter any agricultural land to ascertain 'whether the farm owner or occupier has complied with the 10 % farm forestry' requirement.[69] If the inspector finds that the requirement has not been complied with, the inspector is empowered to issue a farm forestry establishment order requiring the farmer to institute measures to put the farm forest in place.[70]

Apart from the otherwise clear legal responsibility on land owners to maintain a 10 % minimum tree cover on their farms, the rules offer no tools or mechanism to guide decision making by land owners on tree or land husbandry. This is evident because the inspector is only required to ascertain there is a 10 % tree cover, without more, such as checking on the quality of farm forest, or the husbandry practices. This is unlike the Forest Act, which explicitly states that a land owner is entitled to receive technical assistance from forest officers,[71] which ideally could help land owners enhance the ecological benefits of farm forests. Equally, the rules are silent on techniques of how farmers may enhance their economic returns for instance by combining indigenous trees that preserve biodiversity with fast growing exotic trees suitable for timber. Instead, the farm forestry rules just state that 'the species of trees or varieties planted shall not have adverse effects on water sources, crops, livestock, or soil fertility' and should not be invasive species.[72]

Farm forestry is a mechanism that can assist to halt, and reverse land degradation and to enhance soil fertility on farmland. If the socio-economic and environmental benefits are made apparent to land owners, and the land owners receive useful skills and technical assistance, their attitudes are bound to change in favour of farm-level forestry activities. In this sense, the legislative framework for farm forestry could serve as a mechanism that enhances the responsibility of land owners towards the land, beyond the mere economic utility or value.

---

[69]Rule 5.

[70]Rule 5.

[71]Section 25(3).

[72]Rules 5.

6 Conclusion

Kenya continues to face a regulatory challenge. Agriculture is extremely important to the economy and livelihoods of many citizens. Yet, land degradation levels are high, and the diminishing fertility of land adversely impacts agriculture productivity. There is a sound constitutional grounding of the duty to pursue sustainability and stewardship in land management and utilization. However, tenure rights present an insufficient legal structure to enforce the adoption of stewardship by land owners in making land use choices, such as agriculture. This dilemma is problematic as Kenya is part of the global agreement to pursue food security, and protection environmental quality of terrestrial resources, through implementation of the sustainable development goals. The paper has suggested that tenure rights can be utilized as a basis for requiring land owners, and occupiers, to implement stewardship based land use practices. A helpful example is the proposed land preservation rules under the AFFA legislation, to the extent that this are prospective (developed to guide land owners on addressing degradation before it occurs), and are coupled with effective agriculture extension. A similar approach can be taken with the further application of agroforestry, to the extent that it ceases to be fully command and control, and offers guidance on how land owners can pursue environmental quality of the land.

References

African Union (undated) Implementing the Comprehensive Africa Agriculture Development Programme and Restoring Food Security in Africa "The Roadmap", New Partnership for African Development (NEPAD) Secretariat, Midrand. Online: ftp://ftp.fao.org/TC/TCA/CAADP%20TT/Background%20%20documents/Main%20document/Roadmap%20of%20implementing%20CAADP.pdf. Accessed 11 May 2016

Agriculture, Fisheries and Food Authority Act (AFFA)

Attwood D (1990) Land registration in Africa: the impact on agricultural production. World Dev 18(5):659–67, 668–669

Bruce JW, Migot-Adholla S (1993) Introduction: are indigenous African tenure systems insecure? In: Bruce JW, Migot-Adholla S (eds) Searching for land tenure security in Africa. Kendall Hunt Publishing/World Bank, Iowa

Climate Change Act, 2016

Colony and Protectorate of Kenya (1955) A plan to intensify the development of African Agriculture in Kenya. Government Printer, Nairobi

Constitution of Kenya, 2010

Environmental Management and Coordination Act, Act No. 8 of 1999, as amended in 2015

FAO (2002) Trees outside forests: towards a better awareness. FAO, Rome

FAO (2012) Voluntary guidelines on the responsible governance of tenure of land, fisheries and forests in the context of national food security. FAO, Rome

Food and Agriculture Organization [FAO] (2002) Land tenure and rural development. FAO Land Tenure Studies 3, Rome

Kameri-Mbote P (2005) The land has its owners! gender issues in land tenure under customary law. Paper presented at the UNDP-International Land Coalition Workshop: Land rights for African development: from knowledge to action, Nairobi, 31 October–3 November 2005

Kenya (2004) Strategy for revitalizing agriculture: 2004 – 2014. Ministry of Agriculture, Nairobi

Kenya (2009) Sessional Paper No. 3 of 2009 on National Land Policy. Ministry of Lands, Nairobi

Kenya (2010a) Agricultural sector development strategy 2010–2020. Ministry of Agriculture, Nairobi

Kenya (2010b) The Kenya CAADP compact: to be implemented through the agriculture sector development strategy. Ministry of Agriculture, Nairobi

Kenya (2011) Draft National Land Reclamation Policy. Ministry of Water and Irrigation, Nairobi

Kenya (2012) Sessional Paper No. 10 of 2012 on Kenya Vision 2030. Government Printer, Nairobi

Kenya (2015) Economic review of agriculture. Ministry of Agriculture, Nairobi

Lastarria-Cornhiel S (1997) Impact of privatization on gender and property rights in Africa. World Dev 8(25):1317, 1318

Meinzen-Dick R, Brown LR, Feldstein HS, Quisumbing AR (1997) Gender, property rights and natural resources. World Dev 8(25):1303

Physical Planning Act, Cap 286 Laws of Kenya

Roose E (1996) Land husbandry: components and strategy. FAO, Rome

Scott A (2008) The evolution of resource property rights. Oxford University Press, New York

Secretary of State for the Colonies (1955) East Africa Royal Commission 1953–1955 Report. Her Majesty's Stationery Office, Cmd 9475, London

Shaxson TF (1993) Conservation-effectiveness of farmers' actions: a criterion of good land husbandry. In: Baum E, Wolff P, Zobisch MA (eds) Acceptance of soil and water conservation: strategies and technologies. Vol. 3 in Topics in applied resource management in the tropics. German Institute for Tropical and Subtropical Agriculture, Witzenhausen

Shaxson F, Tiffen M, Wood A, Turton C (1997) Better land husbandry: rethinking approaches to land improvement and the conservation of water and soil. Natural Resource Perspectives No. 19

UNCCD (2012) Zero net land degradation, UNCCD Secretariat Policy Brief, Bonn

UNGA (2012) The Future We Want GA Res 66/288, UN GAOR, 66th sess, Agenda Item 19

UNGA (2015) Transforming our world: the 2030 Agenda for Sustainable Development GA Res 70/1, UN GAOR, 70th sess, Agenda Item 15 and 116, 25 Sept 2015

United Nations Convention to Combat Desertification [UNCCD] (2014) Land degradation neutrality: resilience at local, national and regional levels United Nations Convention, Secretariat, Bonn

The Human Right to Food and Sustainable Soil Management: Linking Voluntary Agricultural Sustainability Standards with Food Security

Tina Beuchelt, Anna Mohr, and Rafaël Schneider

1 Introduction

Sustainable soil and land management both contribute to achieving several of the proposed Sustainable Development Goals (SDG), such as food security, land degradation neutrality, climate action and conservation of life on land.[1] The SDG 2—to end all forms of hunger by 2030—specifically includes sustainable soil management and agriculture. The unification of these two targets in one goal underlines the dependency of the realization of the Human Right to adequate Food on sustainable agriculture and soil management. Continued and accelerated soil degradation and loss of access to soil pose a major threat to food security as the global food production systems are almost completely reliant on soil.

With governments shifting from fossil-fuel-based to green or bio-based economies, international demand and trade for agricultural commodities are growing. In the context of limited availability of arable land, water and energy, the increasing demand for non-food biomass[2] will compete with the rising food demand for these resources at local and global levels.[3] This competition will also increase the

[1]Töpfer (2015).

[2]With the term biomass, we specifically refer to biological material derived from plants and animals in the agricultural or forestry sector that is used as food for human consumption or for non-food purposes such as animal feed, energy feedstock, fibre and industrial raw materials.

[3]OECD and FAO (2014).

T. Beuchelt (✉) • A. Mohr
Center for Development Research (ZEF), University of Bonn, Bonn, Germany
e-mail: tbeuchel@uni-bonn.de; annamohr@posteo.de

R. Schneider
Policy Department, Welthungerhilfe, Bonn, Germany
e-mail: Rafael.Schneider@welthungerhilfe.de

H. Ginzky et al. (eds.), *International Yearbook of Soil Law and Policy 2016*, International Yearbook of Soil Law and Policy,
DOI 10.1007/978-3-319-42508-5_17

pressure on soils.[4] Thus, the maintenance or enhancement of global soil resources is essential to meet the actual and future needs for food, water, and energy security.[5] Alternative and sustainable land and soil management practices and suitable governance mechanisms are needed for the biomass production sector. Good agricultural practices protect the soil, maintain and enhance soil fertility, hence avoiding erosion and land degradation. Governmental regulations and control are limited regarding agricultural practices in most low- and middle- but also in high-income countries. Sometimes subsidies are linked to cross-compliance with certain environmental and agricultural management practices. This is, for example, the case in Europe.[6] As a response, private governance systems such as voluntary sustainability standards with corresponding certification schemes have emerged in the last two decades covering environmental aspects and agricultural management practices.[7] The standards differ greatly in their aims, scope and concepts of sustainability and feedstock types (e.g. addressing biomass in general or only a specific commodity such as palm oil or coffee). They usually include soil and water management aspects, e.g. erosion prevention, fertility enhancement or soil contamination. Some standards also cover social aspects. Private sustainability standards which focus exclusively on soils and soil management do not exist.

The voluntary sustainability standards determine what is done or not done on a farm. A corresponding certification scheme sets up a system to control and monitor the implementation of the standard of an individual or a company through a third-party verification process. This requires exactly defined and measurable criteria and indicators which are controlled during on-site audits.[8] In international supply chains, voluntary sustainability standards are also expected to succeed where governmental and intergovernmental processes have failed to establish regulations and control mechanisms.

Voluntary sustainability standards for non-food biomass uses gained support with the introduction of the Renewable Energy Directive of the European Union, which includes a set of mandatory sustainability criteria for bioenergy to avoid negative external environmental effects of this policy.[9] In addition to private sustainability standards, different guidelines emerged at the international level to guide and regulate sustainability aspects of biomass production and agricultural investments on a voluntary basis. These guidelines target the responsibility of investors in the agricultural sector such as the 'Principles for Responsible Investment in Agriculture and Food Systems' (RAI) of the Committee on World Food

[4]Koch et al. (2013).

[5]FAO (2015), Koch et al. (2013), and McBratney et al. (2014).

[6]http://ec.europa.eu/agriculture/envir/cross-compliance/index_en.htm or https://www.gov.uk/guidance/soil-management-standards-for-farmers.

[7]Charnovitz et al. (2008), von Geibler (2013), Klooster (2010), Mohr and Bausch (2013), Scarlat and Dallemand (2011), and van Dam (2009).

[8]Albersmeier et al. (2009).

[9]EC (2009).

Security in 2014[10] and the 'Principles for Responsible Agricultural Investment that Respects Rights, Livelihoods and Resources'[11]. The FAO also developed the 'Voluntary Guidelines on the Responsible Governance of Tenure of Land, Fisheries and Forests in the Context of National Food Security' (VGGT)[12]. These international guidelines remain quite abstract. Certification systems, individual producers, processors or enterprises often have difficulty in translating the principles and guidelines into concrete actions in the field.

Research in low- and middle-income countries shows that producing biomass for export or non-food purposes, whether certified or not, can directly compete with local food production for available land, water and human resources. Studies show both neutral or positive effects of biomass production for non-food uses on local food security as well as negative effects such as lower food supplies and rising food prices.[13,14] The impacts of producing (non-food) biomass depend on site-specific parameters such as the feedstock and technology utilized, the scale of production and the level of local job creation, whether the feedstock competes with traditional crops or is a co-product, and whether indirect price effects of non-food biomass tend to increase or decrease food prices.[15]

However, only few attempts have been made to include food security aspects in private sustainability standards for biomass. Existing attempts vary in rigor.[16] A study assessing the sustainability performance of different biomass certification schemes reveals a lack of methods to monitor and assess impacts on local food security.[17] So far, only one voluntary sustainability standard developed a comprehensive guideline to assess food security.[18] However, its complex assessment method proves ineffective due to the need for intensive data collection and analysis.

The linkage between soil management, biomass production and the Human Right to adequate Food (hereafter called Right to Food) has also not been addressed and discussed in any depth in connection with the SDGs. While Müller et al.[19] look at the role of biomass in the SDGs, FIAN[20] addresses the access to land in combination with

---

[10]CFS (2014).

[11]FAO et al. (2010).

[12]FAO (2012).

[13]de Schutter (2011), Gasparatos et al. (2015), German et al. (2011), Horta Nogueira and Silva Capaz (2013), Locke and Henley (2013), Schut and Florin (2015), van Eijck et al. (2014), and Obidzinski et al. (2012).

[14]Non-food biomass production can also be a potential chance for smallholders and producers in developing countries to generate income, yet whether benefits materialize it is highly context specific (Thompson 2012; van Eijck et al. 2014; Virchow et al. 2016).

[15]Horta Nogueira and Silva Capaz (2013), Thompson and Meyer (2013).

[16]Scarlat and Dallemand (2011), Schut and Florin (2015).

[17]WWF (2013).

[18]The standard is the RSB—Roundtable on Sustainable Biomaterials standard and certification scheme.

[19]Müller et al. (2015).

[20]FIAN (2011).

the Right to Food. However, as Töpfer points out, soil protection and rehabilitation policies need to be based on a human rights framework, and relevant measures have to contribute to the progressive realization of the Right to Food.[21]

Therefore, in this chapter we first discuss how the Right to Food, which is applicable in over 100 countries, relates to the sustainable management of soils and the implications of this connection. Then we show how food security and the Right to Food can be ensured in local biomass production and in certification systems in food-insecure regions. We first present a conceptual framework to integrate the Right to Food in biomass production, processing and trade. Then we present and discuss criteria for voluntary sustainability standards to ensure that the Right to Food is not violated by certified biomass operators.

The following section describes the methods, Sect. 3 the linkage between the Right to Food, sustainable soil management and the SDGs. Section 4 presents the conceptual framework. In Sect. 5 we present the developed rights-based food security criteria, and in Sect. 6 conclusions and recommendations.

2 Methods

We started with a comprehensive review of available literature on land degradation, soil management, sustainability standards, the Right to Food, food security and methods for measuring food security. This review helped to design the conceptual framework, which is described in detail in the fourth section. Based on the conceptual framework, the relevant themes and content for the rights-based food security criteria were identified.

In a scientific workshop with food security experts, ideas were discussed as to how a valid assessment of the local food situation could be conducted and how causality with biomass production could be established given the typical situation of audits facing limited resources, i.e. audits are done at relatively low costs, within a short time and without specific expert knowledge. This discussion was complemented by consultations with food security experts from the FAO as well as the World Food Program.

We concentrate on voluntary sustainability standards and certification schemes in the agricultural and biomass sector. Sustainability standards for biomass were assessed to gain an overview of already existing criteria and indicators based on the themes and elements of the conceptual framework, i.e. the criteria were grouped according to the selected Right to Food guidelines (see also Sect. 4). The screening process included the following ten standards: Forest Stewardship Council (FSC), Round Table on Responsible Soy (RTRS), Roundtable for Sustainable Palm Oil (RSPO), Roundtable on Sustainable Biomaterials (RSB), International Sustainability & Carbon Certification Standard (ISCC), Bonsucro, UTZ Certified, REDCert, the German Initiative on Sustainable Supply of Raw Materials for the Industrial Use

[21]Töpfer (2015), p. 3.

of Biomass (INRO), and the Global Bioenergy Partnership Sustainability Indicators of the FAO. This overview resulted in a list of social and environmental criteria and indicators that already cover the themes of the conceptual framework. The list formed the basis for the selection and specification of criteria for the rights-based food security criteria. The list also helped to identify themes which were not already addressed by existing criteria and indicators. For these themes we suggested new criteria (see Sect. 5).

After finishing the first draft of the rights-based food security criteria, a further stakeholder consultation process was initiated. Interviews and consultations took place with ten experts from the ISCC Standard, the standard 'Cotton made in Africa', experts from the FAO, the World Food Programme and German Agency for International Cooperation/Forum for Sustainable Palm Oil. Two multi-stakeholder workshops with 27 participants from certification bodies, standard initiatives, NGOs, ministries, research institutions, and enterprises were held to discuss and revise the developed criteria.

3 The Linkage of the Right to Food, Sustainable Soil Management and the SDGs

3.1 The Linkage of the Right to Food to Soil and Agricultural Management

Sustainable agriculture requires "technologies and practices that do not have adverse effects on environmental goods and services, are accessible to and effective for farmers, and lead to improvements in food productivity".[22] Sustainable agriculture[23] incorporates the concept of resilience, i.e. the capacity of systems to buffer shocks and stresses, and the concept of persistence, i.e. the capacity of systems to continue over long periods, and hence, it addresses many wider economic, social and environmental outcomes.[24]

---

[22]Pretty (2008, p. 447).

[23]According to Pretty and Bharucha (2014, p. 1577), key attributes of sustainable agricultural systems are that they: "(1) utilize crop varieties and livestock breeds with a high ratio of productivity to use of externally and internally derived inputs; (2) avoid the unnecessary use of external inputs; (3) harness agroecological processes such as nutrient cycling, biological nitrogen fixation, allelopathy, predation and parasitism; (4) minimize use of technologies or practices that have adverse impacts on the environment and human health; (5) make productive use of human capital in the form of knowledge and capacity to adapt and innovate and of social capital to resolve common landscape-scale or system-wide problems (such as water, pest or soil management); and (6) minimize the impacts of system management on externalities such as GHG emissions, clean water, carbon sequestration, biodiversity, and dispersal of pests, pathogens and weeds."

[24]Pretty (2008).

Soil management aims to protect soils and enhance their performance, and this includes the maintenance or increase in soil fertility, and reduction in soil erosion and land degradation. It is an integral part of sustainable agricultural and land management, so that farming remains profitable while the environment quality is preserved and short- and long-term food and non-food biomass availability ensured. Specific soil management practices, such as conservation tillage or agroforestry systems, protect and conserve the soil resources more than others. For more details on sustainable agriculture and soil management, please see the introductory Chapter by Boer et al. (2016). We focus here exclusively on soil management as an agricultural and natural resource issue. Soil management as a social topic related to land tenure, access to land and land use is discussed in Chapter by Windfuhr (2016).

A strong linkage exists between the Right to Food, sustainable agriculture and sustainable soil management. The Right to Food has been formally recognized and enshrined in the adoption of the United Nations Universal Declaration of Human Rights in 1948 and subsequent ratification by nearly all states of the world. A further recommitment to the Right to Food came with the International Covenant on Economic, Social, and Cultural Rights (hereafter, the International Covenant). This International Covenant[25] is a multilateral treaty adopted by the United Nations General Assembly in 1966, and is part of the International Bill of Human Rights, along with the Universal Declaration of Human Rights. The International Covenant explains in more detail the concept of the Right to Food and mentions in its Article 11.2.a that state parties should to take the necessary measures to improve food production by, amongst others, "developing or reforming agrarian systems in such a way as to achieve the most efficient development and utilization of natural resources".[26] In 1999, the General Comment 12 on the Right to Food was introduced by the UN to be able to implement the International Covenant in this regard.[27] The General Comment 12 refers directly or indirectly to soil management in the following paragraphs:

- §7: "...The notion of *sustainability* is intrinsically linked to the notion of adequate food or food *security*, implying food being accessible for both present and future generations."
- §10: "*Free from adverse substances* sets requirements for food safety and for a range of protective measures by both public and private means to prevent contamination of foodstuffs through adulteration and/or through bad environmental hygiene..."
- §12: "*Availability* refers to the possibilities either for feeding oneself directly from productive land or other natural resources, or for well-functioning

[25]ICESCR (1966).

[26]ICESCR (1966), p. 4.

[27]UN-CESCR (1999).

distribution, processing and market systems that can move food from the site of production to where it is needed in accordance with demand."

- §25: "The strategy should address critical issues and measures in regard to all aspects of the food system, including the production, processing, distribution, marketing and consumption of safe food [. . .]. Care should be taken to ensure the most sustainable management and use of natural and other resources for food at the national, regional, local and household levels."
- §36: . . . "States parties should take steps to respect the enjoyment of the right to food in other countries, to protect that right, to facilitate access to food and to provide the necessary aid when required. States parties should, in international agreements whenever relevant, ensure that the right to adequate food is given due attention and consider the development of further international legal instruments to that end."

Paragraph 36 is especially important for international trade and the obligations for importing countries to protect the Right to Food in all its different dimensions, including soil management and agricultural practices. This is relevant for certified biomass that is produced and traded with low- and middle-income countries.

Since even the General Comment 12 still contains only few concrete recommendations for implementation, the *'Voluntary Guidelines to Support the Progressive Realization of the Right to Adequate Food in the Context of National Food Security'* were developed.[28] These Right to Food guidelines, which is how we refer to the voluntary guidelines here, further explain and provide concrete recommendations on how the Right to Food can be implemented to guarantee food security.

The Right to Food guidelines not only embrace many strategies and political measures that address the sustainable use of land and other natural resources but also target the access to land and to productive resources for marginalized groups (e.g. Guideline 8.1, 8.7). Many guidelines are relevant for soil management though some less directly than others. They are also important for biomass sustainability standards (see Sect. 5). The following Right to Food Guidelines and their respective sub-items refer more directly to soil management:

- Guideline 2.5 highlights that "States should pursue inclusive, non-discriminatory and sound economic, agriculture, fisheries, forestry, land-use [. . .] and encourage conservation and sustainable management of natural resources, including in marginal areas."
- Guideline 2.6 emphasizes that in areas where "poverty and hunger are predominantly rural, States should focus on sustainable agricultural and rural development through measures to improve access to land, water, appropriate and affordable technologies, productive and financial resources, [. . .] conserve and protect natural resources [. . .]"

[28]FAO (2005).

- Guideline 8.10 on *Land* addresses the question of land tenure and suggests that states establish mechanisms that "promote conservation and sustainable use of land".
- Guideline 8.13 on *Sustainability* requires states "to protect ecological sustainability and the carrying capacity of ecosystems to ensure the possibility for increased, sustainable food production for present and future generations, prevent water pollution, protect the fertility of the soil [. . .]".
- Guideline 9.5 emphasizes that states should assist farmers and other primary producers to follow good agricultural practices where necessary.

3.2 Sustainable Development Goals, Soil Management and the Right to Food

With the 2030 Agenda for Sustainable Development,[29] the United Nations confirmed their determination "to end poverty and hunger, in all their forms and dimensions" and "to protect the planet from degradation, including through [. . .] sustainably managing its natural resources". The 2030 Agenda is expected to promote a "world of universal respect for human rights and human dignity". With the 17 Sustainable Development Goals and associated 169 targets, the countries agreed on pathways which "seek to realize the human rights of all".

The dependence of the target 2.1 to end all forms of hunger by 2030 on sustainable soil management is clearly set out in Goal 2: "End hunger, achieve food security and improved nutrition and promote sustainable agriculture". The associated target 2.4 reiterates the necessity to "ensure sustainable food production systems and implement resilient agricultural practices that [. . .] help maintain ecosystems [. . .] and that progressively improve land and soil quality." These two targets combined in one goal highlight the dependency between the Right to Food and sustainable land use. Sustainable land management is not only crucial for attaining food security, but also for developing sustainable bio-economies. The relevance of soils for a sustainable development is clearly set out in target 15.3: "By 2030, combat desertification, restore degraded land and soil, including land affected by desertification, drought and floods, and strive to achieve a land degradation-neutral world".

Two other targets (3.9 and 12.4) refer to soil pollution and the related adverse impacts on human health, which is also relevant with respect to the production of adequate and healthy food. Regarding biomass production and use, the following SDGs are also important: SDG 7 on (sustainable) energy, SDG 9 on promotion of sustainable industrialization, SDG 12 with sustainable consumption and production patterns, and also SDG 13 with actions to combat climate change and its impacts.[30]

[29] United Nations (2015).

[30] Müller et al. (2015).

4 Conceptual Framework

4.1 Definitions of Food Security and the Right to Food

Many food security concepts exist. Definitions and thinking around food security has changed over the last decades from a rather production-/supply-oriented definition to a broad one including aspects of access and nutrition.[31] In 1996, 185 government leaders met at the World Food Summit in response to the continued existence of widespread undernutrition and the growing concern about meeting future food needs. At this summit, food security was defined, and since then only slightly revised to its current form at the 2009 World Food Summit[32]: "Food security exists when all people, at all times, have physical, social and economic access to sufficient, safe and nutritious food to meet their dietary needs and food preferences for an active and healthy life. The four pillars of food security are availability, access, utilization and stability. The nutritional dimension is integral to the concept of food security."

The four pillars or dimensions of food security each have several determining factors[33]:

- Food availability refers to the availability of sufficient quantities of food of appropriate quality.
- Food access refers to physical and economic access to food, including access by individuals to adequate resources (entitlements) for acquiring appropriate foods for a nutritious diet.
- Food utilization refers to an adequate diet, also in regard to quality and diversity, food safety, clean water, sanitation and health care to reach a state of nutritional well-being where all physiological needs are met. It emphasizes the importance of non-food inputs in food security.
- Food stability means that a population, household or individual has access to adequate food at all times and should not risk losing access to food as a consequence of sudden shocks, seasonal food insecurity or other events.[34] Stability is also relevant for all other dimensions.

Our understanding of the Right to Food is based on Article 25 of the Universal Declaration of Human Rights in 1948, Article 11, of the International Covenant and the General Comment 12. The Right to Food guidelines are the main document we use in this regard. These guidelines mostly address states and list duties of governments to ensure and protect the Right to Food. In total, 19 voluntary guidelines, each with several sub-items, address important elements of food security such as

[31]Maxwell (1996), Pinstrup-Andersen (2009).

[32]FAO (2009).

[33]FAO (2008), FAO (2006), United Nations (2004).

[34]FAO (2006).

good governance, market systems, legal frameworks, economic development policies and access to resources and assets. Guideline 8 (Access to Resources and Assets) is further detailed in six sub-guidelines (labour, land, water, genetic resources, sustainability, services). Each guideline is considered of equal importance to ensure food security, i.e. if one guideline is completely ignored it is likely that food security is not achieved in the way the Right to Food would require.

4.2 The Framework

To guide the selection of relevant food security criteria, there is a need for a robust conceptual framework which provides the normative basis for the selection decision. We decided to use two concepts for our food security criteria—the Right to Food and the food security definition of the 2009 World Food Summit. We focus on the Right to Food and not on the food sovereignty concept, as the Right to Food is internationally accepted and endorsed by many countries worldwide and is relevant for national and international agricultural, trade and development policies.[35] The food security definition of the World Food Summit presents a globally agreed and very encompassing definition and concept which is of advantage compared to other concepts. Hence, the conceptual framework combines the Right to Food, the Right to Food guidelines and the four dimensions of food security with their respective determinants (Fig. 1).

We complement the four dimensions of food security with a fifth dimension covering cross-cutting aspects which are important elements of the Right to Food but are not that apparent in the four dimensions. The cross-cutting dimension covers aspects such as education, participation in processes, accountability, non-discrimination, transparency, human dignity, empowerment, and rule of law, women rights and gender equity. The selection of decisive factors and determinants for the cross-cutting dimension is based on those Right to Food guidelines that cannot be categorized within the other four dimensions. Without this additional dimension, not all aspects of the Right to Food would be sufficiently and adequately reflected when selecting relevant food security criteria.

In Fig. 1, under each dimension of food security the determinants for that dimension and the relevant Right to Food guidelines that match the determinants are listed. To attribute a Right to Food guideline directly to one food security dimension is not always a clear-cut decision, as the guidelines are formulated broadly and encompassingly. Therefore, a Right to Food guideline is sometimes classified in more than one dimension. In total, we selected 14 guidelines and 5 sub-guidelines out of the original 19 voluntary Right to Food guidelines. They form the five pillars for the rights-based food security criteria. From this

[35] Beuchelt and Virchow (2012).

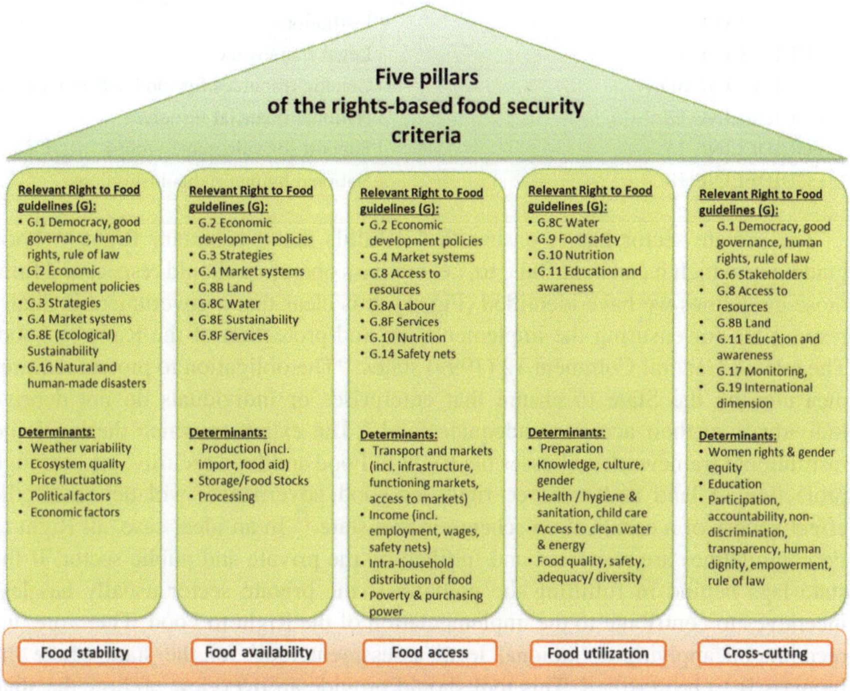

Fig. 1 The five pillars of the rights-based food security criteria with their determinants and relevant Right to Food guidelines.

framework, the criteria are derived to address food security in biomass sustainability standards.

The selection of relevant Right to Food guidelines was done from the viewpoint of biomass production for trade and export. While the trade direction is generally not of importance, we especially address the situation of biomass exports from food- or income-insecure countries to high-income countries. In choosing the guidelines, we always considered potential effects caused by a biomass investment/trade in food-insecure regions. Though the Right to Food guidelines were originally written for states, we distinguish in our framework between guidelines which we consider to be applicable and relevant also for private enterprises from those guidelines that exclusively imply state obligations and can only be fulfilled by a state. The following Right to Food guidelines are considered to be only implementable at state level[36]:

[36]Guideline 15 'International food aid' is not used at all because we found no direct relation to investments/trade in the biomass sector.

• GUIDELINE 5	Institutions
• GUIDELINE 7	Legal framework
• GUIDELINE 8D	Genetic resources for food and agriculture
• GUIDELINE 12	National financial resources
• GUIDELINE 13	Support for vulnerable groups
• GUIDELINE 18	National human rights institutions

The private sector has to ensure that it fulfils those voluntary Right to Food guidelines which it can contribute to, i.e. biomass operators should respect and fulfil those guidelines we have identified (Fig. 1). It is clear that a government is finally responsible for ensuring the implementation and protection of the Right to Food. The related General Comment 12 (1999) states: "The obligation to protect requires measures by the State to ensure that enterprises or individuals do not deprive individuals of their access to adequate food." The extent to which the legal and institutional framework addresses the Right to Food and food security, e.g., through protection of land and resource rights or good governance, will determine the effectiveness of a certification scheme in this state.[37] In an ideal case, all Right to Food guidelines are respected and fulfilled by the private and public sector. If the state lags behind in fulfilling the guidelines, the private sector usually has less incentives to contribute to the implementation of the Right to Food. Therefore, we recommend applying a 'national level assessment tool' for the state where the biomass is to be certified. This tool should provide an overview on how the state performs concerning the Right to Food guidelines. The assessment tool supports companies and certification bodies before their investing in certification, and will indicate states where the risk of non-compliance is higher, and where thus a more intensive auditing and verification process must be conducted. The tool still needs to be developed, but existing tools such as the *'Screen state action against hunger!'* guide by FIAN and Welthungerhilfe[38] could also be used or adapted.

In cases where many Right to Food guidelines are ignored or even violated by states, verification of the rights-based food security criteria applied by a certified operator will be more difficult because trustful information to check the compliance may not be (easily) obtained. Costs for verification will increase as more time is needed for the field assessment, and more stakeholders might need to be consulted in a more pro-active manner (see also Sect. 5.1). More research over longer time frames is needed to analyse whether certification schemes and especially the rights-based food security criteria can actually work, be correctly verified, and lead to the desired benefits in states with weak governance or in failed states.

[37]Bäckstrand (2006), Bracco (2015), and Mutersbaugh (2005).

[38]Suarez-Franco et al. (2007).

4.3 The Responsibility of Biomass Operators Regarding Food Security

Due to their historical development, the Right to Food guidelines are predominately directed at states and less at the private sector, yet they are still relevant for the business sector (e.g. UN-CESCR,[39] §20). The obligation of a company to respect human rights and the Right to Food are also part of many international agreements and guidelines such as the ILO conventions, the VGGT Principles, the agricultural investment principles (e.g. RAI) or the UN Guiding Principles on Business and Human Rights.[40] Although companies are faced with complex value chains, they should conduct a due diligence[41] of their business activities' impacts on human rights, and should be held responsible for impacts on food security within their scope of action.[42] The UN principles on Business and Human Rights address in Principle 13 the accountability of an operator for own activities and for those impacts that arise from activities "directly linked to their operations, products or services by their business relationships, even if they have not contributed to those impacts".[43]

The responsibilities of the private sector, state and the individual concerning food security differ widely (Table 1). For a voluntary sustainability standard, the responsibility of a biomass operator to ensure the Right to Food in the locality where it acts must be directly verifiable by a third-party audit during the certification process. The food security situation of a household or an individual is not part of the operator's responsibility, as the individual has the freedom to decide whether and what to eat and how the obtained income should be spent. For example, the operator pays a living wage but the individual prefers an unhealthy diet or to buy other products such as a TV instead of healthy food. A food security measurement may then detect food and nutrition insecurity, yet the operator cannot influence this private decision and also has no right to impose certain food consumption patterns on individuals. However, the operator has to provide all means to enable an individual and her/his household to be food secure and to fulfil the Right to Food. The operator is responsible for providing the conditions for all its employees, including casual workers and resettled communities, which lead to the fulfilment of their Right to Food. We include resettled communities in this responsibility, as they often are food insecure after a resettlement triggered by the operator's activities. The operator must follow laws, pay living wages or fair[44] prices for the crops purchased (for example, in contract farming/outgrower schemes), provide access to

[39]UN-CESCR (1999).

[40]OECD (2011), United Nations (2011).

[41]Investigation/appraisal of a business.

[42]Bettzieche et al. (2015), de Schutter et al. (2012).

[43]United Nations (2011), p. 14.

[44]Acknowledging that it is very difficult to define what a fair price is.

Table 1 Responsibility of the private sector, the state and the individual for food security

Private sector	State	Individual
Responsible for providing the means for food security to staff, resettled communities and those within the property	Responsible for providing the institutional framework for food security	Responsible for ensuring own nutrition
• Compliance with the law • Corporate responsibility for respecting human rights • Living wages + fair crop prices • Access to remedy • Local value creation • Training	• Duty to protect human rights • Establishment and control of legal framework • Policies supporting the Right to Food • Access to remedy • Education	• Use of capacities to work • Exertion of rights • Balanced nutrition

Adapted from Mohr et al. (2016)

remedy and support local value creation. The individual is responsible for using her or his capacities to work and/or produce food (especially relevant for resettled communities) and for being well nourished. The state must provide the necessary institutional and legal framework, which includes policies that support the implementation of the Right to Food, education or access to remedy.

5 The Rights-Based Food Security Criteria

Key consideration for the selection of the rights-based food security criteria were five questions: (i) What falls under the responsibility of the local operator[45]? (ii) What is desirable from a food security/Right to Food perspective? (iii) What is possible and realistic for an operator (including small investors/farmers) to implement? (iv) What is verifiable/measurable at adequate costs in the field? (v) Can a sound causality between the investment and changes in local food security be established?

Identifying the relationship between the operator's activities and the impacts on a community involves the challenge to retrace the food security outcomes directly to the activities of one local operator. The local food security situation can also be negatively affected by, for example, unfavourable weather events like droughts or floods, food price hikes at global and thus also local level, other biomass operators and enterprises using and polluting land and water resources, or there may be interactions with national policies and other markets. To establish causality in these environments, large data sets including panel data combined with rigorous quantitative (econometric) impact assessment methodology are necessary—an

[45]With operator, we mean a biomass producer or processor holding a biomass sustainability certification. An operator can be, for example, a large estate or plantation, company, public enterprise, cooperative, an individual farmer or a family farmer.

activity done by scientists in lengthy studies and far beyond the scope of an audit and of any auditor's capacities.

Our proposed criteria do not directly measure impacts on local food security due to costs, practicability, problems with causality and the freedom of an individual to forego food or reduce food quality, for example, in order to purchase luxury goods (see Sect. 4.1). We decided to instead use an approach which seeks to ensure the capabilities to secure food and nutrition at the individual level. This approach was also supported in the stakeholder workshops. To protect local communities against adverse impacts on their Right to Food that might occur through an operator, we define criteria which lie directly in the area of responsibility of an operator. Through this approach, the operator can be directly held accountable for non-compliance.

Based on the above-described conceptual framework, we derived the rights-based food security criteria, i.e. a set of 45 criteria regarding food security and the Right to Food (see Annex). The short title for each criteria group originates from the Right to Food guidelines.[46] As described in Sect. 2, not all criteria needed to be developed from scratch, as we identified already existing criteria through the screening of certification systems. From the screened standards, diverse suggestions for the wording of the criteria could be derived. Nine criteria were completely newly defined.[47] The other criteria are already implemented in one way or the other in some sustainability standards, though phrasing or comprehensiveness may differ greatly and not every standard covers the same aspects. As several criteria are already part of sustainability standards for biomass, a broad range of experience on their applicability, practicability and justification exists.[48]

Local communities may experience positive or negative impacts of the investment of an operator in biomass production depending on the business size and production model of the operator.[49] This must also be reflected in the requirements of the certification system. We distinguish between (i) family farmers,[50] (ii) operators with at least one permanent employee, which therefore do not fall under the definition of a family farmer, and (iii) companies with a certain size or production model with a high risk of negative impacts on local communities based on previous comparable experiences documented, for example, in scientific literature.[51] We distinguish between criteria with immediate application and criteria where an implementation period between 1 and 3 years is defined. Ideally, the

[46]FAO (2005).

[47]These are the criteria 2.1, 3.1, 3.2, 4.1, 4.2, 6.1, 13.1, 15.1 and 17.2.

[48]A detailed explanation of all criteria can be found in Mohr et al. (2015, 2016).

[49]For references and examples, please refer to footnote 10.

[50]Family farming is defined as: "a means of organizing agricultural, forestry, fisheries, pastoral and aquaculture production which is managed and operated by a family and predominantly reliant on family labour, both women's and men's. The family and the farm are linked, coevolve and combine economic, environmental, reproductive, social and cultural functions" (Garner and O Campos 2014).

[51]For examples, please refer to footnote 10.

whole set of criteria should be ensured from the beginning of the operation and/or certification process. For marginalized family farmers, a set of key criteria to be met immediately combined with a scheme of 'continuous improvement' is recommended to enable their participation and not overburden them at the beginning with too many requirements (see Annex). A certification does not automatically create added value for the certified operator, which is often a problem for poor family farmers. Certified family farmers may be even poorer than non-certified farmers and food security effects on family farmers being organized in group certification can be difficult to detect.[52]

In countries where the undernourishment level is below 5 % based on national or FAO data, the relevance of the application of the rights-based food security criteria may be questionable. In these countries, the certification of biomass may not need to include the audit of the food security criteria. However, as many middle-income countries often still have regional hotspots with higher levels of food insecurity (e.g. Brazil, Mexico), local and regional data need to be cross-checked. If a biomass operator is to be certified in a location/region with more than 5 % food insecurity/undernourishment, all criteria of the principle have to be checked. They always have to be checked in countries where the Global Hunger Index[53] is defined as moderate, serious, alarming or extremely alarming.

We are concerned that the food security of local communities might be negatively affected by large-scale biomass operations given reported, also scientifically documented, negative impacts including increases in food insecurity, limited employment effects, low wages, expropriation, and loss of access to land and water resources.[54] Although we assume that a full compliance with the rights-based food security criteria would avoid negative impacts through an operation, an additional monitoring of food security impacts must be implemented to gain certainty about this assumption.

For that reason, we inserted a criterion that requires (i) a Right to Food ex-ante impact assessment, and (ii) an ex-post monitoring procedure. The ex-ante impact assessment identifies possible food and nutrition security impacts of an operation before an investment is started and takes necessary action to avoid negative impacts from the very beginning. The ex-post monitoring should address possible impacts on (i) communities inside the operator's property, (ii) resettled communities due to the operator's activities, and (iii) communities surrounding the operator's property. While we understand that this kind of monitoring is a challenging demand for an operator and certification system, operators and politicians promoting non-food biomass uses "must accept the burden of acquiring detailed understanding how

[52]Beuchelt et al. (2010), Beuchelt (2012), and Beuchelt and Zeller (2013).

[53]The Global Hunger Index (GHI) is designed to comprehensively measure and track hunger globally and by country and region, and is calculated each year by the International Food Policy Research Institute (IFPRI), see also: https://www.ifpri.org/topic/global-hunger-index.

[54]Anseeuw et al. (2012), de Schutter (2011), Deininger and Byerlee (2011), Harvey and Pilgrim (2011), and Yengoh and Armah (2015); see also references from footnote 10.

implementation of the technology will affect food insecure people" (Thompson 2012, p. 341).

In general, we expect a positive impact of the suggested criteria on the food security of the directly concerned stakeholders in the long term. In the short term, we expect that the food security and Right to Food situation will not (further) deteriorate. However, it is clear that the possibilities of a certification to improve food security are limited to the directly concerned stakeholder under the direct influence of an operator. Accumulative effects of several operators (even when certified) producing, for example, all non-food biomass and hence possibly leading to price increases on local food markets cannot be solved by introducing the food security criteria in a sustainability standard. The existence of rights-based food security criteria can prevent violations of the Right to Food by certified operators on a local level, but it does not replace the implementation of sound development strategies. Likewise, the rights-based food security criteria will have little positive effects on the production systems in food-insecure regions if the criteria are only integrated in a few sustainability standards, and if there was little or no demand from consumers at the end of the supply chain or political pressure.

The criteria ensuring the Right to Food are designed as a best-practice set which can be added as a whole to the existing criteria and indicators of any biomass sustainability standard. The interest of sustainability standard setters for feasible food security criteria is growing, so the developed criteria are a major step towards a better integration of food security aspects in sustainable biomass certifications. The screening of the sustainability standards showed that there is already some overlap with our criteria, which increases its adoptability. Sound experience already exists among many sustainability standards regarding implementation of some of our suggested criteria, while for the newly proposed criteria experience regarding the verifiability is lacking. Although being discussed with auditors, practitioners and standard setters, we see the need for a field-testing phase of the whole criteria set.

Political interest and support as well as civil society pressure is necessary to foster the likelihood to adopt the proposed food security criteria in standards and certification systems. However, also the demand for products with this type of certification must significantly increase - especially at the end market of the respective biomass value chains. An existing demand makes it more attractive for many standards and certification systems to fully integrate the rights-based food security criteria in their system.

The proposed criteria should also become part of mandatory regulations which are not uncommon as, for example, the obligatory environmental criteria for bioenergy requested by the European Renewable Energy Directive (EU-RED). The rights-based food security criteria could be easily integrated in the EU-RED adding the social dimension of sustainability to the already addressed ecological dimension. Governmental regulations can hence support the creation of a market demand for biomass products where the Right to Food was respected in production, processing and trade.

6 Conclusions and Recommendations

A strong linkage exists between the realization of the Right to Food and sustainable soil management. The interlinkage is addressed in the Right to Food guidelines and the SDGs, but rarely reflected in agricultural sustainability standards. Rising concerns about negative impacts on local food security through growing biomass production in food-insecure regions and international trade prompted us to address the question how the Right to Food can be ensured in biomass production and certification systems. For this, a conceptual framework was developed to integrate the Right to Food in biomass production and trade and to guide the choice of food security criteria for sustainability standards.

Following the conceptual framework, 45 rights-based food security criteria categorized under 17 themes were developed. The criteria are applicable to all biomass types and uses, and to different biomass standards, farm sizes and business types. For the first time, practicable and measurable criteria exist. The criteria ensuring the Right to Food are designed as a best-practice set which can be added as a whole to the existing criteria and indicators of any biomass sustainability standard. These standards also include soil management and good agricultural practices and hence, our criteria set would complement the link to food security and contribute further to reach several SDGs related to food security, soils and biomass.

The voluntary standards and certification schemes are based on the assumption that the implementation of these systems would lead to the desired positive impacts. However, 'compliance' with criteria must not necessarily lead to 'impact'. There is a risk that, despite adhering to the rights-based food security criteria, the desired impact on soil management and local food security of those involved or affected by the biomass operator may not materialize. Since impacts often take time to materialize, rigorous impact assessments are recommended a few years after the rights-based food security criteria have been implemented.

We recommend the inclusion of the rights-based food security criteria in all biomass sustainability standards. Given the limited demand and willingness to pay for sustainably produced goods of consumers besides political statements regarding "food first" when using biomass in the bioeconomy, we see the need to make the rights-based food security criteria part of governmental regulations. Meanwhile, governments could already support the criteria by using their substantial purchasing power to buy goods that have been certified according to the criteria. A change in procurement policies will increase the demand and be a step towards the desired impacts.

Acknowledgements The chapter was developed within the project "Developing food and nutrition security criteria for biomass standards and certifications" (ESSZert), which is funded by the German Federal Ministry for Food and Agriculture (BMEL) based on the decision of the Parliament of the Federal Republic of Germany (FKZ 22013714) and the project "Improving food security in Africa through increased system productivity of biomass-based value webs" (BiomassWeb), which is funded by the German Federal Ministry of Education and Research (BMBF) based on the decision of the Parliament of the Federal Republic of Germany (FKZ 031A258A). Both funding sources are gratefully acknowledged.

Annex: The Developed 45 Rights-Based Food Security Criteria

Food security dimension	#	Criteria
Stability	**1**	**Democracy, good governance, human rights and the rule of law (RtaF-G. 1)**
	1.1	The operator must demonstrate compliance with all applicable national, regional and local laws and regulations.
	1.2	The operator holds a written policy committing to the UN "Guiding Principles on Business and Human Rights" in all operations and transactions. The implementation of the policy must be documented and communicated to all levels of the workforce and operations.
	2	**Strategies (RtaF-G. 3)**
	2.1	The operator endorses existing national strategies with regard to food security and does not contradict them by any of its business activities.
	3	**Sustainability (RtaF-G. 8E)**
	3.1	The operator has to apply Good Agricultural Practices (concerning soil management, chemical application and use, water management, fertilizer application).
	4	**Natural and man-made disasters (RtaF-G. 16)**
	4.1	The operator recognizes all national and /or international natural disaster risk assessments, strategies and maps in the business plan/ strategy.
	4.2	The operator informs suppliers and communities in the concerned region about natural risks and provides support in case of strong adverse natural and man-made disasters.
Access	**5**	**Economic development policies (RtaF-G. 2)**
	5.1	Provision of a business plan showing evidence of long-term economic viability of the operation.
	5.2	The operator has to provide fair, legal and transparent arrangements with suppliers. Agreed payments shall be made in a timely manner.
	6	**Market systems (RtaF-G. 4)**
	6.1	The operator adopts an implementation plan supporting local value creation.
	6.2	The operator must not reduce the access to markets for local communities through its operations.

(continued)

Food security dimension	#	Criteria
	7	**Labour (RtaF-G. 8A)**
	7.1	The operator shows full compliance with the ILO Core Conventions and the Declaration on Fundamental Principles and Rights at Work.
	7.2	The operator pays wages for all workers according at least to the calculated national adequate Living Wages.
	7.3	If payment for piecework is applied, the pay rate, based on an eight-hour workday, allows workers to earn at least the adequate Living Wage.
	7.4	Men and women earn equal pay for equal work.
	7.5	The workers are not subjected to any form on discrimination in hiring, remuneration, benefits, access to training, promotion, termination, retirement or any other aspect of employment, based on race, colour, gender, religion, political opinion, national extraction, social origin, sexual orientation, family responsibilities, marital status, union membership, age or any other condition that could give rise to discrimination. Their awareness is for discriminatory practices is trained.
	7.6	The workers confirm that no deductions from wages as a result of disciplinary measures are made.
	7.7	The operator provides all employees with fair, legal, written contracts, signed by both the employee and the employer.
	7.8	The operator endorses a health and safety policy where the main health and safety risks are assessed. An implementation plan addressing measures for mitigation of these risks is in place. The policy and plan applies to all workers, including contractors, workers and suppliers. The implementation is regularly monitored and improved.
	7.9	All workers involved in the operation shall be adequately trained in safe working, using adequate and appropriate protective equipment.
	7.10	An adequate share of workers must be trained in first aid.
	7.11	Occupational injuries shall be recorded using Lost Time Accident (LTA) metrics.
	8	**Services (RtaF-G. 8 F)**
	8.1	The operator provides agricultural services and capacity building for suppliers and communities inside the property (plantation). A plan and prove of activities has to be available.
	9	**Safety nets (RtaF-G. 14)**
	9.1	Workers are provided with medical care in case of accidents or work-related diseases. Additionally, workers are covered with a public accident and medical insurance, if existent in the respective country. Sick leaves are paid according to the law.
	9.2	All permanent workers are provided with an occupational pension fund according to the national law.

(continued)

Food security dimension	#	Criteria
Availability	**10**	**Land (RtaF-G. 8B)**
	10.1	The operator respects all human rights and legitimate tenure rights, and conducts an appropriate assessment to prevent any adverse impacts on these.
	10.2	All decisions regarding land rights and land-use rights, such as buying, selling or valuing related to the operator are based on the free, prior, and informed consent of all stakeholders involved.
	10.3	There has been no forced or involuntary physical or economic displacement, resettlement or relinquishment of land rights for the purpose of the production.
	10.4	Land used by the operator may not be under dispute, contested and/or under conflict.
	11	**Water (RtaF-G. 8C)**
	11.1	The operator implements a water management plan and monitoring system. If communities rely on the same water source, the plan must be agreed with free, prior, and informed consent of all stakeholders involved, and may not be under dispute.
	11.2	Water used by the operator does not negatively affect availability, quality and access to the water supply to communities which rely on the same water resources. There has to be a continuous monitoring of the availability, quality and access to the water resources.
Utilization	**12**	**Food safety and consumer protection (RtaF-G. 9)**
	12.1	The operator must not use pesticides and chemicals that are categorised as World Health Organisation Class 1A, 1B, or 2 and/or that are listed by the Stockholm or Rotterdam Conventions. Any use of pesticides and other chemicals must be documented.
	12.2	The operator uses integrated pest management (IPM) and supports scheme suppliers with training in IPM.
	12.3	Workers have always access to safe drinking water.
	13	**Nutrition (RtaF-G. 10)**
	13.1	The operator shall make demonstrable efforts to improve workers' access to adequate, safe, sufficient and affordable food.
	13.2	Breastfeeding women have two additional 30-min breaks per day to nurture the child.
Cross-cutting	**14**	**Stakeholders (RtaF-G. 6)**
	14.1	The operator has to establish an internal grievance mechanism for workers and an external grievance mechanism for stakeholders. The mechanism has been made known and is accessible to the communities. All grievances are documented and monitored. A responsible person for grievances is named and known to the workers and communities around the farm.
	14.2	The affected persons and communities at large support the project before the operator starts the process.

(continued)

Food security dimension	#	Criteria
	15	**Women's rights and gender equity (RtaF-G. 8)**
	15.1	Women should not be discriminated, and their rights have to be respected.
	15.2	No work with pesticides must be undertaken by pregnant or breastfeeding women.
	16	**Education and awareness raising (RtaF-G. 11)**
	16.1	The operator implements a formal training programme that covers all agricultural activities of the company (e.g. use and application of chemicals and fertilizers). This includes regular assessments of training needs and documentation of the programme.
	16.2	All children living on the operation have access to quality primary school education which does not exceed local school fees.
	17	**Monitoring, indicators and benchmarks (RtaF-G. 17)**
	17.1	The operator has to conduct an ex-ante impact assessment on food security and the Right to Food of concerned communities (on the operator's property, within its operating scale (e.g. outgrower schemes) and nearby surrounding communities). The availability, access, quality and stability of food must not be negatively affected by the planned operator investments and activities. This applies only to new investments.
	17.2	The operator is responsible for ensuring food security for the inhabitants (communities) within the operator's property and administrative boundaries, even when the inhabitants are not employees of the operator. The food security situation must be monitored by a food security screening.
	17.3	In communities resettled according to the FPIC, the operator has to monitor the food security situation through a food security screening and, for example, a continuous dialogue, and ensure their food security.
	17.4	Operations where a large impact is expected (e.g. operations above 1000ha) have to conduct a food security screening also in the surrounding communities of the operator's property and administrative boundaries. In the surrounding communities the availability, access, quality and stability of food must not be reduced by the operator's activities. The operator is responsible for ensuring that the investment does not create or exacerbate local or national food insecurity.

Adapted from Mohr et al. (2016)

References

Albersmeier F, Schulze H, Jahn G, Spiller A (2009) The reliability of third-party certification in the food chain: from checklists to risk-oriented auditing. Food Control 20(10):927–935. doi:10.1016/j.foodcont.2009.01.010

Anseeuw W, Alden Wily L, Cotula L, Taylor M (2012) Land rights and the rush for land: findings of the global commercial pressures on land research project. ILC, Rome

Bäckstrand K (2006) Multi-stakeholder partnerships for sustainable development: rethinking legitimacy, accountability and effectiveness. Eur Environ 16(5):290–306. doi:10.1002/eet.425

Bettzieche L, Schuller C, Utlu D, Windfuhr M (2015) National Baseline Assessment: Umsetzung der UN-Leitprinzipien für Wirtschaft und Menschenrechte. Deutsches Institut für Menschenrechte, Berlin

Beuchelt T (2012) Analyzing organic and fairtrade certification schemes: participation and welfare effects on small-scale farmers in coffee value chains. Cuvillier, Göttingen

Beuchelt T, Virchow D (2012) Food sovereignty or the human right to adequate food: which concept serves better as international development policy for global hunger and poverty reduction? Agric Hum Values 29(2):259–273. doi:10.1007/s10460-012-9355-0

Beuchelt T, Zeller M (2013) The role of cooperative business models for the success of smallholder coffee certification in Nicaragua: a comparison of conventional, organic and Organic-Fairtrade certified cooperatives. Renew Agric Food Syst 28(03):195–211. doi:10.1017/S1742170512000087

Beuchelt T, Kiemen A, Zeller M (2010) Value adding through certification? Insights from the coffee sector in Nicaragua. In: van Trijp H, Ingenbleek P, van Tilburg A (eds) Markets, marketing and developing countries: where we stand and where we are heading. Wageningen Academic Publishers, Wageningen, pp 115–120

Boer B, Ginzky H, Heuser IL (2016) International soil protection law –history, concepts and latest developments. In: Ginzky H, Heuser IL, Qin T, Ruppel OC, Wegerdt P (eds) International yearbook of soil law and policy 2016. Springer, Heidelberg (in this volume). doi:10.1007/978-3-319-42508-5

Bracco S (2015) Effectiveness of EU biofuels sustainability criteria in the context of land acquisitions in Africa. Renew Sust Energ Rev 50:130–143. doi:10.1016/j.rser.2015.05.006

CFS (2014) Principles for responsible investment in agriculture and food systems. Committee on World Food Security, Rome. Available via FAO http://www.fao.org/3/a-au866e.pdf. Accessed 16 Sept 2016

Charnovitz S, Earley J, Howse R (2008) An examination of social standards in biofuels sustainability criteria. IPC Discussion Paper – Standards Series December 2008, International Food & Agricultural Trade Policy Council,Washington, DC

de Schutter O (2011) How not to think of land-grabbing: three critiques of large-scale investments in farmland. J Peasant Stud 38(2):249–279. doi:10.1080/03066150.2011.559008

de Schutter O, Ramasastry A, Taylor MB, Thompson RC (2012) Human Rights due diligence: the role of states. International Corporate Accountability Roundtable. http://icar.ngo/wp-content/uploads/2012/12/Human-Rights-Due-Diligence-The-Role-of-States.pdf. Accessed 2 Oct 2015

Deininger K, Byerlee D (2011) Rising global interest in farmland: can it yield sustainable and equitable benefits? Agriculture and development series. The World Bank, Washington

EC (2009) Directive 2009/28/EC of the European Parliament and of the Council of 23 April 2009 on the promotion of the use of energy from renewable sources, European Parliament and the council of the European Union, Brussels

FAO (2005) Voluntary guidelines to support the progressive realization of the right to adequate food in the context of national food security, adopted by the 127th Session of the FAO Council, November 2004. Food and Agriculture Organization of the United Nations, Rome

FAO (2006) Food security. Policy brief issue 2. Available at FAO. http://www.fao.org/forestry/13128-0e6f36f27e0091055bec28ebe830f46b3.pdf. Accessed 22 Feb 2016

FAO (2008) Food security information for action. Practical guides: an introduction to the basic concepts of food security. EC - FAO Food Security Programme. Available at FAO. http://www.fao.org/docrep/013/al936e/al936e00.pdf. Accessed 22 Feb 2016

FAO (2009). Declaration of the World Summit on Food Security, WSFS 2009/2. World Food Summit, 16-18 November 2009, Rome: FAO. Available via FAO. http://www.fao.org/fileadmin/templates/wsfs/Summit/Docs/Final_Declaration/WSFS09_Declaration.pdf. Accessed 26 April 2016

FAO, IFAD, UNCTAD, World Bank (2010) Principles for responsible agricultural investment that respects rights, livelihoods and resources (PRAI). Available via FAO. http://www.fao.org/fileadmin/templates/est/INTERNATIONAL-TRADE/FDIs/RAI_Principles_Synoptic.pdf. Accessed 16 Sept 2016

FAO (2012) Voluntary guidelines on the responsible governance of tenure of land, fisheries and forests in the context of national food security. Food and Agriculture Organization of the United Nations, Rome

FAO (2015) Revised world soil charter: FAO conference revised charter adoption. Available at FAO. http://www.fao.org/3/a-i4965e.pdf. Accessed 22 Feb 2016

FIAN (2011) Zugang zu Land und das Recht auf Nahrung: Ein Beitrag zur Klärung menschenrechtlicher Anforderungen in Zusammenhang mit dem Zugang zu Land und natürlichen Ressourcen. FIAN, Köln

Garner E, de la O Campos AP (2014) Identifying the "family farm". An informal discussion of the concepts and definitions. ESA Working Paper 14–10. FAO, Rome

Gasparatos A, von Maltitz GP, Johnson FX, Lee L, Mathai M, Puppim de Oliveira JA, Willis KJ (2015) Biofuels in sub-Sahara Africa: drivers, impacts and priority policy areas. Renew Sust Energ Rev 45:879–901. doi:10.1016/j.rser.2015.02.006

German L, Schoneveld GC, Pacheco P (2011) Local social and environmental impacts of biofuels: global comparative assessment and implications for governance. E&S 16(4). doi: 10.5751/ES-04516-160429

Harvey M, Pilgrim S (2011) The new competition for land: food, energy, and climate change. Food Policy 36:40–51. doi:10.1016/j.foodpol.2010.11.009

Horta Nogueira LA, Silva Capaz R (2013) Biofuels in Brazil: evolution, achievements and perspectives on food security. Glob Food Secur 2:117–125. doi:10.1016/j.gfs.2013.04.001

ICESCR (1966) International covenant on economic, social and cultural rights. Available via OHCHR. http://www.ohchr.org/Documents/ProfessionalInterest/cescr.pdf. Accessed 22 Feb 2016

Klooster D (2010) Standardizing sustainable development? The Forest Stewardship Council's plantation policy review process as neoliberal environmental governance. Geoforum 41 (1):117–129. doi:10.1016/j.geoforum.2009.02.006

Koch A, McBratney A, Adams M, Field D, Hill R, Crawford J, Minasny B, Lal R, Abbott L, O'Donnell A, Angers D, Baldock J, Barbier E, Binkley D, Parton W, Wall DH, Bird M, Bouma J, Chenu C, Flora CB, Goulding K, Grunwald S, Hempel J, Jastrow J, Lehmann J, Lorenz K, Morgan CL, Rice CW, Whitehead D, Young I, Zimmermann M (2013) Soil security: solving the global soil crisis. Glob Policy 4(4):434–441. doi:10.1111/1758-5899.12096

Locke, A, Henley, G (2013) A review of the literature on biofuels and food security at a local level. Assessing the state of the evidence. Overseas Development Institute, London. https://www.odi.org/sites/odi.org.uk/files/odi-assets/publications-opinion-files/8869.pdf. Accessed 22 Feb 2016

Maxwell S (1996) Food security: a post-modern perspective. Food Policy 21(2):155–170. doi:10.1016/0306-9192(95)00074-7

McBratney A, Field DJ, Koch A (2014) The dimensions of soil security. Geoderma 213:203–213. doi:10.1016/j.geoderma.2013.08.013

Mohr A, Bausch L (2013) Social sustainability in certification schemes for biofuel production: an explorative analysis against the background of land use constraints in Brazil. Energ Sustain Soc 3(1):1–14. doi:10.1186/2192-0567-3-6

Mohr A, Beuchelt T, Schneider R, Virchow D (2015) A rights-based food security principle for biomass sustainability standards and certification systems. ZEF Working Paper No 143, Centre for Development Research (ZEF), Bonn. Available via ZEF. http://www.zef.de/uploads/tx_zefportal/Publications/ZEF_Working_Paper_143.pdf. Accessed 16 Sept 2016

Mohr A, Beuchelt T, Schneider R, Virchow D (2016) Food security criteria for voluntary biomass sustainability standards and certifications. Biomass Bioenerg 89:133–145. doi:10.1016/j.biombioe.2016.02.019

Müller A, Weigelt J, Götz A, Schmidt O, Lobos Alva I, Matuschke I, Ehling U, Beringer T (2015) The Role of Biomass in the Sustainable Development Goals: A Reality Check and Governance Implications. IASS Working Paper, IASS, Potsdam

Mutersbaugh T (2005) Just-in-space: certified rural products, labor of quality, and regulatory spaces. J Rural Stud 21(4):389–402. doi:10.1016/j.jrurstud.2005.08.003

Obidzinski K, Andriani R, Komarudin H, Andrianto A (2012) Environmental and social impacts of oil palm plantations and their implications for biofuel production in Indonesia. Ecol Soc 17 (1):25. doi:10.5751/ES-04775-170125

OECD (2011) OECD guidelines for multinational enterprises, 2011 edition. OECD Publishing, Paris. Available via OECD. doi:10.1787/9789264115415-en. Accessed 16 Sept 2016

OECD, FAO (2014) OECD-FAO agricultural outlook 2014–2023. OECD Publishing, Paris. doi:10.1787/agr_outlook-2014-en. Accessed 16 Sept 2016

Pinstrup-Andersen P (2009) Food security: definition and measurement. Food Sec 1(1):5–7. doi:10.1007/s12571-008-0002-y

Pretty J (2008) Agricultural sustainability: concepts, principles and evidence. Philos Trans R Soc Lond Ser B Biol Sci 363:447–465. doi:10.1098/rstb.2007.2163

Pretty J, Bharucha ZP (2014) Sustainable intensification in agricultural systems. Ann Bot 114:1571–1596. doi:10.1093/aob/mcu205

Scarlat N, Dallemand J (2011) Recent developments of biofuels/bioenergy sustainability certification: a global overview. Energy Policy 39(3):1630–1646. doi:10.1016/j.enpol.2010.12.039

Schut M, Florin MJ (2015) The policy and practice of sustainable biofuels: between global frameworks and local heterogeneity. the case of food security in Mozambique. Biomass Bioenerg 72:123–135. doi:10.1016/j.biombioe.2014.11.009

Suarez-Franco A, Ratjen S, Schneider R (2007) Screen state action against hunger! How to use the Voluntary Guidelines on the Right to Food to monitor public policies? FIAN International, Heidelberg

Thompson P (2012) The agricultural ethics of biofuels: the food vs fuel debate. Agriculture 2:339–358. doi:10.3390/agriculture2040339

Thompson W, Meyer S (2013) Second generation biofuels and food crops: co-products or competitors? Glob Food Secur 2:89–96. doi:10.1016/j.gfs.2013.03.001

Töpfer K (2015) Global soil week 2015: soil. The substance of transformation. Chairman's conclusions: transformation is needed – transformation is possible. http://globalsoilweek.org/wp-content/uploads/2015/04/150422_Chairman_Conclusions.pdf. Accessed 22 Feb 2016

UN-CESCR (1999) Substantive issues arising in the implementation of the international covenant on economic, social and cultural rights: general comment 12: the right to adequate food (Art. 11). Available via FAO. http://www.fao.org/fileadmin/templates/righttofood/documents/RTF_publications/EN/General_Comment_12_EN.pdf. Accessed 22 Feb 2016

United Nations (2004) Food security and its determinant factors. United Nations Albania. Available via UNICEF. http://www.unicef.org/albania/Food_Security_ANG.pdf. Accessed 22 Feb 2016

United Nations (2011) Guiding principles on business and human rights: implementing the United Nations "Protect, Respect and Remedy" Framework, New York/Geneva

United Nations (2015) Transforming our world: the 2030 agenda for sustainable development. Resolution adopted by the General Assembly on 25 September 2015, A/RES/70/1, New York, US. http://www.un.org/ga/search/view_doc.asp?symbol=A/RES/70/1&Lang=E. Accessed 22 Feb 2016

van Dam JM (2009) Sustainability of bioenergy chains: the result is in the details. Dissertation, Utrecht University

van Eijck J, Romijn H, Balkema A, Faaij A (2014) Global experience with Jatropha cultivation for bioenergy: an assessment of socio-economic and environmental aspects. Renew Sust Energ Rev 32:869–889. doi:10.1016/j.rser.2014.01.028

Virchow D, Beuchelt TD, Kuhn A, Denich M (2016) Biomass-based value webs: a novel perspective for emerging bioeconomies in Sub-Saharan Africa. In: Gatzweiler FW, von

Braun J (eds) Technological and institutional innovations for marginalized smallholders in agricultural development. Springer, Cham/Heidelberg/New York/Dordrecht/London, pp 225–238. doi:10.1007/978-3-319-25718-1

von Geibler J (2013) Market-based governance for sustainability in value chains: conditions for successful standard setting in the palm oil sector. J Clean Prod 56:39–53. doi:10.1016/j.jclepro.2012.08.027

Windfuhr M (2016) FAO - Voluntary guidelines on responsible governance of tenure of land, forests and fisheries – relevance, reception and first experiences in implementation. In: Ginzky H, Heuser IL, Qin T, Ruppel OC, Wegerdt P (eds) International yearbook of soil law and policy 2016. Springer, Heidelberg (in this volume). doi:10.1007/978-3-319-42508-5

WWF (2013) Searching for sustainability: comparative analysis of certification schemes for biomass used for the production of biofuels. WWF Germany, Düsseldorf

Yengoh GT, Armah FA (2015) Effects of large-scale acquisition on food insecurity in Sierra Leone. Sustainability (Switzerland) 7:9505–9539. doi:10.3390/su7079505

Economics of Land Degradation: Achievements and Next Steps

Hannes Etter, Tobias Gerhartsreiter, and Naomi Stewart

1 Introduction

Desertification, land degradation, and drought are widely occurring global threats, placing immense pressure on the future development and sustainability of our earth. This relates particularly to food, water and energy security, but also to climate change mitigation and adaptation, and livelihood resilience. Negative ecosystem impacts human activities, especially land use and management. This in turn has the biggest impact disproportionately on the most vulnerable parts of the human population, the rural poor. Estimations by the ELD Initiative have shown that up to 1.4 billion people are currently living on less favored land and agricultural areas.[1]

However, looking back on past efforts to combat land degradation and desertification, it is clear piecemeal approaches by different parties are not successful. What is needed are integrated approaches which interlink biophysical, technological, and social management strategies and factors. A further and critical obstacle has been that one of the core drivers is not sufficiently addressed so far: the focus on short-term economic gains from land, which actually yield long-term costs. Activities focusing on the intermediate results of natural resources use neglect the actual value of land by degrading possible long-term resource production. In order to reach this goal, the real economic value and potential of land needs to be identified and communicated practically and in a way that this information is useful for all

[1]Barbier and Hochard (2014).

H. Etter (✉) • T. Gerhartsreiter
Deutsche Gesellschaft für Internationale Zusammenarbeit (GIZ) GmbH, Bonn, Germany
e-mail: tobias.gerhartsreiter@giz.de; hannes.etter@giz.de

N. Stewart
Freelance Consultant, London, UK
e-mail: naomi.m.stewart@gmail.com

H. Ginzky et al. (eds.), *International Yearbook of Soil Law and Policy 2016*,
International Yearbook of Soil Law and Policy,
DOI 10.1007/978-3-319-42508-5_18

decision-makers especially policy makers and the private sector, who have a great say in how land is used and managed.[2]

In order to bridge existing silos and unite different stakeholders towards a more sustainable development path for land management, a common vision and goal is necessary, which requires common metrics and terms.

2 The Value of Land

Out of this understanding, the international and collaborative Economics of Land Degradation (ELD) Initiative was incepted in 2011. The Initiative resulted of a joint decision by an international consortium of donors, including the German Ministry for Economic Cooperation and Development (BMZ), the European Commission (EC) and the Korea Forest Service (KFS) to strengthen the political significance of land and its sustainable management. Hence, the ELD Initiative aims at bringing awareness of the economics of land degradation and sustainable land management into the scientific, political, public, and private spheres. Through research, capacity-building, and active knowledge exchange, it seeks to ensure that the economics of sustainable land management are understood and useful.

In 2015, 4 years of research effort undertaken by the Initiative, including academics, scientists, policy-/decision-makers, and practitioners culminated with the publication of reports on the value of land targeting specific groups. Within the international year of soils, the ELD Initiative published a range of tailored reports for political decision makers and scientist to raise awareness and provide tools for such stakeholders.

The report,[3] is a comprehensive assessment of land degradation and sustainable land management. It was presented during a high level launch event in Brussels as well as at UNGA in New York City and UNCCD COP12 in Ankara and experienced great attention by more than 150 media outlets in 15 different languages world-wide.

The ELD Initiative's reports describe the huge opportunities in sustainable land management—for private and public investors alike.

The reports prove that land, as an undervalued global asset, offers huge opportunities for sustainable development, with investments in sustainable land management as a significant opportunity for recovering these losses. By increasing awareness of the costs and benefits of different land management techniques, policy-/decision-makers can take more strategic decisions at national, regional, and local levels.

[2]ELD (2015c).

[3]ELD (2015d).

2.1 Importance of Sustainable Land Management

There are many under-appreciated values of land, which are often not considered when we make land management decisions, like, air purification, water retention, and storm protection. Fully informed economic valuations are key in sustainable land management. They provide a neutral and equalising language for different groups to discuss their various needs, costs, and benefits, and come to optimal, mutually acceptable sustainable land management strategies. Accomplishing sustainable land management will be critical for humanity, as it is an essential aspect of a productive biophysical future that involves both environmental and human well-being.

Answering the economic questions of land degradation and providing integrated frameworks for informed action are particularly important in the context of increasing land scarcity. Globally, the human population has reached a stage where cultivated areas can no longer be expanded, except in limited areas of South America and Sub-Saharan Africa. Even in these few cases, the geographical extent of exploitable land may be over-estimated, and we thus need economics to find equitable, optimal paths forward into a sustainable future.[4]

The ELD Initiative highlights and facilitates discussions on the economic value of sustainable land management. Land is an underperforming asset in the global economy and land degradation threatens the growing need and availability of fertile lands and land-based ecosystems, and the benefits humanity derives from it. Competition for access to increasingly scarce resources such as healthy and productive soils and water is accelerating, and is further intensified by population pressures, climate change, and resource demands.

The aim of the ELD Initiative is to integrate the economic assessment of land use into public and private sector decision-making. For example, the latest ELD case study highlights the policy of sustainable agroforestry measures adopted in Mali, where every single Euro invested is estimated to create 12 Euros benefit in return through increased crop yields in the long-term.[5]

Box 1: Key Facts and Figures from the ELD Research

- The loss of ecosystem service values from land degradation ranges from USD 6.3 to 10.6 trillion every single year. This works out to about USD 900–1450 per person, and USD 47,000–78,000 per square kilometer.[6]

(continued)

[4]Lambin et al. (2013).

[5]Sidibé et al. (2014).

[6]ELD Initiative (2015d).

Box 1 (continued)

- Globally, a large portion of the vulnerable human populations—the rural poor—live on degrading and less favoured agricultural lands without market access, leading to increases in overall poverty rate.[7]
- However, up to USD 75.6 trillion can be gained annually from transforming global policies to enable sustainable land management.[8]
- Regionally, taking action against soil erosion induced nutrient loss in 42 African countries by improving grain crops alone could generate up to almost USD 63 billion, creating an estimated total economic annual growth rate of 5.31% for all of the countries.[9]
- At the national level, studies in Jordan show that adopting traditional pastoralism in just one part of the country could generate USD 192–438 million in benefits.[10] Further, studies in Sudan showed it was possible to reduce carbon sequestration costs by USD 867 per hectare with sustainable land management.[11] In Mali, studies on implementing agroforestry showed up to USD 500 in economic returns over a 25 year horizon.[12]

2.2 *Establishing a Sound Scientific Foundation to Assess the Economics of Land*

Calculating the value of land is no easy feat, and so the ELD Initiative provides ground-truthed tools and assessments to facilitate cost-benefit analyses of land and land uses through total economic valuation. To mainstream a harmonized assessment methodology—which encourages a robust approach with comparable results—the ELD Initiative has adopted a framework from Noel and Soussan (2010). This stepwise approach is useful in bridging the requirements for a guiding framework, which can ensure that there is conformity between different assessments, while leaving sufficient space to adopt necessary evaluation methodologies towards the eco- and land use systems under consideration.

Table 1 outlines the ELD's 6 + 1 step methodology, which has been applied through a wide range of different case studies globally. The steps often overlap when being applied, but have been teased apart theoretically to aim for the consideration of all relevant aspects of an economic assessment.

[7]Barbier and Hochard (2014).

[8]ELD Initiative (2015d).

[9]ELD Initiative (2015d).

[10]Myint and Westerberg (2015).

[11]Aymeric et al. (2015).

[12]Sidibé et al. (2014).

Table 1 The ELD 6 + 1 methodology to assess the economics of sustainable land management[a]

1. Inception	Identification of the scope, location, spatial scale, and strategic focus of the study, based on stakeholder consultation. Preparation of background materials on the socio-economic and environmental context of the assessment.
2. Geographical characteristics	Establishment of the geographic and ecological boundaries of the study area identified in Step 1, following an assessment of quantity, spatial distribution, and ecological characteristics of land cover types that are categorised into agroecological zones and analysed through a Geographical Information System (GIS).
3. Types of Ecosystem Services	For each land cover category identified in Step 2, identification and analysis of stocks and flows of ecosystem services for classification along the four categories of the ecosystem service framework (provisioning, regulating, cultural, and supporting services).
4. Roles of ecosystem services and economic valuation	Establishment of the link between the role of ecosystem services in the livelihoods of communities living in each land cover area and in overall economic development in the study zone. Estimation of the total economic value for each ecosystem service.
5. Pressures and patterns	Identification of land degradation patterns and drivers, pressures on sustainable management of land resources and drivers of adoption of sustainable land management (including determining the role of property rights and legal systems), and their spatial distribution to inform the establishment of global scenarios. Revision of previous steps if needed, to ensure the assessment is as comprehensive as possible.
6. Cost-benefit	Cost-benefit analysis (CBA) comparing costs and benefits of an 'action' scenario to that of a 'business-as-usual' scenario to assess whether the proposed land management changes lead to net benefits. ('Action' scenarios include land management changes that can reduce or remove degradation pressures). Mapping of net benefits for identification of the locations for which land management changes are suitable from an economic perspective. This will lead to the identification of "on-the-ground" actions that are economically desirable.
+1 Take action	

[a]ELD (2015c), p. 7f

For each step, multiple approaches and individual methodologies exist, which are not exclusive. The underlying concept is based on ecosystem services, which are generally defined as "the benefits people obtain from ecosystems",[13]

[13]MEA (2005).

			Benefits		Costs		Net benefits	Net benefits of action
			Action	Inaction	Action	Inaction		
Business-as-usual		Rice	N/A	8,522	N/A	6,804	1,717	N/A
		Mango	N/A	10,513	N/A	4,563	5,959	N/A
Sustainable land management	Rice	Scenario 1a. horizontal desalination	11,589	N/A	11,304	N/A	285	-1,432
		Scenario 2a. replacing rice by quinoa production	30,000	N/A	10,000	N/A	20,000	18,282
	Mango	Scenario 1b. organic production	8,655	N/A	1,205	N/A	7,450	1,491
		Scenario 2b. production as part of an agro-forestry system	27,049	N/A	2,074	N/A	24,974	19,015

All figures in Peruvian nuevo sol (PEN). Exchange rate PEN/USD = 0.31

Fig. 1 Example of cost-benefit scenario. ELD Initiative (2015d), p. 43

encompassing a clear anthropocentric perspective. The ELD Initiative takes this up and transfers the expression of such benefits towards an economic realm, based on monetary terms. These are expressed by indicating the ecosystem's overall use for people as the Total Economic Value (TEV). However, it must be noted that the TEV also encloses benefits such as diversity and ecosystem health, thereby being inclusive of the ecological sustainability. For those interested in further discussions, the ELD Interim Report (2013a) highlights a wide range of valuation options that can be applied based on the system's characteristics.

As a result of such an assessment, policy-/decision-makers will have a range of economic land management options. These choices are derived from stakeholder consultations and developed during the research through iterative interactions with local population, administration and national decision makers. As outlined in Fig. 1, options are determined with specific cost-benefit ratios towards relevant stakeholders.

It is important to point out that while economic information present universally understandable arguments that allow for a comparative decision-making process, such numbers cannot be the only factor. Sound economic assessments reflect the values of preferences of different stakeholder groups, but the developed land-use scenarios and options can only reflect these to a certain degree.[14]

[14]ELD Initiative (2015d).

"Movement towards sustainable land management requires a multi-sector approach at national and sub-national levels", highlighting the need to widen the group of engaged persons beyond the participants during the research. Therefore it remains important to validate the results and incorporate different perspectives again when proceeding to the most important step, which is framed under the '+1'-component of the ELD methodology: the *'taking action'*.

Figure 1 represents a typical outcome of the first six steps, providing the range of benefits for each option. Once this is assessed, it is then up to the respective policy-/ decision-makers to decide, which option delivers the desired outcomes.

2.3 +1 Step from Valuation: Taking Action Towards Sustainable Land Management

Sustainable land management should be compatible with the local ecological environment while being economically viable, and also comply with the social and political setting.[15] Although the ELD's six steps yields applicable scenarios, further emphasis must be made on creating an enabling environment that mainstreams these into the political sphere. The political system consists of wider range of sectors, groups, and social configurations in which "sustainable land management is an important cross-cutting issue of concern to a range of different stakeholders. [. . .] This diversity requires effective integration of perspectives in order to deliver sustainable land management actions that are feasible to implement, and which also consider and serve the varied needs and scales of operation of different stakeholders".[16]

"Approaches involving stakeholders should ensure that the most economically desirable option is compatible with existing economic mechanisms, and is also technically and legally feasible, and environmentally and socially acceptable. Additionally, physical and monetary resources to achieve the practical implementation of sustainable land management should be accessible and available. Comprehensive (re-) design of portfolios of options, including current, revised, and new measures, can help make sure that there is convergence and that action is taken based on assessment results."[17]

In order to facilitate this, the ELD Initiative has developed different approaches to engage with stakeholders and develop solutions based on local conditions. This includes an array of tools for a successful science-policy dialogue, which is necessary to actually put the chosen scenarios into place. These tools and methods are summarized in the ELD Practitioner's Guide.[18]

[15] ELD Initiative (2015d).

[16] ELD Initiative (2015d), p. 92.

[17] ibid, p. 102.

[18] ELD Initiative (2015b).

The ultimate aim of stakeholder consultation is an enabling environment where decisions are made with all relevant stakeholders. This should include consideration for incentives that will change the way land is used under detrimental 'business-as-usual' scenarios. A non-exhaustive list of potential incentives and tools to reach this goal is provided in Box 2, and often a mix of different ones are required. It is critical to remember however, that their selection must be relevant in local contexts. Additionally, participants of sustainable land management stakeholder roundtables should reach consensus on the following issues:

- Sustainable land management transmitting the sublime perception of the resource-system
  Human-land interactions are a social-ecological system where anthropological perceptions are reflected in the way the natural system is configured and vice-versa. Land management approaches thus represent a manifestation of the human perception of nature. As this is subjective, it is often difficult to reframe land management when such a present system already presents a palimpsest of the dominating understanding
- Multiplicity of valuations
  The economic value of ecosystems and land is determined by social perceptions. Neglecting differences across stakeholders may lead to imbalanced values that prioritize some over others. The inclusion of all groups through the valuation and implementation process is thus important in ensuring the valuation is truly representative.[19]
- Contextual framing of and identification of local particularities
  Management of social-ecological systems remains a complex exercise, even with a range of management options (see Box 2): Degradation processes such as soil erosion might differ locally in their causes and impacts or optional incentive mechanisms might not be socially adequate. This creates the need to review findings of assessments and the potential interventions by the respective groups before applying them as blueprints.[20]

Box 2: Examples of Policy Instruments to Enable the Adoption of Sustainable Land Management[21]

The ELD Main Report indicates a wide range of potential options, which can be instrumentalized to facilitate the uptake of sustainable land management:

- Bans: Bans restrict the use of products proven to be harmful for the environment or public health, such as certain pesticides.

(continued)

[19]ELD Initiative (2015d).

[20]Ibid.

[21]ELD Initiative (2015c), p. 15f.

Box 2 (continued)

- Contract farmland set-asides: Land owners abandon the right to use parts or all of their farmland to foster the delivery of environmental benefits, and receive a payment in return.
- Eco-labels and certification: Eco-labels are a form of sustainability measurement for food and consumer products with the aim to facilitate the purchase of eco-sensitive commodities. Eco-labels result from a standardised certification process controlled by bodies such as the International Organization for Standardization (ISO), FairTrade® Foundation, or Forest Stewardship Council (FSC).
- Insurance schemes: In the US, Canada, and India, the governments provide insurance against crop losses due to weather extremes or declines in global commodity prices. If crop yields at the end of a cropping season are lower than a pre-established reference amount, farmers receive compensation.
- Microfinance: Microfinance is a specific form of credits that support the establishment of local, small-scale businesses. Micro-credits are provided at a lower interest rate than those offered by traditional banks and have helped to reduce poverty at the individual and village levels in many developing countries such as Bangladesh. In providing for easily accessible start-up capital, micro-credits are a particularly well suited tool to facilitate livelihood diversification.
- Payments for conservation investments: Certain investments into sustainable land management are financially rewarded by the government. Agri-environmental measures by the EU are one example.
- Payments for ecosystem services (PES): Land owners are rewarded for the provision of certain ecosystem services by the beneficiaries of these services. To this end, managers of the service-providing ecosystems close a deal either with a private company, the government, or a nongovernment organisation. Globally, the REDD scheme has gained wide attention in its effort to compensate developing countries for the preservation of forests and the carbon stored therein, as well as for the enhancement of forest carbon stocks ("REDD+"). The PES mechanism itself values the provided ecosystem services by linking the demand by the beneficiaries to the supply side of the land managers. The agreed payments for the services therefore result from two—sometimes differing—values assigned to land, which must be met in such a PES scheme.
- Permanent conservation easements: Permanent conservation easements are voluntary, legally binding agreements by which certain land usages are prohibited. They serve to protect the ecological or aesthetic values of land. National parks are one example.

(continued)

Box 2 (continued)

- Taxes and environmental fees: Environmental taxes and fees aim to raise the cost of production or consumption of environmentally damaging goods so as to limit their demand. One example is the eco-tax on plastic-based products in Europe through which the recycling of plastic is being funded.
- Trading of emission reductions: A pollution goal or allowance is set and pollution permits are distributed which can thereafter be traded. Several emissions trading schemes have been established globally (e.g., EU Emissions Trading System), yet with limited success so far.
- Transferable development rights: These allow for the development of a certain area of land on the condition that land of a comparable type and quality is restored as a compensation measure.
- Conservation banking or offsets: Conservation offsets aim at compensating for environmental damage caused by land development. Developers can source conservation credits through a market mechanism to offset the loss of ecosystem services at one site, with conservation gains elsewhere.
- Voluntary carbon offsets: On a voluntary basis, individuals, governments or companies can purchase carbon offsets to compensate for greenhouse gas emissions caused by electricity use or transportation (e.g., personal air travel).

3 Achieving Land Degradation Neutrality (LDN) and the Sustainable Development Goals

The activities of the Initiative will also contribute to the international debate on land, including in the context of a land degradation neutral world under the Sustainable Development Goal (SDG) 15 set in 2015, to be reached by 2030 (see Box 3).

Box 3: Land Degradation Neutrality in the Context of the SDG 15

September 2015 represented a milestone in the international debate on sustainable development as the 70th session General Assembly of the United Nations had ratified the SDGs in the context of the Post-2015 Development Agenda. While all 17 goals focus on mainstreaming human and environmental health, goal 15 specifically aims to "protect, restore and promote sustainable use of terrestrial ecosystems, sustainably manage forests, combat desertification, and halt and reverse land degradation and halt biodiversity

(continued)

Box 3 (continued)
loss".[22] It thereby is an essential step to include sustainable resource of land resources in international and national policy agendas.

As part of this goal, SDG 15.3 specifically addresses land degradation under a concept coined land degradation neutrality. Land degradation neutrality is the aim of working to maintain or even improve the amount of healthy and productive land resources over time, and in line with national sustainable development priorities.[23] However, it is not just about rehabilitating a few degraded zones or planting trees, but requires the reorientation and reform of various sectors and policy areas, and will also impact the economic framework of each country. In order to adequately reflect the overall significance of the natural resource base included in land within the realm of political decision making, a good understanding of the economic dimension of the LDN target is necessary. This creates a need to compare the total economic value of achieving the LDN target to the often short-term oriented economic interest observable globally.

LDN is country specific, and therefore will need to be implemented at national levels. This is where the ELD Initiative can offer tools and support to help countries work towards it.

Ultimately, reaching land degradation neutrality is a highly political issue, and understanding the total economic value of land is a precondition for political processes around land. The perspectives, methods, approaches, and knowledge developed by the ELD Initiative and its partners offer the necessary tools to create such understanding facilitate decision-making, making it a central component of any land degradation neutrality processes. The Initiative will further contribute to this global vision by assessing the benefits to humanity from protecting and restoring land based ecosystems in economic terms. Through sound economic assessments, it has already proven that sustainable land management directly contributes to LDN, and also enhances the monetary revenues of land users.[24]

3.1 Creating Knowledge Across Scales and Institutions

Resulting from a disproportion in research capacity between the global south and global north, the ELD Initiative has openly emphasized a need for human capacity development in addition to supporting localised implementation and adaption of the ELD methodology.

[22]UNDP (2016).
[23]UNCCD (2016).
[24]Etter (2015).

"Once rigorous cost-benefit analyses have been performed, the ELD Initiative thus endeavors to provide scientific evidence, support networks, and capacity building for all stakeholders to choose informed and sustainable land management options, towards shared prosperity."[25]

Creating sound understanding of the inter-linkages between local livelihood structures, and the implications for sustainable land management often builds upon pre-existing knowledge which must be used and accessed through capacity building measures tailored to the respective contexts.

With respect to the Initiative's post 2015 implementation phase, the ELD Initiative will continue and expand the development and implementation of several capacity building activities. These aim at supporting learning processes on the economics of land degradation, sustainable land management and help create shared understanding of these topics. The initiative aims to meet this need for instance by offering web-based trainings such as Massive Open Online Courses (MOOC) or through on-site training workshops.

3.2 Massive Open Online Courses (MOOCs)

The ELD Initiative launched Massive Open Online Courses (MOOCs) in 2014 and 2015 with over 1500 registrants each. Each course also resulted in practitioners guides developed from the course outcomes and participant contributions, which are freely available for use and research.

The first ELD MOOC 'The Economics of Land Degradation—Assessing the socio-economic benefits of land ecosystems' brought together policy-/decision-makers, managers, supervisors, seniors as well as educational and research professionals from international organizations, governments, the private sector, and NGOs. The course demonstrated how to assess the economic benefits of land as a first step towards preventing degradation. It introduced methods of economic analysis and valuation to help manage land sustainably and efficiently. Course participants learned problem-solving using economics approaches and methods.

In continuing to support the ELD Initiative capacity-building endeavours, a second ELD MOOC '*Options and pathways for action: Stakeholder Engagement*' was offered by the Initiative in 2015. Building off the first MOOC, this course addressed the importance of working with relevant stakeholders to implement sustainable land management choices effectively and sustainably. It introduced approaches to identify options and pathways for action, methods to identify stakeholders, and how to establish a basis for initiating engagement and discussion.

In the courses, participants gained introductory knowledge and understanding of the economics of land degradation and the ELD Initiative approach. They also had the opportunity to apply these by conducting economic valuations for specific case

[25]Etter et al. (2015), p. 4.

studies of their own choosing. The international participants, organised themselves in working groups and delivered a wide range of assignments and case studies. Their work generated knowledge about land degradation and sustainable land management and gave provided nuanced insights and research at varying scales. Participants from Mexico for example have applied the ELD approach presented in the MOOC and could identify a clear need to reshape the local water management in a local watershed based on the assigned water and land values by stakeholders.[26]

The Initiative will continue with the successful outreach of MOOCs beyond 2015, centralising the ELD Initiative approach as the recognised methodology for economic valuations of land for interested actors and agencies. A third ELD MOOC will be offered in 2016, and as with the others, will be freely open and accessible. The ELD Initiative is also collaborating on external courses, supporting the efforts of both online and on-site land management and related courses that include economic components. Additionally, an ELD module is being developed for the World Overview of Conservation Approaches and Techniques (WOCAT), and the Initiative is contributing content to curricula and online trainings for partner institutions such as the UNCCD or Commonland.

3.3 Training Workshops

The ELD Initiative also supports capacity building through training workshops in an iterative and multi-scale dialogue between relevant parties involved in land management. This is done in order to ensure that results from economic land assessments are applied and implemented on a wider scale, as well as being upscaled.

Based on scientific case studies on the economic benefits of sustainable and conservative land management approaches, trained policy-/decision-makers develop solutions for the implementation of identified feasible approaches together with networks of other stakeholders. Integrative local networks are then developed and negotiations on the appropriate management are facilitated by a common language—economics.

4 From Knowledge to Action: The Relevance of the Economics of Land Degradation to Land Degradation Neutrality

In 2016 and beyond, the focus of the ELD Initiative will shift towards actualising lessons learned from the previous research-intensive phase, including capacity-building at different scales, and openly sharing ELD findings. Following the post-

[26]ELD (2014b).

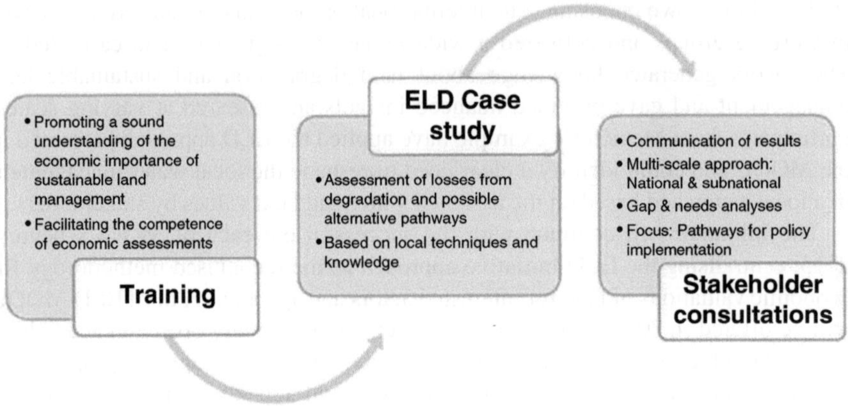

Fig. 2 Implementing the economics through iterative science-policy dialogue. Etter (2015), p. 315

2015 development agenda expressed in the SDGs, the economic value of land must be reflected in discussions on steering the world towards the state of LDN. Enhanced provision of ecosystem services can yield important financial and economic returns, which are crucial to meet financial gaps related to the LDN vision. The ELD Initiative has pledged itself to embrace this vision and support governments in their efforts, as well as to NGO and research partners in providing the necessary tools, approaches, and data to initiate and steer related discussions. More specifically, the Initiative will break down the on-going work and perspectives of the ELD Initiative community to the regional, national, and sub-national levels, link sound scientific results with policy dialogues, and provide tailored solutions and advice. On-going examples of these workstreams include work currently being undertaken in Namibia, a country characterised by strikingly high rates of land degradation and requires a full set of economic tools to address the problem at a regional scale amongst different countries. The following section will outline this approach with key examples.

4.1 Iterative Stepwise ELD Approach

In order to ensure the application and implementation of assessment results on a broader scale, as well as facilitate upscaling, the ELD Initiative promotes an iterative, multi-scaled dialogue between relevant parties involved in land management. Land management can sufficiently influence the practice of resource use, including three target groups relevant for land management (scientific community, private sector, and policy-/decision-makers). Figure 2 presents the schematic process, in which economic arguments can be mainstreamed in decision making consisting of three core components: (i) an inaugural training workshop, (ii) an assessment following the above described methodology, and (iii) subsequent stakeholder consultations and guided transformative advocacy for reframing land management.[27]

Training of relevant policy-/decision-makers on the ministerial level or members of a company's supervisory board can develop a sound understanding of the importance of land and relate this discussion to relevant visions, such as the SDGs or greening the company's brand. Further, sound understanding of economic assessments and the relevance of the resource itself can also raise the awareness of the importance of sound land management. This also sets the frame and focus for subsequent ELD assessments, ensuring results that ultimately meet the demands and needs of policy-/decision-makers and their constituents by contributing to the relevant vision.[28]

Results from assessments of the benefits from action can subsequently feed into stakeholder consultations at different administrative scales. These enable the uptake of sustainable land management can further provide the foundation to consultations are currently being used to communicate the potential of sustainable land management, as well as identify feasible pathways and tools for the implementation of desired management options. For instance, successful consultations by the ELD Working Group on Options and Pathways to Action in Kenya concluded with clear recommendations to enable the implementation of sustainable land management on a national and subnational level.[29] Potential tools to facilitate the development of financing mechanisms include as payments for ecosystem services, taxes, subsidies, and the establishment and use of new markets such as carbon offsets.[30] These must be adapted to context-specific needs and opportunities for each system. While pathways to action are mostly informed by the scientific community and steered by political actors, they also often have a strong impact on and are influenced by the private sector. This calls for connecting stakeholder realms and initiating intersectoral discourse. As such, and also in light of their influence on the use of natural resources, it is prudent to include the private sector as policy-/decision-makers and relevant stakeholders when performing cost-benefit analyses of sustainable land management.[31] Even though the interest of such business-oriented stakeholders might differ from the logic under which the public sector works, there remains considerable overlap in agendas: The ELD Business Brief[32] proves especially productive industrial sectors are facing high risks for economic losses when facing degradation. Further, the public sector can influence the businesses through incentivizing more sustainable sourcing.

[27]Etter (2015).

[28]ibid.

[29]ELD Initiative (2014a).

[30]ELD Initiative (2013a).

[31]Etter (2015).

[32]ELD (2013b).

Box 4: Implementation of the ELD Approach in the Context of the SDGs: The Namibian Case

Namibia as one of the driest countries in the world is suffering from a specific form of degradation known as bush encroachment. This severely limits agricultural or livestock productivity and also reduces available freshwater sources. High densities of invasive species are often linked to mismanagement of the rangelands, especially overgrazing and suppression of natural bush fires. Facing this challenge, Namibia has included the LDN vision within its recently approved National Action Programme (NAP), making this SDG target an intersectoral priority.

In order to combat this form of degradation, a local GIZ project is cooperating with the ELD Initiative in order to reduce degradation and enable Namibia to reach a land degradation neutral state. Together with a consortium consisting of relevant ministries, NGOs and private sector representatives, an integrated regional land use planning (IRLUP) process was identified as a feasible vehicle to realize the LDN vision by integrating potential land management approaches together with economic values. Understanding the total economic values of land in the consideration of LDN will also require measuring these values against short-term economic interests that currently drive many land management decisions. The ELD study will assist policy-/decision-makers to determine the necessary amount of public funds to be directed towards supporting/incentivizing bush control measures in Namibia, taking into consideration the different types of land use. Further, good land management/rehabilitation practices will be elaborated based on already available experience and matched with their potential to meet the LDN target. The compilation of good practices will be integrated in the IRLUP as management recommendations for the respective land use type. Having an economic valuation of identified best practices for land management under consideration of the planned land use zones will make the implications of the ILRUP process much clearer to the stakeholders involved in the planning process. Through this planning process, the LDN vision can directly contribute to the LDN vision and fulfill Namibia's commitment to the SDG process. As a replicable best-practice, this can then be scaled-out throughout the country.[33]

In parallel to the implementing work if the Initiative, strategic research will also continue to be performed, keeping the state of knowledge on the economics of land degradation current and applicable. Emerging areas of study that have already been requested for the ELD Initiative to undertake and support include: migration and land degradation, economic incentives for land degradation neutrality and

[33]More information can be obtained from the authors.

sustainable land management, and large scale land rehabilitation processes. New insights from ELD research is envisioned to contribute to and enhance the options and alternatives in countries. It is important to also integrate insights from consultations and work with stakeholders in the growing compendium of scientific knowledge.

5 Conclusion

The ELD Initiative has developed an assessment approach for the economic value of land based on the current state of knowledge on the economics of land degradation and demonstrated the benefits of its application on multiple scales. This includes an online case-study database, as well as commissioned research projects to fill knowledge gaps at global, national and subnational scales. Stakeholder consultations were undertaken by ELD partners in 15 countries across the world, to explicitly understand the different needs and perspectives on the ground. Two successful massive open online courses (MOOCs) were run in 2014 and 2015 respectively, which included more than 3000 registered participants learning about undertaking cost-benefit analyses and stakeholder engagement. On-site training workshops for policy-/decision-makers have taken place in countries like Tunisia to communicate scientific results clearly, and these will be an integral part in future efforts to upscale sustainable land management. In terms of the private sector, numerous workshops and partnerships have taken place, and a Land Materiality Screening Toolkit is currently being launched to help businesses assess their risk from and impact on land and land-based ecosystems.

For the past several years, the Initiative has realised its many outputs based on the strength and collaborative efforts of its partners, and seeks to amplify this success in the future. Thus, the core partners of the ELD Initiative have created avenues for interested partners and stakeholders to participate and benefit from the efforts of this global team. This is particularly crucial for the upcoming next phase of the ELD Initiative in 2016, which will see the focus shift from scientific research to implementation work on regional, national, and sub-national levels. Here, relationships will be critical in establishing a global network of land valuation as a means to achieve sustainable land management and land degradation neutrality as described in the UN Sustainable Development Goals. Potential partners are invited to interlink with the ELD Initiative for an exchange of best practices, knowledge, and networks in order to financially benefit from securing and realising sustainable land management on the ground.

All activities of the ELD Initiative are intended to be part of a holistic global set of solutions in the desertification, land degradation, and drought crisis. This is part of contributing to a rebalance of physical and social imbalances that are driven by resource mismanagement worldwide. The ELD Initiative focuses on efforts to achieve a world in which environmental and human well-being exist and are maintained for the good of humanity and the physical earth. Understanding the deeper implications of economic understandings of land, and seeing its full value is part of this necessary future. In recognition of this, the ELD Initiative will ensure that relevant knowledge and tools remain available and practical.

References

Aymeric R, Myint MM, Westerberg V (2015) An economic valuation of sustainable land management through agroforestry in eastern Sudan. Report for the ELD Initiative by International Union for Conservation of Nature, Nairobi, Kenya. Available at: www.eld-initiative.org

Barbier E, Hochard JP (2014) Land degradation, less favored lands and the rural poor: a spatial and economic analysis. A report for the economics of land degradation initiative. Available at: www.eldinitiative.org

Etter H (2015) Land degradation. An economic perspective. In: Chabay I, Frick M, Helgeson J (eds) Land restoration. Reclaiming landscapes for a sustainable future. Academic Press, pp 311–317

Etter H, Quillérou E, Stewart N, Gerhartsreiter T, Thomas R, Schauer M (2015) The economics of land degradation initiative: economics as a tool for prosperous, shared, and sustainable land management

ELD Initiative (2013a) The rewards of investing in sustainable land management. Interim report for the economics of land degradation initiative: a global strategy for sustainable land management. Available from, www.eld-initiative.org

ELD Initiative (2013b) Opportunity lost: mitigating risk and making the most of your land assets. An assessment of the exposure of business to land degradation risk and the opportunities inherent in sustainable land management

ELD Initiative (2014a) Support towards the economics of land degradation initiative. Report on the ELD Kenya Stakeholders Consultations, April 2014. Available from www.eld-initiative.org

ELD Initiative (2014b) Practitioner's guide 'the economics of land degradation'. Principles of economic valuation for sustainable land management based on the Massive Open Online Course "the economics of land degradation". Available from: www.eld-initiative.org

ELD Initiative (2015a) ELD initiative user guide: a 6+1 step approach to assess the economics of land management. GIZ, Bonn, Germany. Available from www.eld-initiative.org

ELD Initiative (2015b) Pathways and options for action and stakeholder engagement, based on the 2015 ELD Massive Open Online Course "Stakeholder Engagement". Practitioner's guide. Available from: www.eld-initiative.org

ELD Initiative (2015c) Report for policy and decision makers: reaping economic and environmental benefits from sustainable land management. Available from www.eld-initiative.org

ELD Initiative (2015d) The value of land: prosperous lands and positive rewards through sustainable land management. Available at: www.eld-initiative.org

Lambin E, Gibbs H, Ferreira L, Grau R, Mayaux P, Meyfroidt P, Morton DC, Rudel TK, Gasparri I, Munger J (2013) Estimating the world's potentially available cropland using a bottom-up approach. Glob Environ Chang 23:892–901

MEA (2005) Ecosystems and human well-being: synthesis. Island Press, Washington

Myint MM, Westerberg V (2015) An economic valuation of a large-scale rangeland restoration project through the Hima system in Jordan. Report for the ELD Initiative by International Union for Conservation of Nature, Nairobi, Kenya. Available at: www.eld-initiative.org

Noel S, Soussan J (2010) Economics of land degradation: supporting evidence-based decision making. Methodologies for assessing costs of degradation and benefits of sustainable land management. Paper commissioned by the Global Mechanism of the UNCCD to the Stockholm Environment Institute (SEI)

Sidibé Y, Myint M, Westerberg V (2014) An economic valuation of agroforestry and land restoration in the Kelka Forest, Mali. Assessing the socio-economic and environmental dimensions of land degradation. Report for the Economics of Land Degradation Initiative, by International Union for Conservation of Nature, Nairobi, Kenya. Available from: www.eld-initiative.org

UNCCD (2016) Land degradation neutrality - frequently asked questions. Available from: http://www.unccd.int/en/programmes/RioConventions/RioPlus20/Pages/LDNFAQ.aspx

UNDP (2016) Goal 15: Life on land. Available from: http://www.undp.org/content/undp/en/home/sdgoverview/post-2015-development-agenda/goal-15.html

Part IV
National and Regional Soil Legislation

National Developments in Soil Protection in Mongolia

Ian Hannam

1 Introduction

The fundamental philosophy of soil protection is that the conservation and enhancement of the quality of soil is a common concern of all humanity. Inherent in this philosophy is the responsibility of humanity to maintain and enhance the productivity of soil, taking effective measures to prevent soil degradation, including legislative measures to meet the needs of land use, society, and nature.[1] At present, the soils of Mongolia are vulnerable to the effects of climate change and anthropogenic impact. Severe degradation, including widespread contamination, compaction, accelerated erosion, severe depletion of carbon and nutrients, pressures from urbanization and frequent hazards threaten their sustainable use, ecosystem services and the essential quality of human life. Mongolia has taken legislative and strategic steps to improve the management of its soil resources including the introduction of the Law for Soil Protection and Prevention of Desertification 2012 and the drafting of a Pastureland Law. In addition, there are several other existing laws related to soil protection and prevention of desertification which together make up the total framework of soil protection law. Together, these laws provide part of the platform needed by Mongolia to help satisfy Goal 15 of the Sustainable Development Goals agreed by the countries at the United Nations Sustainable Development Summit

[1]Boer and Hannam (2015).

I. Hannam (✉)
Australian Centre for Agriculture and Law, University of New England, Armidale 2351, NSW, Australia
e-mail: ian.hannam@ozemail.com.au

H. Ginzky et al. (eds.), *International Yearbook of Soil Law and Policy 2016*,
International Yearbook of Soil Law and Policy,
DOI 10.1007/978-3-319-42508-5_19

in 2015.[2] Goal 15 is [to] "Protect, restore and promote sustainable use of terrestrial ecosystems, sustainably manage forests, combat desertification, and halt and reverse land degradation and halt biodiversity loss" (SDG 15). More specifically, Target 15.3 specifies [that] "By 2030, combat desertification, restore degraded land and soil, including land affected by desertification, drought and floods, and strive to achieve a land degradation-neutral world."[3]

Since the 1950s Mongolia had been transformed by rapid economic development and industrialization from what was predominantly an agricultural economy based on nomadic animal husbandry towards an industrial economy.[4] Today over half the population lives in urban areas. As was the case with many rapidly industrializing countries, environmental and resource management considerations were generally absent from land planning and utilization. Land that is productive is under developmental pressures that are leading to environmental deterioration. Land available for agricultural production is decreasing: grazing land was 141 million hectares in the 1960s but has now dropped to 117 million hectares while the number of livestock has risen to 43 million head.[5] This number has increased because Mongolia has not had a *dzud* for several years.[6]

2 Land Resources

Mongolia's climate is harsh with extremes of heat and cold. Availability of water and other renewable natural resources is very limited, and natural ecosystems are fragile, highly susceptible to degradation and slow to recover. Low precipitation is

[2]http://www.undp.org/content/undp/en/home/presscenter/pressreleases/2015/09/24/undp-welcomes-adoption-of-sustainable-development-goals-by-world-leaders.html. Accessed 10 Feb 2016; United Nations General Assembly resolution A/RES/70/1, "Transforming our world: the 2030 Agenda for Sustainable Development"; see also http://countryoffice.unfpa.org/mongolia/2015/09/30/12840/social_good_summit_2015_in_mongolia_focuses_on_the_new_sustainable_development_goals/. Accessed 17 Feb 2016.

[3]https://sustainabledevelopment.un.org/?menu=1300. Accessed 10 Feb 2016; Dooley et al. (2015); UNCCD COP 12 Decision COP (12)/L.4 "invites Parties" to develop voluntary targets.

[4]Mongolia is divided into 21 administrative units (provinces) called Aimags. The population of cities range in size from 50,000 to over 600,000 in the capital of Ulaanbaatar. The cities are divided into districts. Urban population accounts for about 51 % of the country's total population and there is a strong tendency for migration from the rural to urban areas. The Aimag populations range from 40,000 to approximately 100,000. They are divided into Aimag centers and rural Soums or districts around the centers. Soums are comprised of Soum centers and Bags. The Bags are the lowest level rural administrative. There are a total of 333 Soums and 1564 Bags in the country.

[5]Includes camels, yaks, horses, cattle, sheep, goats.

[6]A dzud is a winter disaster that covers pastures with ice and causes mass livestock starvation. The most recent dzuds affected Mongolia in 1996–1997, 1999–2000 and 2000–2001—the last of which killed 13 % of the country's total livestock.

unequally distributed both temporally (majority occurs during 3-month period in summer) and geographically (north of the country receives most of the precipitation; southern part is semi arid or desert).[7] Climatological data indicate that the Mongolian climate is gradually growing drier, which indicates that its environmental laws have to be adaptable to the effects of climate change. Separated from the moderating influences of the oceans, Mongolia experiences great extremes in temperature; wide temperature fluctuations in both diurnal and annual temperatures are common. These climatic factors combine to significantly limit the availability of renewable natural resources. More specifically, current land resource issues include[8]:

- *Soil erosion*—during the past 40 years around 50 % of cultivated land has suffered high levels of erosion; 7 million hectares of grazing grassland has been destroyed; productive capacity of overgrazed grasslands has decreased by almost 5 times and the number of crop species has reduced.
- *Threatened forests*—forest reserves occupy 17.5 thousand hectares; over the last 20 years the forested area has decreased by 1.4 million hectares from forest fires and exploitation.
- *Increasing desertification*—Mongolia's desert occupies 42 % of the total territory; over the last 40 years the desert area has increased by 38,000 ha, of which 88 % is newly desertified land mainly in the Gobi region, with 12 % in the northern part of the country; around 95 % of total land is considered to be highly vulnerable to desertification; droughts covering as much as 25 % of the country occur every 2–3 years, and drought reaching half the land occurs every 4–5 years; due to climatic changes desertification has spread even more widely, which is reflected in the increased numbers of lakes and rivers that are drying up, the worsened quality and growth levels of plants, soil and water salinization, loss of productive capacity, and increasing sand movement.
- *Threats to biodiversity*—Around 200 species of flora and fauna are at the brink of extinction; 100 species of animals classified as extremely rare and rare, and 128 species of plants require immediate protection.
- *Threats to water resources*—annual water reserves available total 34.6 cubic km; although this is relatively low in absolute volume compared to other countries, given Mongolia's small population the amount of water per person is substantial; this is deceiving given climatic trends and the increased process of desertification; the area under constant snow cover will diminish with the result that in winter time the herders and their herds will suffer water shortages; this will be coupled with an increase in evaporation followed by higher rates of drying up of springs, oasis, and well water supplies.
- *Severe threats from natural disasters*—loss of life and extensive damage are caused by natural disasters in Mongolia every year; almost half of the

[7]Batima (2006).

[8]Hannam (2010).

atmospheric related disasters, totaling between 25 and 40 occurrences a year, result in substantial economic loss and some loss of human life; fires, heavy rains and snowfall, river flooding, sand and dust storms, drought, and blizzards are among the primary natural disasters.[9]

3 Social Transformation and Soil Protection

Since the 1950s Mongolia had been transformed by rapid economic development and industrialization from what was predominantly an agricultural economy based on nomadic animal husbandry towards an industrial economy. Mongolia was a socialist state with a centrally planned economy until 1990 when it became independent of the USSR and adopted a democratic form of government and free market economy.[10] As was the case with many rapidly industrializing countries, environmental and resource management considerations were generally absent from land planning and utilization. Many environmental, social, economic, political and legislative factors interact that affect the goal of effective soil protection. The success of Mongolia's soil protection law, especially in terms of helping to achieve SDG 15 is challenged by its harsh climate and its extremes of heat and cold. Climatological data indicate that the Mongolian climate is gradually growing drier, which indicates that its environmental laws have to be adaptable to manage the effects of climate change. In this regard, the legal framework for soil protection and desertification prevention in Mongolia also assists with meeting Sustainable Development Goal 13 to take urgent action to combat climate change and its impacts, in particular Target 13.2 "Integrate climate change measures into national policies, strategies and planning" (SDG 13). [11]

These climatic factors combine to significantly limit the availability of renewable natural resources. The rates of humus production, vegetative regeneration and growth, and livestock productivity are very low in comparison to other countries in the region. In view of the strong dependence of Mongolia's economy, food supply, and other sectors on land resources, halting and reversing the ongoing processes of soil degradation, desertification and other forms of environmental degradation are high priority.[12] The critical need to strengthen the environmental legal and institutional framework, and build institutional capacity is seen as essential to achieving effective soil management and environmentally sustainable development.[13] The

---

[9]Mongolia (2008).

[10]UNDP (2005).

[11]https://sustainabledevelopment.un.org/?menu=1300. Accessed 10 Feb 2016; see also Adoption of the Paris Agreement, http://unfccc.int/resource/docs/2015/cop21/eng/l09.pdf. Accessed 10 Feb 2016.

[12]Hannam (2010).

[13]Nature-Future Fund Non-Governmental Organization (2008).

distribution of agricultural resources and uses are not even and are adversely affected by concentrations along transportation routes, around market centres, water sources, and winter grazing grounds. These factors have led to severe overgrazing and desertification in marginal areas and rapidly declining productivity of grazing lands and elements of poverty in the agricultural and urban sectors.[14]

One quarter of the population and more than half of Mongolia's industry have become concentrated in the capital city (Ulaanbaatar) due mainly to the lack of policies on population settlement and city planning and the inability to provide sufficient employment and social and cultural services to retain residents.[15] Establishing an appropriate legal framework that can more efficiently and rationally deliver better goods and services, including environmental services of all kinds to the dispersed rural population, is essential to raise the social standards of living of provincial regions and to protect and utilize soil resources sustainably.[16] The agricultural sector has important implications throughout the nation's manufacturing sectors, providing the essential material inputs into many processing industries. A privatization program for agriculture began over a decade ago and is now essentially complete, with more than 95 % of livestock in private hands and most of the formerly state owned farms now joint-stock companies. The results of the changes have been uneven and serious management and financial shortfalls exist in privately held agricultural operations. Agricultural production is subject to the harshness and unpredictable nature of the climate which places significant pressure on the soil resources.[17]

4 Law for Soil Protection

Investigations carried out by United Nations Development Program (UNDP) and the Mongolian Government in 2008–2009 established various legislative, policy and institutional reforms to improve overall natural resources management including soil protection.[18] The reforms focused on the institutional structure at central and local levels and six elements formed the basis of the reform, including:

- Reorganizing the system for managing legislation.
- Using an ecological approach as the basis for environmental law and to re-structure institutional arrangements.

[14]World Bank (2002), ADB (2013).

[15]Hannam (2015).

[16]Mongolia (2006).

[17]Hannam (2010).

[18]Tortell et al. (2008), Hannam (2009).

- Introducing a management system with specialized institutions/agencies to manage "clusters" of environmental legislation associated with principal problem areas.
- Reforming many existing natural resources laws and developing new types of environmental law.
- Accepting a long term plan to undertake and implement the legal, policy and institutional reforms.

Following the approval of the 68th decree for the elimination of overlaps, contradictions and gaps in Mongolia's legal system in 2010, the government carried out a detailed analysis of 18 existing environmental laws.[19] These laws were systematically compacted into eight laws and two new laws were added, including the *Law on Soil Protection and Prevention of Desertification*. The new laws were ratified by the Mongolian Parliament in May 2012 and the following principles were applied in the reform: to enable sustainable national development; to achieve economic efficiency, responsibility, environmentally friendly development; to implement compliance to international standards of auditing; to employ the "polluter pays" principle; to ensure local involvement in environmental protection; and to generate sustainable financial resources for funding environmental protection.[20]

4.1 Law on Soil Protection and Prevention of Desertification 2012 (LSPPD)

The LSPPD is central to Mongolia achieving its obligations to SD15 and to conserving its soil. It is a law to protect soils from the intensification of agriculture, mining, road construction, urban land use as well as climate change. An important consideration in its development was the facilitation of a safe and healthy environment for humans The purpose of the LSPPD (Article 1) is to regulate the protection of soils from erosion, degradation and pollution and implement soil rehabilitation and desertification prevention measures. Article 2.1 specifies that the LSPPD

[19]See UNEP Governing Council of the United Nations Environment Program UNEP/GC/25/INF/15, Fourth Program for the Development and Periodic Review of Environmental Law (Montevideo Program IV), II Conservation, management and sustainable use of natural resources, C "Soils", where the objective is "To improve national and international principles and standards and to support efforts under the United Nations Convention to Combat Desertification for the further development of legal approaches for the conservation, restoration and sustainable use of soils", http://www.unep.org/delc/Portals/119/MontevideoIV.pdf. Accessed 11 Feb 2016.

[20]Erdenebayasgalan and Sainbayar (2013).

consists of the Constitution of Mongolia,[21] the Law on Environmental Protection 1995, the Law on Land 2002, the Law on Subsoil 1995, the Minerals Law 1998, the Law on Water 2004, the LSPPD and other relevant legislative acts.

Soil protection is defined under Article 3.1.6 as the measures to protect soil from erosion, degradation and desertification caused by human and climatic change. Desertification prevention is defined under Article 3.1.8 as the measures to prevent and mitigate soil degradation in dry, semi-dry and moisture lacking areas through soil protection and improvement activities, in order to ensure sustainable development. Several other key terms are defined in the LSPPD including: soil, soil fertility, soil erosion, degradation and pollution, desertification, and soil rehabilitation (Chapter Three). The LSPPD provides for the classification, assessment and analysis of soil erosion, degradation, and soil pollution and desertification, soil protection, desertification prevention measures, and refers to the rights and responsibilities of public institutions, individuals, and economic entities. It has a mechanism to resolve disputes associated with soil protection and desertification prevention, annul illegal decisions and to compensate for damage.

There is a provision (Chapter Four) to classify soil erosion, degradation, pollution and desertification based on comparison with conditions from previous years (Articles 4.1.1–4.1.4). Soil pollution is classified (Article 4.2) as chemical polluting and hazardous elements, from biological elements and from physical or radioactive elements (Articles 4.2.1–4.2.3). The main role of the government (Article 5.1) is to enforce and implement a national program on soil protection, combat desertification and ensure ecological safety (Article 5.1.1). It also has to establish a national committee to implement the United Nations Convention to Combat Desertification,[22] to facilitate desertification prevention (Article 5.1.2). The responsibility of the Department of Environment and Green Development (Article 5.2), includes implementing government policies and national programs on soil protection and desertification prevention (Article 5.2.1), adopting and enforcing the rules, regulations, methodology and guidelines relating to soil protection and desertification prevention measures (Article 5.2.2), and limiting certain operations in particular areas for a specified period if the activities of individuals, economic entities or organizations is causing soil erosion, degradation, pollution or desertification (Article 5.2.3).

[21]Article 6 of the Constitution of Mongolia 1992 states that "The land, its subsoil, forests, water, fauna and flora and other natural resources in Mongolia shall belong exclusively to the people and be under the State protection"; Article 16 specifies the citizens of Mongolia shall be guaranteed the privilege to enjoy the following right and freedom: [the] 2) "Right to healthy and safe environment, and to be protected against environmental pollution and ecological imbalance"; Article 38 specifies that government shall implement State laws .. 4) "to undertake measures on the protection of the environment, rational use and restoration of natural resources".

[22]UNEP (1994).

4.1.1 Rights and Responsibilities

The rights and responsibilities of citizens and legal bodies with regard to soil protection and desertification prevention include requesting authorities to impose liability on individuals and legal bodies who are causing soil pollution or desertification, to ban their activities and to implement soil protection and desertification prevention activities and measures (Article 6.1). There is also a responsibility to retrieve and provide information on soil erosion, degradation, pollution and desertification to and from the related authorities (Article 6.2), and to eliminate damage and rehabilitate the soil if their illegal acts are deemed to have caused soil pollution or degradation (Article 6.3).

4.1.2 Soil Protection and Desertification Prevention Measures

Effective soil protection and desertification prevention measures are critical to achieving SDG 15. Under the LSPPD they consist of prevention, combating, limiting certain activities and rehabilitation (Article 7.1). The LSPPD provides for soil protection measures to be implemented (Article 7.2), including development of a list of areas of soil pollution (Article 7.2.1), conducting risk assessment in areas of soil pollution in compliance with the Law on Environmental Impact Assessment Law 1998 and the Law on Hazardous and Toxic Chemicals 2006 (Article 7.2.2). The assessment is to specify the type of soil pollution, the area affected and the impact and soil pollution mitigating measures (Article 7.2.3). The risk assessment will decide on the measures for securing polluted areas, neutralizing, burying or stripping the polluted soil or other related measures in line with conclusions made by the environment institution (Article 7.2.4).

The LSPPD requires pasture users and possessors to maintain rotational use of their grazing areas to ensure soil protection and desertification prevention, in compliance with Article 5.4.3. Crop field users have a responsibility to implement soil protection and desertification prevention activities on these lands (Article 7.4) including technologies to increase soil fertility and protect the soil (Article 7.4.1), build windbreaks around crop fields (Article 7.4.2), and use crop fields in strips of 50–100 m width and 30–50 m length (Article 7.4.3). Prohibited acts include reducing soil fertility from its initial level (Article 7.5.1) and the construction of canals and ditches around a crop field unless the field is irrigated (Article 7.5.2).

4.1.3 Incentives for Soil Protection and Desertification Prevention Measures

The government can approve various incentives to individuals and economic entities that have introduced up to date technologies for soil protection and desertification prevention, or have implemented desertification prevention activities

(Article 8.1). Incentives are provided to crop field users to improve soil fertility (Article 8.2.1), to construct windbreaks surrounding a minimum of 50 % of the crop field area (Article 8.2.2), and for using minimum tillage technology during seeding, green fertilizers and organic fertilizers to prevent soil erosion (Article 8.2.3).

4.1.4 Assessment of Soil Erosion, Degradation, Pollution and Desertification

It is the responsibility of the environment institute to assess and analyse soil erosion, degradation, pollution and desertification (Article 9.1). Responsibilities include identifying the causes and sources of these soil problems (Article 9.2.1), defining indicators for soil erosion, degradation, pollution and desertification (Article 9.2.2.), and defining characteristics, severity level and coverage of soil erosion, degradation, pollution and desertification (Article 9.2.3). There is also a responsibility to identify soil quality (Article 9.2.4), estimate direct and indirect damage (Article 9.2.5), estimate ecological and economic value of eroded, degraded and polluted soil (Article 9.2.6), and to implement planning measures for reducing and mitigating soil erosion, degradation, pollution and desertification (Article 9.2.7).

Other aspects of the LSPPD include preparation of a database on soil protection and desertification prevention measures (Chapter 10), monitoring and ratification of soil protection and desertification prevention measures (Chapter 11), payment of compensation for soil pollution (Chapter 12), and the liabilities for a breach of the LSPPD (Chapter 13).

4.2 Related Laws for Soil Protection

In addition to the laws already mentioned above a number of other laws are critical to the objective of soil protection and achieving SDG 15.

4.2.1 The Law on Forests 2007 (LF)

The LF is important for soil protection and desertification control as well managed forest ecosystems will reduce soil hazards.[23] The purpose of the LF (Article 1.1) is to regulate the protection, sustainable use and the reproduction of forests. More broadly, the LF provides for the restoration, forestation, tenure, and use of forests and the control of fires in forest and steppe areas. Under the LF, "forest" means (Article 3.1.1) "a complex of diverse wildlife, plants, and other biological micro-organisms, natural and planted trees, bushes and shrubs that form an ecosystem of

[23]Crisp et al. (2004).

natural resources". Forest protection (Article 3.1.10) includes the measures to prevent forest deterioration, to regenerate the forest and to monitor these activities.

The LF regulates the ownership (Chapter Four) of planted forests and tenure forests. It also provides for a Forest Fund (Article 7; 12.1.11) based on two conservation and utilisation regimes: conservation forest zones and production forest zones. The amendment of the LF in 2012 is important for soil protection, [24] as it recognizes the use of forests for greenhouse gas (GHG) sequestration and defines forest inventory and forest taxation as comprehensive measures to protect forests, their sustainable use and restoration. It provides for the assessment of the forest resources, the distribution, composition, quality, conditions, and estimation of GHG sequestration and the study of forest biomass.[25] This is an important aspect of SDG 13—"to take urgent action to combat climate change and its impacts".[26]

4.2.2 Law on Disaster Prevention 2003 (LDP)

The LDP is important for soil protection and desertification prevention as it regulates matters relating to the principles and powers of disaster protection agencies, their activities, as well as the rights and duties of the State, local authorities, enterprises, entities and individuals in relation to disaster protection (Article 1.1). Disaster Protection is defined as the measures to prevent, protect and rescue people, livestock, properties and environment from the impact of disasters and to limit their consequences. It also acts to facilitate rapid recovery and to train the public in these activities (Article 4.1.1). The term "disaster", includes the loss of human lives and livestock, damage to property and the environment caused by hazardous phenomena, including droughts and extreme weather events, e.g., dzud (Article 4.1.2).

The state organisation in charge of disaster protection is responsible for the implementation of the state disaster protection policy and the legislation on disaster protection and the organisation of nation-wide disaster protection activities (Article 22). It is the responsibility of the government to submit the draft of the disaster protection programme to the Parliament for its consideration (Article 23.1), and to take measures to implement it.

4.2.3 Law on Environmental Protection 1995 (LEP)

The purpose of the LEP is important for the protection of soil resources because it regulates relations between the state, citizens, economic entities and organizations in order to guarantee the human right to live in a healthy and safe environment, have

[24]Erdenebayasgalan and Sainbayar (2013).

[25]Tsogtbaatar (2004).

[26]Footnote 12.

ecologically balanced social and economic development, and to protect the environment for present and future generations. The LEP regulates the use and restoration of natural resources (Article 1.1). It protects a variety of conservation resources (under Article 3.1) from adverse effects in order to prevent ecological imbalance, including to the "land and its soil".

The definition of "environment" (Article 3.2.1) includes the "geosphere" and "biosphere"...."within the territory of Mongolia, directly or indirectly affecting the lives of human beings, and the relationship between them". The term "environmental protection" (Article 3.2.7) applies to the proper use and rehabilitation of natural resources.

4.2.4 Law on Subsoil 1988 (LS)

The purpose of the LS is to regulate relations concerning the use and protection of subsoil for present and future generations (Article 1.1). The subsoil is the property of the State (Article 3). Under the LS (Article 4.1), "subsoil" means "the space that spreads into the depth beneath the soil, as well as all material objects in this space (all types of gravel, minerals and other geological objects…"). Although the LS provides for the use of geologic material the basic right and duties of users of subsoil (Article 20) includes the protection of the land, forests and water, specially protected areas and items of importance to national, historical and cultural studies. In this regard, the LS together with the LSPPD provides for the management of the entire soil medium.

4.2.5 Law on Land 2002 (LL)

The concept of land, land ownership and land rights has received a lot of attention in Mongolia.[27] The purpose of the LL (Article 1.1) includes the regulation of possession and use of land by a citizen, entity and organization. Land means (Article 3.1.1) "a piece of space including the land surface, its soil, forests, water and plants". To "use land" (Article 3.1.4), which has important implications for Mongolia to meet the objectives of SDGs 13 and 15, is to undertake a legitimate and concrete activity to make use of some of the land's specific qualities in accordance with contracts made with owners and possessors of land. Under the LL, "land" is classified into various uses, including "agricultural land" (Article 10.1.1), where "agricultural land" includes (Article 11.1) pastureland, hay fields, cultivation land, fallow land, land under agricultural construction, and other land for agricultural production.

The fundamental characteristics of agricultural livelihood strategies have not changed greatly over the years. Mobile and flexible grazing strategies adapted to

[27] Myers and Hetz (2004), SDC (2015).

cope with harsh and variable production conditions remain the cornerstone of Mongolian agricultural land use. Although land tenure regimes have evolved towards increasingly individuated tenure over pastoral resources, the pasturelands continue to be held and managed as common property resources in most locations, although these institutions have been greatly weakened in the past half century. The recent changes in Mongolia's political economy threaten the continued sustainability of Mongolian pastoral systems due to changes in both pastoral land use and land tenure.[28]

4.3 Pastureland and Soil Protection

A law to protect pastureland is an important part of the legal framework to protect the soil and control desertification. It will also be an important law to help meet the objectives of SDGs 13 and 15. Under the LL "pastureland" means rural agricultural land covered with natural and cultivated vegetation for grazing of livestock and animals (Article 3.1.6). With pastureland covering 70 % of Mongolia (112 million ha) its sustainable management is vitally important for conservation of soil and land resources. The climate is harsh with extremes of heat and cold. Availability of water and other renewable natural resources is very limited, and pastureland ecosystems are fragile, highly susceptible to degradation and slow to recover. With the climate gradually growing drier, this signals changes to the way Mongolia must manage the pastureland resources.[29] A significant portion of pastureland is degraded from overgrazing, deforestation, soil erosion and desertification. Moreover, the soil of pastureland is degraded because herders are unable to apply sustainable grazing practices. The pastureland is not valued so its regulation and management has been avoided in the past. Herders continue to graze their livestock on public land unrestrained, where there is high competition for good pasture. They use public pasture and water free of charge and without initiative to protect and properly use it.[30] Climate change is a contributing factor to desertification and wind erosion on the pastureland.[31] As relatively intact terrestrial ecosystems, the Mongolian pasturelands play a significant role in sequestrating atmospheric carbon dioxide, conserving biodiversity, and especially in providing livelihood benefits to local herders. Over the past 50 years, pastureland degradation has undermined the ecosystem services they generate.[32]

---

[28]Fernandez-Gimenez (2006).

[29]Heiner et al. (2011).

[30]Swift (2007).

[31]Mongolian Government (2008).

[32]ADB (2013).

4.3.1 Draft Pastureland Law (PL)

A legal framework for pastureland use and management was drafted in 2015 to provide tenure for livestock producers, herder groups, cooperatives and herding households, to establish remote pasture reserves, and for the operation of pastureland use committees as a basis to overcome many of the problems that stem from lack of land ownership.[33] The State Policy on Herders and Herding Households 2009 will provide strategic support to the PL. The policy aims to create a favourable legal, economic and business environment and produce better living conditions for herders, prevention of poverty in herders and employment and social security.[34]

The concept of a PL has been through extensive community and parliamentary discussion since 2007 and it accommodates the general objectives of SDGs 13 and 15. The main purpose is to provide a legal method for the transition from an unplanned pastureland user system to a system characterized by: secure possession of pastureland for herders and legal entities; a pastureland planning and management system; the development of information; distinguishing the duties and responsibilities between the different levels of pastureland administration; mechanisms to improve the management of problems associated with soil degradation and the effects of climate change. The draft PL includes procedures to classify pastureland on an ecological basis and provide for the agricultural and economic needs of traditional herding communities and the livestock husbandry industry. It will support economically productive pastoral agriculture while managing pressures on the ecological environment from climate change, desertification and natural disasters.[35]

Importantly, the draft PL includes a procedure to allocate land for grazing and for its management and protection, which constitutes the basis of land tenure and this will help overcome many problems that stem from the traditional pasture usage system.[36] The procedure includes the identification and classification of pastureland, an application for pastureland usage for the purpose of livestock husbandry and the issue of a certificate for usage. This procedure will help promote the sustainable use of pastureland and development of a stewardship ethic which is important for a stable long term tenure system. These are important procedures and should increase the capability of Mongolia to manage its pastureland resources more effectively in the face of the increasing effects of climate change and other natural events that cause or exacerbate soil degradation.[37] The draft PL contains the legal elements considered essential for a successful pastureland management system including support systems to enable herder communities and legal entities to

---

[33]Hannam and Sereenov (2015); an initial draft law was prepared in 2007 and a further draft in 2010 and both were publically debated.

[34]Mongolian Government (2009).

[35]Mongolian Government (2007, 2010).

[36]Fernandez-Gimenez (2006), Hannam and Sereenov (2015).

[37]Batima (2006).

achieve a sustainable livelihood and for the State to achieve its national goals for pastureland management.[38] Important elements in the PL include: development of operational policy; development of a national strategy for pastureland management; development of pastureland management use plans; formation of herder pastureland use groups; developing a comprehensive education and training program; and a procedure for stakeholders to access information and knowledge.

4.3.2 User Pays

Under the Mongolian Constitution 1992, all land is State-owned. Since the mid 1990s, under a special provision of the LEP and various individual natural resource 'use fee' laws that followed, citizens have had access to many natural resources for the payment of a fee.[39] Although this system is similar in concept to the worldwide "principle of user pays", there are various limitations in the way it is applied in Mongolia, where the fee amount is very low (and not indexed), is poorly administered, is not accompanied by natural resource incentives and support schemes, and a substantial amount of fee collected is siphoned off by the central budget and only a small amount of revenue is reinvested locally for conservation and land improvement purposes—which is its principal object. A further limitation for conservation of soil resources has been the absence of a fee paying system over Mongolia's natural pastureland. The government has investigated the application of the user pays system over the pastureland where it is proposed to charge a fee on livestock (based on sheep head equivalent) and herders would receive various inducements to improve their land stewardship under this scheme.[40] This scheme is backed by the Policy on Herders and Herding Households which identifies 36 priority actions to improve the management of pasturelands, including such critical issues as: valuation and pricing of pastureland and establishing a market mechanism for pastureland; studies into the economics of grazing; investigating national and international markets and trade of pastureland products; development of a resettlement plan for herders displaced by financial constraints and natural disasters, and for those who voluntarily seek an alternative livelihood. Given the high incidence of poverty among the herder community there is some scepticism over the ability of people to pay and of the overall financial viability of the scheme.[41]

---

[38] Hannam (2007), Mongolian Government (2009), and Hannam and Sereenov (2015).

[39] Including forest, water, land, natural plants and hunting native animals.

[40] Mongolia (2008), Hannam and Adiyasuren (2009).

[41] Mongolia (2009).

5 Community Organizations

Local pasture user organizations have been promoted as an important part of the process to achieve sustainable use of the land resources and they are crucial for the effective implementation of the LSPPD and related laws.[42] Over time herders in Mongolia have been organized into different institutions, both formally and informally.[43] Moreover, since 1999 Mongolia has seen a number of community-based pastureland management projects aimed to improve herders' livelihoods and pasture management practices. This has been facilitated by various donor pasture-management programs and international NGOs. In recent years, special attention has been given to two donor-funded projects which take a different approach to herder organization, namely 'Herder Groups' (through a UNDP/Sustainable Grasslands Project) and 'Pasture User Groups' (PUGs). Since 2004, the Swiss Government funded Green Gold project has facilitated the formation of hundreds of such groups and will, in the years to come continue to do so by scaling up this collective action approach and broadening its geographical reach.[44]

Under the two main laws concerning herders' relations to land management in Mongolia, namely the LL (Art. 52.2) and the Civil Code of Mongolia 2002 (Article 481), pastureland can be managed by groups of people for a specific period of time. As a form of organisation, PUGs are entitled to enter into a 'pastureland management contract' or 'pasture use agreement' with a local government organization. As a pastureland management tool at the local level, the contract establishes the management requirements for a defined area of land. Each PUG is thus covered by the contract for pasture use which is signed by the *Soum* and *Bagh* governors, the head of the PUG and the PUG leader.[45] Despite the PUG system having the benefit of the combination of a given territory and a contract system which regulates relations within that territory, the issue of pasture use rights still has to be resolved to ensure all legal and practical aspects of the contractual arrangements can be properly realized, in particular: roles, rights and responsibilities; access/use and flexibility; boundaries, flexibility and permeability/leakage; monitoring, control and enforcement; and security of tenure. The contract can form the centerpiece in linking the legal, policy and institutional framework in the management of pastureland at the local level.[46] In some PUGs pasture utilization is covered all year round with the pastures of all the four seasons being registered on maps which form a part of the contract. The contracts are made in a similar legislative format to that

[42]Fernandez-Gimenez et al. (2008), Blattler (2012), and Hannam (2012).

[43]Fernandez-Gimenez (2006).

[44]MSRM (2012); Herder Groups, are usually fairly small with up to 20 households, are organized primarily on the basis of a combination of kinship and shared use of key resources (e.g. wells, winter pastures). By contrast, PUGs involve up to 150+ households and are organized on the basis of territory where local areas are divided into a number of territorial units.

[45]Blattler (2012).

[46]Blattler (2012), Hannam and Sereenov (2015).

provided for in the LL, and can specify the size of the grassland area, its location, seasonal use, and stock carrying capacity, procedures for dispute settlement and conflict resolution, and specific tasks for the PUG. The contract specifies the limit on animal numbers in a prescribed area and in this form it becomes an integrated instrument for pastureland management.[47]

The establishment of PUGs generally works from a bottom-up approach in an attempt to reach out and involve all herders in a given area in the decision making process. Experience shows that land use systems that enable stakeholders to formally participate in the decision-making process generally provide more satisfactory and balanced outcomes to all parties and will help promote the sustainable use of pastureland and development of a pastureland stewardship ethic which is important for a stable long term tenure system. For the PUG concept to be successful there needs to be a good relationship between PUGs and local government. In this regard, the draft PL is an appropriate framework to formally establish PUGs.[48]

Many Forest User Groups (FUGs), comprising groups of herders, have been formed in the northern forest area of Mongolia to improve livelihoods, the ecological status of the forests, reverse land degradation and aid climate adaptation.[49] The FUGs have many benefits including strengthening the national policy on co-management, improving forest planning and management, and improving biodiversity conservation of pastureland and the forest ecosystems. The FUGs operate under a contract with the local Soum and a management plan is prepared with an operational period of 10 years. According to Article 4(5) of the LF, FUGs can operate under a contract with local Soum for a period up to 60 years.

6 Climate Change

In recent years Mongolia has joined 14 environment-related UN conventions and treaties, including the United Nations Framework Convention for Climate Change (UNFCCC).[50] Since 1992, the Parliament has passed several laws and regulations on environmental protection which have an important role in managing the effects of climate change on soil resources, such as the Law on Environmental Protection (1995, amended in 2007), the Water Law 2004, the Law on Forests 2012, the Law on Air 1995 (amended in 2012), the Law on Energy 2001 and the Disaster Prevention Law of 2003. Climate change is having a severe impact on food security

[47]Hannam (2012).

[48]Hannam and Sereenov (2015).

[49]FUGs are formed under Chapter Four of the Law on Forest 2012; see Project "Mainstreaming biodiversity conservation", SFM and carbon sink enhancement into Mongolia's productive forest landscapes, GCP/MON/008/GF, Ulaanbaatar 2015.

[50]UNEP (1995).

and sustainable livelihood of rural people in Mongolia.[51] The government has also introduced a number of action plans which contain soil protection mechanisms, including the Mongolian Environmental Action Plan, the Mongolian Action Programme for the 21st Century (MAP 21), National Action Plan to Combat Desertification, and the National Biodiversity Action Plan. MAP 21 includes concrete considerations and recommendations related to adaptation to climate change and mitigation of GHGs emissions.

Mongolia is committed to the UNFCCC negotiation process towards adopting at COP21 a legal instrument or an agreed outcome with legal force under the Convention, applicable to all Parties, in line with confining global warming below 2 °C. It communicated its Intended Nationally Determined Contributions (INDC) to reiterate its determination to support the multilateral negotiation process and international community efforts in fulfilling the objectives of the UNFCCC. Importantly for soil, in its INDC, Mongolia outlined a series of policies and measures that the country commits to implement up to 2030, in the energy, industry, and agriculture and waste sectors.[52] These commitments are contingent upon gaining access to new technologies and sources of finance through internationally agreed mechanisms and instruments under the auspices of the UNFCCC.[53]

6.1 *Climate Change Strategies and Soil Protection*

A number of Mongolian programs on climate change management are important for the conservation of soil resources,[54] while contributing to the achievement of SDG 13.

6.1.1 The National Action Programme on Climate Change (NAPCC)

The NAPCC is the most relevant policy document addressing climate change.[55] Approved by Parliament in 2011 it intends to meet UNFCCC obligations and commitments and establishes a national policy and strategy to tackle the adverse impacts of climate change and to mitigate GHG emissions. The NAPCC aims to create a sustainable environment for development by promoting capacities and

[51]Narantuya (2006).

[52]Further to the negotiations under the Ad Hoc Working Group on the Durban Platform for Enhanced Action (ADP), the Conference of the Parties (COP), by its decision 1/CP.19, invited all Parties to initiate or intensify domestic preparations for their INDCs towards achieving the objective of the Convention as set out in its Article 2.

[53]Mongolia (2015); http://www4.unfccc.int/submissions/INDC/Published%20Documents/Mongolia/1/150924_INDCs%20of%20Mongolia.pdf. Accessed 15 Feb 2016.

[54]Schulze (2009).

[55]Mongolian Government (2010).

measures on adaptation to climate change, halting decline in ecosystems and protecting them. The implementation strategies in the NAPCC include institutional, legislative, financial, human, education and public awareness, and research programs, as well as co-ordination with other national and sectoral development plans. The goal is to ensure ecological balance, development of socio-economic sectors adapted to climate change, reduce vulnerabilities and risks, mitigate GHG emissions and promote economic effectiveness and efficiencies and implementation of green development goals. The NAPCC also includes adaptation and mitigation strategies and measures for key socio-economic sectors of the country. Importantly for soil protection, it establishes a foundation for green economic growth and development.

The NAPCC is implemented in two phases. The first phase (2011–2016) aimed to strengthen national mitigation and adaptation capacity by setting up the legal environment, structure, institutional and management system, and improving community and public awareness and participation in climate change activities. The second phase (2017–2021) aims to implement climate change adaptation and mitigation measures.

6.1.2 Climate Change Coordination Office (CCCO)

The CCCO was formed to carry out the activities necessary to implement the commitments under the UNFCCC and the Kyoto Protocol,[56] and to integrate climate change-related issues in various sectors. These are important functions in relation to SDG 13. The inter-disciplinary and inter-sectoral National Climate Committee (NCC), now led by the Ministry of Environment and Green Development, co-ordinates and guides national activities and measures to adapt to climate change and mitigate GHG emissions. The National Agency for Meteorology, Hydrology and Environment Monitoring (NAMHEM) monitors the environment and climate carrying out a range of climate change studies and research.[57] In addition, the government is working to raise public awareness on climate change and its impacts on land resources.[58]

6.1.3 Green Development Policy (GDP)

The Parliament approved the GDP in June 2014.[59] The Green Development concept supports the global commitment to a transition to a socially inclusive, low carbon

---

[56] Climate Change Secretariat (1997).

[57] SDC (2015).

[58] ADB (2013).

[59] http://www.mn.undp.org/content/dam/mongolia/DevelopmentDialogues/DD2013/DD4/dialogue4eng.pdf. Accessed 6 Jan 2016.

and low waste development model by changing current development trends and conserving natural resources and ecosystem values along with increased human well-being and reduced poverty. The GDP aims to improve human well-being by safeguarding environmentally friendly, inclusive economic growth, enhancing natural resources utilization and achieving ecosystem sustainability, all which will benefit soil resources. It is based on the concept of sustainable development, and recognizes the need for a change in socio-economic development trends and patterns. Several of the criteria which define green development support the objective for a sustainable approach to Mongolia's pastoral lands. In general, the criteria that relate to soils and pastureland include waste management, energy resources utilization and methods of managing air and soil pollution.

The GDP establishes various strategic objectives to ensure green development all of which benefit the conservation of land and soil resources, including: preserving ecosystem balance, promoting living in harmony with nature, development of a population settlement plan in accordance with climate change, availability of natural and other resources in the regions and restoration capacity. For each of the strategic objectives the GDP establishes specific and measurable targets of which three are important in soil protection, including:

- Processing of agricultural raw materials through the promotion of sustainable agriculture development and development of green-tech industry.
- Protecting at least 60 % of Mongolia's water flow and spring water area and the expansion of protected areas to 25 and 30 % by 2020 and 2030, respectively, and creation of sustainable financing mechanisms.
- Enhancing forest absorption of carbon by intensifying reforestation and expanding forest areas to 9 % of the country by 2030.

6.1.4 Partnership for Action on Green Economy (PAGE)

Mongolia will benefit from the PAGE which is a response to the outcome document of the United Nations Conference on Sustainable Development (Rio + 20), which recognised the green economy as a vehicle for sustainable development and poverty eradication.[60] This partnership brings together the UNDP and other global agencies to support Mongolia in refining and implementing its GDP. An initial economic analysis of the country and a green jobs mapping study have been completed and modeling work is underway to quantify the impacts and costs of the country's transition. Moreover, PAGE is working together on the ground with UNDP and other global agencies to provide a coordinated international response to support the country in its ambitions for a greener, more inclusive economy.

[60]http://en.openei.org/wiki/Mongolia-Partnership_for_Action_on_Green_Economy_(PAGE). Accessed 6 Jan 2016.

6.2 Adaptation and Mitigation

Climate change exacerbates existing natural resource concerns, including diminution of water resources, soil degradation and desertification. Desertification has become a national disaster, affecting more than 70 % of pastureland.[61] A comprehensive assessment of the climatic situation, including future climate scenario and vulnerabilities, is found in Mongolia's Second National Communication to the UNFCCC.[62] In recent years, the UNFCCC process has made significant progress in providing a mechanism for developing countries such as Mongolia to access climate change funds to implement adaptation and mitigation activities.[63] In this regard, the development of an effective legal, policy and institutional framework would be an essential component of a national strategy to manage the climate change impacts on land based ecosystems, in particular the soil and pastoral ecosystems.[64] Adaptation strategies are based on reducing land degradation, improving livestock management and improving human livelihood. Mitigation actions include policy development, monitoring and reporting, economic assessment and capacity building.[65]

The National Development Strategy of 2008 defines the country's policy up to the year 2021. The strategy calls for the promotion of Mongolia's "capacity to adapt to climate change and desertification, to reduce their negative impacts", with adaptation activities and measures identified. The Second National Communication to the UNFCCC also lists a number of priority adaptation options and measures in different sectors vulnerable to climate change: animal husbandry, arable farming, water resources, human health and forestry. Within these sectors, strategies and measures to combat land degradation and desertification have been identified both in the Second National Communication and in the national programme for climate change.[66] The government has also approved a National Livestock Programme (NLP) which includes directives relevant for climate change adaptation in animal husbandry and the pastoral industry.[67] A study for the National Climate Risk Management Strategy was prepared in 2009 and the NLP seeks to build climate resilience at the community level through reducing risk, and facilitating adaptation.

[61] ADB (2013).
[62] Mongolian Government (2010).
[63] Streck and Guimaraes (2011).
[64] Batima (2006), Hannam (2012).
[65] Wilkes et al. (2011).
[66] Mongolian Government (2010).
[67] Neely et al. (2009), Hannam (2012).

7 Conclusion

One of the greatest challenges Mongolia faces now and into the future is the sustainable management of its soil resources in particular with changing circumstances from climate change, especially if it is to meet the objectives and targets of Sustainable Development Goals 13 and 15. It must ensure that balance is maintained between economic development and environmental protection given the country's high dependence of people's livelihoods on the soil resources, and the vulnerability of the environment to natural and human induced disasters. Engaging and empowering people and local communities through knowledge of the environmental capacity of soil resources is a vital component of this process. Many citizens and government officials, particularly those at local and provincial levels, are under-informed about the rights and obligations that are set down in environmental legislation in general. Raising awareness of the country's legal provisions and the rights and obligations of citizens and the government through the Law on Soil Protection and Prevention of Desertification and other legislation related to protecting soil resources will go a long way towards promoting responsible use of the soil, and ensuring the variety of economic, social, environmental and other soil-related interests are best-served. In particular, the main challenge faced by Mongolia includes: improving the capacity and performance of central government organizations towards the protection of soil; improving the coordination of these organizations and local governments on the use of soil resources; creating enabling conditions for civil society to participate in soil resources governance; strengthening partnerships with non-government organizations and the private sector to promote soil resources awareness, ecological education of the public; and enforcing environment standards for soil conservation and desertification control. The implementation of such measures will greatly assist Mongolia achieve its national obligations to the 2015 Sustainable Development Goals for climate change (SDG 13) and land (SDG 15).

References

Batima P (2006) Climate change vulnerability and adaptation in the livestock sector of Mongolia, Final Report Project AS06. Assessments of impact and adaptation to climate change, International START Secretariat, Washington DC, pp 409

Blattler D (2012) Draft sub-study policy: project linking herders to carbon markets, pp 23

Boer BW, Hannam ID (2015) Developing a global soil regime. Int J Rural Law Policy, Special edition 1, Soil Governance, pp 1–15

Climate Change Secretariat (1997) The Kyoto Protocol to the convention on climate change, Bonn

Crisp N, Dick J, Mullins M (2004) Mongolian forest sector review. World Bank Report, Washington DC

ADB (Asian Development Bank) (2013) Making grasslands sustainable in Mongolia: adapting to climate and environmental change. Asian Development Bank, Mandaluyong City, Philippines, pp 54

Dooley E, Roberts E, Wunder S (2015) Land degradation neutrality under the SDGs: national and international implementation of the land degradation neutral world target. Environ Law Netw Int Rev, 1+2/15:2–9

Erdenebayasgalan G, Sainbayar A (2013) (Mongolia's Ministry of Environment and Green Development) Winter, Alum Newsletter

Fernandez-Gimenez ME (2006) Land use and land tenure in Mongolia: a brief history and current issues. USDA Forest Service Proceedings RMRS-P-39, pp 30–36

Fernandez-Gimenez ME, Kamimura A, Batbuyan Batjav (2008) Final report implementing Mongolia's land law: progress and issues. A Research Project of the Centre for Asian Legal Exchange (CALE), Nagoya University, Japan

Mongolia (Mongolian Government) (2006) Agricultural development strategy paper, Vol. 1

Mongolia (Mongolian Government) (2008) Terminal Report, Sustainable Grassland Management Project, Ulaanbaatar

Hannam ID (2007) Report to United Nations Development Program Mongolia on review of draft pastureland law of Mongolia. UNDP Sustainable Land Management for Combating Desertification in Mongolia, Ulaanbaatar

Hannam ID (2009) Compilation report, environmental law and institutional framework Mongolia. UNDP Project Strengthening Environmental Governance in Mongolia

Hannam ID (2010) Human dimensions of soil and water conservation: Mongolia. In Napier TL (ed) Human dimensions of soil and water conservation. Nova Science Publishers, Inc, pp 341–356

Hannam ID (2012) Working Paper 2, legal and policy framework to support the livestock/grassland NAMA. ADB R-CDTA 7534, Strengthening Carbon Financing for Regional Grassland Management in Northeast ASIA, p 89

Hannam ID (2015) Study on the integration of climate change and biodiversity in land-use planning, country study: Mongolia. IUCN Environmental Law Centre, Bonn, p 42

Hannam ID, Adiysuren Ts, Borjigdkhan (2009) Feasibility study on pasture use fee and its reinvestment into local improvement activities for ensuring sustainability of pastoral resources in Mongolia. UNDP Project Sustainable Land Management for Combating Desertification in Mongolia

Hannam ID, Sereenov M (2015) Report on the preparation of a draft pastoral land law in Mongolia. UNFAO Rome, CST-NRCCD 402/08/2015, p 65

Heiner M et al (2011) Identifying conservation priorities in the face of future development: applying development by design in the grasslands of Mongolia. The Nature Conservancy, p 62

Mongolia Government (2007) Draft pastureland law 20 July 2007 and brief introduction to the draft law on pastureland, p 26

Mongolian Government-UNDP (2008) Assessment of environmental laws: Final Report. UNDP Project Strengthening Environmental Governance in Mongolia, p 136

Mongolian Government (2009) Government policy toward Herders, 4 June 2009, Resolution 39, Ulaanbaatar

Mongolian Government (2010) Mongolian Second National Communication under the United Nations Framework Convention on Climate Change, Ministry of Nature, Environment and Tourism, p 158

Mongolian Government (2015) Intended Nationally Determined Contribution (INDC) Submission by Mongolia to the Ad-Hoc Working Group on the Durban Platform for Enhanced Action (ADP)

MSRM (Mongolian Society for Rangeland Management) (2012) Green Gold Pasture Ecosystem Management Project, Annual Report, 1 January to 31 December 2011

Myers G, Hetz P (2004) Property rights and land privatization: issues for success in Mongolia. SAID/Mongolia, p 22

Narantuya T (2006) Comprehensive analysis of climate change impacts on food security and sustainable livelihood of rural people. In Climate change and sustainable livelihoods of

rural people in Mongolia. MNET of Mongolia and Netherlands Climate Change Studies Assistance Programme, Ulaanbaatar, p 12
Nature-Future Fund Non-Governmental Organization (2008) Team report of analyses on Mongolian environmental policies and fiscal legislations, strengthening environmental governance in Mongolia
Neely C, Bunning S, Wilkes A (eds) (2009) Review for evidence on dryland pastoral systems and climate change, implications and opportunities for mitigation and adaptation. Food and Agriculture Organization of the United Nations, Rome, p 37
Schulze A (2009) Land legislation and the possibilities for pastoral risk management and adaptation to climate change – the example of Mongolia. Environment, Climate Change and Bioenergy Division, Natural Resources Management and Environment Department, FAO, Rome, p 41
SDC (Swiss Agency for Development and Cooperation) (2015) National Report on the Rangeland Health of Mongolia. Ulaanbaatar, Mongolia
Streck C, Guimaraes L (2011) Nationally appropriate mitigation actions in developing countries emerging opportunities for private sector engagement. Climate Focus 3/14, Washington DC
Swift JJ (2007) Case study: institutionalizing pastoral risk management in Mongolia: lessons learned, prepared under the overall guidance from the rural institutions and participation service. FAO Project Pastoral Risk Management Strategy, TCP/MON/0066, p 58
Tortell P, Adiyasuren Ts. Borjigdkhan, Erdenesaikhan Naidansuren (2008) Institutional structures for environmental management in Mongolia. UNDP Project Strengthening Environmental Governance in Mongolia
Tsogtbaatar J (2004) Deforestation and reforestation needs in Mongolia. For Ecol Manag 201(1): 57–63
UNDP (United Nations Development Program) (2005) Economic and ecological vulnerabilities and human security in Mongolia
UNEP (1994) United Nations Convention to Combat Desertification, Nairobi
UNEP (1995) United Nations Framework Convention on Climate Change, Nairobi
Wilkes A, Wang S, Tennigkeit T, Feng J (2011) Agricultural monitoring and evaluation systems: what can we learn for the MRV of agricultural NAMAs? ICRAF Working Paper No. 126, World Agroforestry Centre, Beijing, China
World Bank (2002) Mongolia Poverty Assessment, Poverty Reduction and Economic Management East Asia and the Pacific Region, Report No. 35660-MN

The Protection of Soil Under Namibian Law

Oliver C. Ruppel and Anielle von Finckenstein

Namibia, land of the brave
Beloved land of savannahs
Extract from the Namibian National Anthem

1 Background

1.1 Decolonisation

The inhospitable desert constituted a barrier to European colonisation of south-western Africa until the late eighteenth century when traders and missionaries first explored the area. Soon the territory fell under German administration, henceforth to be known as German South West Africa. German domination ended during World War I in the wake of South Africa's military occupation of the German colony. On 17 December 1920, South Africa took over South West Africa in terms of Article 22 of the 1919 Peace Treaty of Versailles. South Africa was from now on mandated with the power of administration and legislation over the territory.[1]

In 1946, the League of Nations was superseded by the newly formed United Nations. When the United Nations requested South Africa to place the territory under a trusteeship agreement it refused. In 1966 the South African mandate was officially revoked by the UN General Assembly.[2] Political and social unrest within Namibia and the Anti-Apartheid movement outside Namibia increased markedly

[1]Ruppel and Ruppel-Schlichting (2016), p. 1.
[2]For further details see Zaire (2014), p. 37.

O.C. Ruppel (✉)
Development and Rule of Law Programme (DROP), University of Stellenbosch, Stellenbosch, South Africa
e-mail: ruppel@sun.ac.za

A. von Finckenstein
Koep & Partners, Windhoek, Namibia
e-mail: aniellevf@gmail.com

H. Ginzky et al. (eds.), *International Yearbook of Soil Law and Policy 2016*,
International Yearbook of Soil Law and Policy,
DOI 10.1007/978-3-319-42508-5_20

during the 1970s.[3] In 1978, the UN Security Council passed Resolution 435 and authorised the creation of a Transition Assistance Group to monitor the country's transition to independence.[4] After more than a century of foreign domination and a long struggle on both diplomatic and military levels, Namibian independence was achieved and officially declared on 21 March 1990.[5]

1.2 Todays Legal Dispensation

Namibian law reflects the country's history and is the product of different sources: Roman law; the fusion of Roman law and Roman Dutch customary law—hence the term Roman Dutch law—which came in the wake of Dutch colonisation at the Cape of Good Hope; from the early nineteenth century onwards English law asserted itself, leaving deep traces in Roman Dutch law, after British hegemony in southern Africa had been established; and indigenous customary law from time immemorial.[6]

Roman-Dutch law is based on Roman law as it was applied by the courts of Holland and other provinces in the Netherlands; it was developed by writers such as Hugo de Groot and Simon van Leeuwen in the seventeenth and eighteenth centuries.[7] Roman-Dutch law came to the Cape of Good Hope, when the Dutch East India Company under its local governor Jan van Riebeeck established a refreshment post—today's Cape Town in 1652. Roman-Dutch Law in South Africa was subject to further developments under the influence of particularly English law.[8] With the effect of Proclamation 21 of 1919, the Roman-Dutch law developed by the South African courts became the common law of the territory, binding on the Namibian courts until Independence. This position was affirmed by Article 66 (1) of the Namibian Constitution of 1990, which provides that "both the customary law and the common law of Namibia in force on the date of Independence shall remain valid to the extent to which such customary or common law does not conflict with this Constitution or any other statutory law."

With few exceptions German legal influence has disappeared completely. Today the Namibian legal system is an object of fascination for comparative lawyers, legal ethnologists and sociologists. The concept of legal pluralism—a situation in which more than one type of law or legal tradition operates simultaneously—is commonplace in Namibia.[9] The sources of law in which they are usually consulted are

[3]Melber (2011), p. 27 ff.

[4]Amoo and Skeffers (2008), p. 17.

[5]Ruppel and Ruppel-Schlichting (2016), p. 1.

[6]With further references Ruppel and Ruppel-Schlichting (2016), p. 4.

[7]Du Plessis (1999), p. 40.

[8]Du Plessis (1999), p. 49.

[9]Ruppel and Ruppel-Schlichting (2012), pp. 33–64.

statute law or legislation; judgements of the courts; international law (Article 144 of the Constitution); common and customary law (Article 66 of the Constitution) and to some extent legal writing.[10]

1.3 Geographic Indicators, Land and Soil Degradation

Namibia borders on Angola in the north, Zambia and Zimbabwe in the north-east, Botswana in the east, South Africa in the south and the Atlantic Ocean to the west. The latest available census dated 2011 enumerated a population of 2,104,900.[11] The population density lies at 2.5 inhabitants per km^2, which makes it the second least densely populated country in the world; about 42 % of the population live in urban areas.[12] The capital of Namibia is Windhoek, with a population estimated to be 368,000.[13]

Namibia's surface area is 824,268 km^2 with three major categories of land tenure: the so-called commercial farmland with freehold tenure (approximately 44 % of the country situated predominantly in the south and centre of Namibia), communal areas which are situated mainly in contiguous blocks in the northern Namibia (approximately 41 % of the country), and the state land including conservation areas (approximately 15 % of the country).

Namibia is a country with two deserts, the Kalahari and the Namib. Moreover, it is—as also the Namibian National Anthem—rightfully states Namibia a "land of savannahs". In fact, Namibia is one of the driest countries in sub-Saharan Africa with a mean annual rainfall of approximately 270 mm with wide regional and seasonal variation. Against the backdrop of variation in climate and aridity in the country, it is explainable that the vegetation cover in Namibia is generally low. The main groups of soils in the country are unconsolidated sand (arenosols) and shallow and weakly developed soils on bedrock (lithosols, xerosols, regosols and yermosols).[14] Owing to very low contents of clay in the soil, the water holding capacity is generally very low. Nonetheless, Namibia has a broad variety of vegetation types including deserts, savannahs (dwarf shrub savannah, various acacia-based tree and shrub savannah associations and the mopane savannah) and dry woodlands. Moreover, Namibia has an abundant dense and diverse mammalian fauna.

To quite some extent, Namibia faces environmental problems that are similar to those experienced in many parts of Africa, including climate change,[15] water stress, land degradation and soil erosion, and deforestation.

[10] With further references Ruppel (2016a), p. 49.

[11] Cf. Government of the Republic of Namibia (2012b), p. 2.

[12] Government of the Republic of Namibia (2012a, b), pp. 42 and 5.

[13] CIA (2015).

[14] Sweet and Burke (2006).

[15] Niang and Ruppel (2014).

The major driving forces of land degradation in Namibia include: poverty in rural areas; population pressure; land management policies; unsustainable use of water; limited capacity and cross sectoral collaborations to effectively prevent land degradation; limited financial and technical resources and climate change.[16]

Land degradation threatens environmental quality and has a negative economic impact, which makes it one of the primary environmental concerns in Namibia. Farming has deep cultural and social significance and about 70 % of the Namibian population depends on agricultural activities for a livelihood.[17] In this light, the conservation of land by legal means is of critical importance for the country.[18]

The effects of land degradation in Namibia, like elsewhere in the world, manifest in different forms and the causes of land degradation are manifold.[19] Climatic variations, especially the high variability of rainfall patterns, and human activities play a primary role. According to the Namibia Household Income and Expenditure Survey 2009/2010,[20] 23 % of Namibian households depend on subsistence farming as the main source of income. Even though this figure has decreased from 38 % in 1993/1994 and from 29 % in 2003/2004, many Namibians depend—directly or indirectly—more heavily on farming than on any other economic activity.[21] Despite the fact that the whole agriculture and forestry sector, which includes processing, only made up 5.1 % of GDP in 2009,[22] most of the land in Namibia is used for agricultural purposes.[23]

Some of the main causes of soil degradation recognised by policy makers and affected persons are overstocking and overgrazing. Rural areas are particularly affected, as poverty forces people to practise unsustainable environmental management, in order to ensure food supply. It is very common that the densities of livestock exceed the carrying capacity of the land, which inevitably places a severe strain on the environment. Unsustainable harvesting of forest resources, wild plants and game, and the clearing of land for farming or housing purposes are further examples of practices which have an adverse impact on soils in Namibia.[24]

The most vulnerable and poor groups in Namibia are disproportionately affected by the adverse effects of land degradation. Not only does land degradation have negative economic consequences in that it reduces the country's resources, it also poses a serious threat to food security and rural livelihoods. Deforestation, soil erosion, bush encroachment, soil salinisation and the decreased availability of

[16]Government of the Republic of Namibia (2014a).

[17]Government of the Republic of Namibia (2007), p. 1.

[18]Cf. Ruppel et al. (2016) and Hinz and Ruppel (2008).

[19]Klintenberg and Seely (2004).

[20]Namibia Statistics Agency (2012), p. 56.

[21]Iyambo, N, then minister of Agriculture, Water and Forestry in his foreword to Mendelsohn (2006).

[22]HSF (2012), p. 15.

[23]Mendelsohn (2006), p. 10.

[24]Ministry of Environment and Tourism (2006), p. 1.

palatable grass species are some of the most distressing aspects of land degradation.[25] Furthermore, the recent experience of persistent periods of drought has exacerbated these concerns, and swift and decisive action is necessary in order to prevent further and irreparable harm.

The preservation of soil as a natural resource in Namibia is imperative in light of the heavy reliance on agriculture by large numbers of the population.[26] Soil degradation can take place in various ways, such as desertification due to climatic influence, soil contamination due to industrial impacts and unsustainable farming practices.[27] In Namibia, the agricultural use of land, commercially and communally, is widespread. Even though many commercial farmers are able to mitigate negative impacts by means of rotational grazing and continuous monitoring, these opportunities do not generally exist for communal farmers.[28] In light of this background, the importance of soil conservation becomes apparent.

2 Institutional Background on Ministerial Level

The Ministry of Agriculture, Water and Forestry (MAWF) is responsible for soil management and promotion and development of sustainable soil management practices in the agriculture, water and forestry sectors through appropriate policy and legal instruments.[29] Among other activities, the Ministry provides extension services to the agricultural sector in the form of information dissemination, training, communications, and advice. A participatory approach is adopted that places responsibilities of the development process in the hands of rural communities. One recent example of activities directed at promoting sustainable farming practices is the Dry Land Crop Production Programme by the MAWF applied in several regions of the country, which has stimulated increasing food production and security in Namibia by assisting farmers with subsidised farm inputs (fertiliser and improved seeds) and agricultural services (ploughing and weeding services).[30] Ongoing research into adaptable crops is a clear indication of active steps taken to promote sustainable farming practices and to adopt a new approach to the usage of soil.[31] The MAWF has also been working in partnership with international and national institutions such as the FAO, the Namibian Organic Association and the

[25]Klintenberg and Seely (2004), p. 7.

[26]Mendelsohn et al. (2009), p. 10.

[27]Government of the Republic of Namibia (2014a), p. 23.

[28]Government of the Republic of Namibia (2014a), p. 28. Communal farmers are generally restricted by the size of the land allocated to them, as well as with regard to technical knowledge and fiscal means.

[29]Ministry of Agriculture, Water and Forestry (2014), p. 8.

[30]Ministry of Agriculture, Water and Forestry (2014), p. 18.

[31]Ministry of Agriculture, Water and Forestry (2014), p. 41.

French National Institute for Agricultural Research, to continue research concerning sustainable farming in Namibia.[32]

Although the MAWF has the primary responsibility for soil management, it must be emphasised that owing to the varied and patchy nature of the legal and policy framework pertaining to issues relating to soil conservation (as will be outlined below) a multitude of government bodies might be involved when it comes to soil conservation in Namibia and overlaps in terms of competence might occur. Whereas some of the laws and policies are developed and administered by the MAWF, others are initiated by or mandate action from the Ministry of Environment and Tourism, the Ministry of Health and Social Services; the Ministry of Works, Transport and Communication; the Ministry of Mines and Energy; the Ministry of Trade and Industry; or the Ministry of Regional and Local Government and Housing. In other cases, local authorities are responsible for enforcing certain laws.

3 Land Tenure in Namibia

The era of colonial reign over Namibia has skewed land ownership of the country in favour of the white minority. After Namibia acquired independence, the government promulgated several laws aimed to implement a comprehensive plan of land reform.[33] Even though there have been shortcomings with regard to the overall success of land redistribution in Namibia, a comprehensive framework for the reform of land tenure, acquisition and ownership exists. In this respect, policies dictate institutionally, socially and economically sustainable resettlement, in additional to environmentally sustainable practices.[34]

The National Land Policy of 1998 is intended to address the social injustices of the past, and encapsulates constitutional principles. It provides for the establishment of urban areas and promotes community involvement, in addition to tax and other financial incentives intended to promote the protection and rehabilitation of the natural environment. The policy goes so far as to state that the failure to practise sustainable environmental practices may be a ground to deny or terminate a title. This approach is in line with Article 95(1) of the Constitution, which promotes environmental sustainability.

The National Land Tenure Policy applies to all urban, communal, commercial (freehold) and resettlement areas, and is intended to regulate land tenure comprehensively. Natural persons, the state and legal entities can hold land in Namibia. Overall, the state holds all communal land in trust for the indigenous tribes who reside on the land, in addition to owning all nature reserves, game parks, military bases and certain urban properties.[35] The types of land tenure are ownership and

---

[32]See http://www.fao.org/africa/news/detail-news/en/c/296655/. Accessed 20 Nov 2015.

[33]USAid (2010), p. 3.

[34]Government of the Republic of Namibia (2001).

[35]USAid (2014), p. 6.

freehold tenure; communal tenure; conservancies; and leaseholds. Regarding tenure in informal settlements, the practice introduced prior to 1990 of Permissions to Occupy (PTO) still applies but is in the process of being phased out and replaced by leaseholds under the Communal Land Act.[36]

Ownership and freehold tenure gives owners the rights to property as developed in common law. Most commercial agricultural and urban land is privately owned, amounting to 44 % of Namibia's land.[37] The Communal Land Reform Act, in addition to common law, regulates leaseholds. The Communal Land Board can grant communal and commercial land leases for a period of 99 years.[38] Land held in such leaseholds may be transferred, inherited, renewed and mortgaged.[39]

Conservancies, once established, become legal entities and for such purposes require identified boundaries, a constitution and defined membership. In addition, it needs to be demonstrated that there is an ability to manage finances.[40] The Nature Conservation Ordinance of 1975 and its Amendment in 1996 form the legal framework for this type of land tenure.

The occupancy in informal settlements is still in a period of major transition. As stated above, PTOs, as issued during the previous administration, are in the process of being phased out. However, these give the holder the right to apply for ownership or leasehold rights once these become available.[41] The new Flexible Land Tenure Act[42] aims to provide secure tenure to the large part of the population residing in informal settlements and envisages an alternative system to the formalisation of land rights in this context.[43]

Overall, these policies promote sustainable use of all land, and other resources, while recognising the arid climate and persistent drought risks facing Namibia.

4 International Regulatory Framework on Soils Pertinent to Namibia

Even though several international conventions recognise the importance of soil conservation, no overarching and transnational framework exists as yet. One of the reasons advanced by opponents of an overarching, global and binding framework is that soil is non-moving and has *locally unique* problems, which should be dealt with locally.[44]

[36]Act No. 5 of 2000.
[37]USAid (2014), p. 6.
[38]Sections 2 and 3 of Act No. 5 of 2000.
[39]Amoo (2014), p. 234.
[40]USAid (2014), p. 6.
[41]USAid (2014), p. 6.
[42]Act No. 4 of 2012.
[43]USAid (2014), p. 6.
[44]Montanarella (2015).

However, 2015 was the International Year of Soils, which has resulted in a wealth of awareness activities across the globe, in addition to putting soils back on the international policy agenda.[45] This has also led to a new international dialogue concerning the protection and rehabilitation of soils and sustainable farming practices in general.[46] The Global Soil Partnership (GSP) is a body established prior to the International Year of Soils and aided in the implementation and coordination of the roll-out of the year-long activities.[47] The GSP further encourages research, plans conferences and establishes local and regional partnerships.[48] However, criticism has been voiced relating to the felt absence of tangible results and calls for specific actions are mounting.[49]

The European Soil Charter of 1972 is held to have been the first international document relating to soil.[50] The World Soil Charter and the World Soils Policy[51] was negotiated by the United Nations Environment Programme (UNEP) in coordination with the United Nations Food and Agricultural Organization (FAO) and was adopted in 1981. Both instruments contain non-binding guidelines and principles relating to soil conservation[52] and were intended to aid states in formulating domestic policies. However, in light of modern environmental practices, these instruments are considered to be outdated.[53]

New scientific knowledge has been gained over the past 3 decades, "especially with respect to new issues that emerged or were exacerbated during the last decades, like soil pollution and its consequences for the environment, climate change adaptation and mitigation and urban sprawl impacts on soil availability and functions".[54] In this respect, the World Soil Charter has been revised and was unanimously endorsed in June 2015, during the course of the International Year of Soils, by the member states of the FAO during the 39th Session of the FAO Conference.[55] The revised guidelines are intended to ensure that "soils are managed sustainably and that degraded soils are rehabilitated or restored."[56] The actions are targeted at individuals and the organised private sector, governments and

[45]FAO (2015).

[46]FAO (2015).

[47]FAO (2016).

[48]Montanarella (2015).

[49]Montanarella (2015).

[50]Alori and Nwapi (2015), p. 105.

[51]Text available at http://www.unep.org/Documents.multilingual/Default.asp?DocumentID=62&ArticleID=553&l=en. Accessed 16 Nov 2015.

[52]Alori and Nwapi (2015), p. 105.

[53]Alori and Nwapi (2015), p. 106.

[54]Introduction to the 2015 Revised World Soil Charter see http://www.fao.org/fileadmin/user_upload/GSP/docs/ITPS_Pillars/annexVII_WSC.pdf. Accessed 16 Nov 2015.

[55]The revised World Soil Charter is organised into a preamble, nine principles, and guidelines for action.

[56]See Section 3 of the Revised World Soil Charter.

international organisations. Furthermore, 2015 has resulted in a wealth of awareness activities across the globe, in addition to putting the protection of soils back on the international policy agenda.[57] A new international dialogue concerning the protection and rehabilitation of soils and sustainable farming practices in general has been initiated.[58]

Further documents which contain provisions relating to soil management are the World Charter on Nature[59] and Agenda 21.[60] However, these instruments have been criticised to be inappropriate to aid in soil conservation, as their wording is too broad to establish clear norms.[61]

The 1994 United Nations Convention to Combat Desertification in those Countries Experiencing Serious Drought and/or Desertification, Particularly in Africa (UNCCD), which is binding upon Namibia, is the main international legal document to combat desertification and mitigate the effects of drought in countries affected through effective action at all levels supported by international cooperation. This is specifically relevant for soil protection in Namibia. The definition of desertification in this convention relates clearly to soil conservation.[62] The tangible effect of the convention however remains limited, as the focus is primarily placed on capacity building, as opposed to creating binding obligations *per se*.[63]

Subsequently the Namibian Program to Combat Desertification (NAPCOD) has been established.[64] With regard to soil protection, NAPCOD concentrates on education and the creation of awareness surrounding this issue.[65] The Regional Awareness Programme aims to enhance the understanding of desertification, soil erosion, deforestation and related issues by local and traditional decision makers. The spreading of information to communities and the initiation of engagement were recognised as crucial to raising awareness. The primary educational activities were centred at the Gobabeb Training and Research Centre, which included programmes to educate teachers. This was intended to allow them, subsequently, to undertake environmental education. Media awareness workshops are also undertaken by NAPCOD and are intended to raise awareness of the extensive implications arising from desertification.[66] NAPCOD further engages in a variety of other programmes

[57] FAO (2015).

[58] FAO (2015).

[59] 28 October 1982, UN GA, A/RES/37/7; text available at http://www.un.org/documents/ga/res/37/a37r007.htm. Accessed 16 Nov 2015.

[60] United Nations Conference on the Environment and Development, Agenda 21, UN Doc a/CONE151/4 (1992).

[61] Alori and Nwapi (2015), p. 106.

[62] "Land degradation in arid, semi-arid and dry sub-humid areas resulting from various factors including climatic variations and human activities."

[63] Alori and Nwapi (2015), p. 107.

[64] Seely and Montgomery (2009).

[65] Seely and Montgomery (2009), p. viii.

[66] Seely and Montgomery (2009), p. 100.

related to education and, more important, practical implementation of efforts to combat desertification.[67]

The Third National Action Programme for Namibia (NAP3) is the framework intended to aid in the implementation of the UNCCD between 2014 and 2024. The focus is placed on illustrating the present obstacles which Namibia faces with regard to the environment, desertification, land degradation and drought processes, and how these pose an immediate danger to Namibia's land-based agricultural sector.[68] The programme also names several problematic issues which need to be addressed in the context of sustainable land management and farming in Namibia. These include water and soil degradation as well as overstocking and overgrazing of land. The document further lists poverty and population growth, unsustainable resource usage, and severe impact of climate change. The inadequate institutional and individual capacity and weak mechanisms of cross-sector collaboration for sustainable land management are also named in this context.[69] The inadequate application of technology is another large obstacle. NAP3 proposes pragmatic steps which should be taken to address these concerns, such as raising awareness and education; ensuring reliable data are available in order to lay the foundation for new policies to be formulated and the implementation of existing policies; and providing for a functional monitoring system.[70]

The objectives as set out in NAP3 aim to address these obstacles. This is done by setting six specific outcome targets and proposing tangible and pragmatic solutions, in order to achieve the chief objective to "prevent and reverse desertification and land degradation in affected areas and to mitigate the effects of drought in Namibia in support of poverty reduction and environmental sustainability".[71] The need to improve cross-sectoral collaboration between government agencies *inter se* is emphasised,[72] as well collaboration between relevant actors and research institutes. This is aimed to encourage research and to allow data to be used more effectively when developing and implementing policies and programmes.[73] Additionally, the NAP3 deliberates policies and programmes currently in place and puts forward suggestions for improvements where deemed necessary.[74]

Relating to this, are the Sustainable Development Goals (SDGs), which were formulated as a successor to the Millennium Development Goals (MDGs) at the UN

[67]These include workshops for farmers and farm managers relating to the prevention of soil erosion and sustainable farming practices, community projects allowing for increased food and social security standards, and general institution and capacity building.

[68]Government of the Republic of Namibia (2014a), p. 3.

[69]This specifically refers to overlapping and contradictory capacity of ministries and departments with opposing goals.

[70]Government of the Republic of Namibia (2014a), p. 15.

[71]Government of the Republic of Namibia (2014a), p. 13.

[72]Government of the Republic of Namibia (2014a), p. 44.

[73]Government of the Republic of Namibia (2014a), p. 46.

[74]Government of the Republic of Namibia (2014a), p. 31.

Conference on Sustainable Development. The SDGs were recently adopted, and Goal 15 pertains to soil protection. The goal is formulated so as to achieve the "[p] rotection and promotion of sustainable use of terrestrial ecosystems, halt desertification, land degradation and biodiversity loss" and further aims to "achieve a land degradation neutral world". In light of this goal, the Land Degradation Neutrality National Report on Namibia was published.[75] The report *inter alia* states, that the NAP3 already incorporates elements relating to Goal 15 and the subsequent aim to achieve land degradation neutrality.[76]

The Convention on Biological Diversity (CBD) is another relevant international instrument, as the diversity within species and ecosystems is closely linked and reliant upon the conservation of soils and ecosystems, specifically in Namibia. It aims at conserving biological diversity, promoting the sustainable use of its components, and encouraging equitable sharing of the benefits arising out of the utilisation of genetic resources.

Namibia also signed and ratified the 1992 United Nations Framework Convention on Climate Change (UNFCCC), which was adopted to regulate levels of greenhouse gas concentration in the atmosphere, so as to for instance avoid the occurrence of climate change on a level that would compromise initiatives in food production. In this regard the convention is relevant together with a variety of other multilateral environmental agreements (MEAs) and international legal instruments that (directly or indirectly) deal with climate change.[77]

Article 144 of the Namibian Constitution incorporates international law explicitly as law of the land. International law is thus integrated into domestic law.[78] Where possible national authorities and the judiciary in particular can, therefore, apply international law directly on the national level, before cases are taken to regional or international judicial or quasi-judicial bodies. International agreements become Namibian law when they come into force for Namibia. The conclusion of or accession to an international agreement is governed by Articles 32(3)(e), 40 (i) and 63(2)(e) of the Namibian Constitution. It is important to mention, that the Constitution does not require the promulgation of an international agreement in order for it to become part of the law of the land.[79]

[75]Government of the Republic of Namibia (2015a).

[76]Government of the Republic of Namibia (2015a), p. 5.

[77]Cf. Ruppel (2013), p. 29 ff.

[78]Article 144 reads as follows: "Unless otherwise provided by this Constitution or Act of Parliament, the general rules of public international law and international agreements binding upon Namibia under this Constitution shall form part of the law of Namibia."

[79]Ruppel (2016b), p. 55 ff.

5 National Regulatory Framework Relevant for the Protection of Soil

The following section shall provide a broad overview of the most relevant regulatory instruments dealing with the protection of soil. As a starting point it can be stated, that Namibia's legal framework dealing with soil protection is fragmented and sectoral rather than integrated. Nonetheless, Namibia has taken many regulatory steps, in order to comply with its obligations under international law and to ensure the conservation of natural resources by regulatory means.

5.1 The Constitution

According to Article 1(6) of the Namibian Constitution of 1990, the latter is the law above all laws. Therefore, all legislation ought to be consistent with the provisions of the Constitution. The Constitution lays the foundation for all policies and legislation in Namibia and contains three key environmental clauses relevant to sustainable use of natural resources.[80]

Article 100 of the Constitution vests all natural resources in the state, unless otherwise legally owned. Thus, unless legal ownership in a specific locality is proven, such natural resources are owned by the state; the provision implies that natural resources can be legally owned as private property. The land (and the soil on the land) belongs to the state in terms of Article 100 of the Constitution, if not otherwise lawfully owned.

Article 95(l) stipulates that the state shall actively promote and maintain the welfare of the people by adopting policies, which include "the maintenance of ecosystems, essential ecological processes and biological diversity of Namibia and utilisation of living natural resources on a sustainable basis for the benefit of all Namibians (...)". Through this particular Article, Namibia is obliged to protect its environment and to promote a sustainable use of its natural resources.[81]

Article 95(l) compels state organs to be directed by the environmental principle of state policy. Constitutional principles of state policy serve as a stimulus for new initiatives or endeavours—especially where existing policy, law or programmes seem inadequate to attain the principles' objectives. The principles must similarly be employed as direction indicators in setting Government priorities. Also, the judiciary should apply the principles of state policy in constitutional interpretation and use them to fill gaps in the legislative framework when and where necessary. These generic features of constitutional principles of state policy arguably also apply to the environmental principle of state policy in the Constitution of Namibia.

[80]Ruppel (2016b), p. 30.

[81]Ruppel (2016b), p. 30.

Furthermore, Article 91(c) stipulates that one of the functions of the Ombudsman is "the duty to investigate complaints concerning the over utilisation of living natural resources, the irrational exploitation of non-renewable resources, the degradation and destruction of ecosystems and failure to protect the beauty and character of Namibia." The environmental mandate of the Ombudsman is constitutionally entrenched and could also become relevant when it comes to the protection of soil.

5.2 *Policy Framework*

Namibia's national policies on land and agriculture are directed at making sustainable farming the rule in Namibia. A number of policies impact on land and agriculture and thus on the protection of soil in general. These policies include the National Agricultural Policy, the National Drought Policy and Strategy, and the Namibia Forest Development Policy. To ensure environmental protection, these policies promote community-based natural resources management (CBNRM). This means that the role of the Government is limited to regulatory functions and the provision of technical support that will enable farmers to improve their capacity to manage resources more effectively. The Government provides the necessary fiscal and administrative support under these policies, while the farmers do the groundwork of managing their land and agricultural resources. However, issues such as bush encroachment require collaborative effort. Besides the national policies, several strategies and action plans are also relevant for the protection of soil in Namibia.[82]

5.2.1 Vision 2030

Namibia's Vision 2030 was launched in June 2004.[83] The vision's rationale is to provide long-term policy scenarios on the future course of development in the country at different points in time up until the target year of 2030. Vision 2030 regards the sequential 5-year National Development Plans (NDPs) as the main vehicles for achieving its long-term objectives.

Considering constraints to sustainable agriculture in the country, namely low land capability, surface water scarcity, variable availability of grazing, and low

[82]These include inter alia the Third National Action Programme for Namibia to Implement the United Nations Convention to Combat Desertification 2014–2024 (NAP3); the National Biodiversity Strategy and Action Plan (NBSAP) 2013–2022; the National Climate Change Strategy and Action Plan (2013–2020); the Forestry Strategic Plan 1996; the National Drought Policy and Strategy 1997; and the Strategic Action Plan for the Implementation of Renewable Energy Policy 2006.

[83]GRN (2004).

livestock carrying capacity, one of the objectives of Vision 2030 is "[t]o ensure that all Namibians have equitable access to land and other natural resources, and that these resources are sustainably and efficiently used, while maximizing Namibia's comparative advantages."[84]

The successive NDPs contain the goals and intermediate targets (milestones) that will eventually lead to the realisation of Vision 2030. The fourth national development plan (NDP4) spanning the period 2012/13—2016/2017 sets forth three overarching goals: high and sustained economic growth; increased income equality; and employment creation. Four sectors, which enjoy priority status within NDP4 are logistics, tourism, manufacturing, and agriculture. Environment and climate change are recognised under the category of values and principles,[85] which Namibia cherishes as a nation: "Our environment is clean, and we will continue to keep it so. We expect all elements of society, and businesses in particular, to support a precautionary approach to environmental challenges and alterations of the natural world contributing to climate change, undertake initiatives to promote greater environmental responsibility, and encourage the development and diffusion of environment-friendly technologies."[86] Furthermore, NDP4 recognises that "environmental management is both an enabler and driver of economic development. The issue of environmental management is firmly anchored in Namibian laws and policies whose roots are in the Namibian Constitution and has earned the country an excellent reputation internationally for prudent environmental management based on innovative and progressive legislative framework."

5.2.2 Land-Use Planning: Towards Sustainable Development

This policy document drafted by the Ministry of Environment and Tourism in 1994 defines five physiographic land forms: Communal state land; privately-owned commercial farmland; proclaimed state land; urban areas; and wetland systems, including their catchment areas. The policy emphasises sustainability of natural resources, biodiversity and essential ecological processes.

5.2.3 The National Land Policy

The National Land Policy drafted in 1998 is based on constitutional principles and on the national commitment to redress the social and economic injustices inherited

---

[84]GRN (2004), p. 145.

[85]The other values and principles which are contained in NDP4 are: National sovereignty and human dignity; upholding the Constitution and good governance; partnership; capacity enhancement; comparative advantage; people-centred economic development; gender equality and the empowerment of women; sustainable development; and peace and security.

[86]Government of the Republic of Namibia (2012a, b), p. 4.

from Namibia's colonial past. The policy calls for the establishment and proclamation of urban areas as townships and municipalities and strives to promote decentralisation and community involvement. This policy proposes financial and tax incentives for the protection and rehabilitation of natural environments (e.g. planting of indigenous trees and using alternative energy to reduce rates of deforestation and pollution). It states that, in accordance with Article 95(1) of the Constitution, the policy will promote environmentally sustainable land use, and goes further to state that failure to demonstrate environmental sustainability may be grounds for the denying or termination of a title.

One of the aims of this policy is to establish a Land Use and Environmental Board (LUEB) to promote environmental protection and contribute towards coordinated planning and management at national and regional levels. This LUEB shall ensure that environmental protection is promoted in order to guarantee environmental, social and economic sustainability.

5.2.4 The National Resettlement Policy

This policy provides for resettlement, which is institutionally, socially, economically and environmentally sustainable and will enable the beneficiaries to become self-supporting, in accordance with the basic objectives of the Government.

5.2.5 The National Land Tenure Policy

The policy covers all land tenure systems in urban, communal, commercial (freehold) and resettlement areas and is intended to guide all land tenure rights in Namibia. The policy promotes sustainable utilisation of the nation's land and other resources, provides a way to regulate different land tenure right systems, provides secure tenure for informal urban settlers, farm workers and occupiers (those who have been employed less than 10 years on a single farm and do not have secure tenure elsewhere), and provides guidelines on compensation for occupiers of expropriated land. In keeping with the National Agricultural Policy (1995), the policy recognises the environmental limitations of a country as dry as Namibia.

5.2.6 The National Agricultural Policy

The National Agricultural Policy of 1995 endorses the sustainable use of Namibia's land and natural resources,[87] as well as demanding the strict implementation of instruments pertaining to soil erosion, which are widely applicable in the context of sustainable farming. The National Agricultural Policy provides an enabling environment for increased food production by smallholder producers, as a means of improving employment opportunities, incomes, household food security and the

[87] Government of the Republic of Namibia (1995), paragraph 21.

nutritional status of all Namibians. In terms of the National Agricultural Policy, long-term or continuing subsidies will be avoided. However, the policy still allows for the possibility that well-targeted subsidies can play an important part in achieving short-term agricultural and socio-economic objectives. There is an apparent need for a well-formulated policy to provide for the management of the savannahs, whether on commercial or communal land. Such a policy has to create a socio-economic environment that provides incentives for farmers to improve the productivity of their pastures by controlling intruder bush and preventing re-infestation in an environmentally sustainable way.[88] At the same time, improved pasture management practices need to be encouraged to minimise the risks of future land degradation.[89]

The National Agricultural Policy regards land degradation as a serious problem and recognises that water resources in Namibia are limited and that growth within the agricultural sector should not be at the expense of the natural environment. Furthermore, it encourages the use of Environmental Impact Assessments (EIAs) for agricultural projects and proposes a review of legislation related to the use of agrochemicals. The aims of the National Agricultural Policy are largely economic and focus on increasing agricultural productivity and real farm income as a contribution to national and household food security. It recognises the limitations imposed by the Namibian climate and soils and seeks to promote sustainable utilisation of the land and other natural resources within the context of a vulnerable ecosystem. Potential problems such as deforestation, soil erosion, bush encroachment and over-grazing are addressed.

5.2.7 The Green Scheme Policy

The Green Scheme Policy of 2003 (GSP)[90] makes provision for several irrigation projects to be commenced in Namibia. "The Green Scheme is an initiative conducted by the Ministry of Agriculture, Water and Rural Development to encourage the development of irrigation based agronomic production in Namibia with the aim of increasing the contribution of agriculture to the country's Gross Domestic Product and to simultaneously achieve the social development and upliftment of communities located within suitable irrigation areas, but to also promote the human resource and skills development within the irrigation sub-sector to possibly enhance cross-border investment and facilitate the exchange of relevant and limited resources with neighbouring countries in this regard."[91]

[88] Groenewaldt (2008).

[89] Groenewaldt (2008).

[90] Text available at http://www.iwrm-namibia.info.na/downloads/green-scheme-policy---final1.pdf. Accessed 16 Nov 2015.

[91] Green Scheme Policy para. 1.1.5.

The policy emphasises environmental impacts assessment requirements and water pricing methods. However, the implementation has been haphazard and marked by several obstacles. These are most markedly the potential loss of biodiversity if the project is expanded as planned. The GSP has further been criticised for over-emphasising the potential behind irrigation schemes to become the driving force behind agricultural production, in spite of the fact that Namibia is one of the driest countries south of the Sahara.[92]

5.2.8 The National Drought Policy and Strategy

The National Drought Policy and Strategy of 1997 includes provisions intended to reduce the long-term vulnerability to drought, by means of improving soil fertility and moisture retention, which is only possible by means of utilising sustainable farming methods. The National Drought Policy and Strategy shifts the onus of drought management from Government aided relief to appropriate farming techniques aimed at empowering farmers to better cope with droughts themselves. Although incentives such as the Forum for Integrated Resource Management (FIRM) promotes this actively in communal areas that participate in the National Programme to Combat Desertification (NAPCOD)[93] recent responses to crop failures in the north and north east have again reverted to relief programmes. Drought preparedness is one of the important aspects of sustainable resource use and strongly advocated in activities of conservancies elsewhere in the country. However, some measures envisaged in the Drought Policy, such as the subsidy on fodder, have in their implementation contradicted this objective and have led to criticism.[94]

5.2.9 Namibia's National Policy on Climate Change

Namibia's National Policy on Climate Change has been prepared and officially launched by the Ministry of Environment and Tourism in October 2011.[95] The policy seeks to outline a coherent, transparent and inclusive framework on climate risk management in accordance with Namibia's national development agenda, legal framework, and in recognition of environmental constraints and vulnerability. The goal of the Policy is to contribute to the attainment of sustainable development in line with Namibia's Vision 2030 through strengthening of national capacities to reduce climate change risk and build resilience for any climate change shocks. The

[92]See Government of the Republic of Namibia (2005), p. 12.

[93]Bethune (2003).

[94]The fodder subsidy has been criticised for leading to unsustainable farming practices since its inception. See Vigne and Whiteside (1997), p. 51.

[95]See Government of the Republic of Namibia (2011), p. iii.

policy also serves to guide "Government on the development and enactment of climate-specific legislation to establish appropriate legal mechanisms for policy implementation."[96] To date, however, no such climate-change specific legislation is on the national radar.[97] Moreover, Namibia is still to prepare its National Adaptation Plan (NAP) and as such has not yet developed an advanced adaptation strategy and plan.

In September 2015 Namibia submitted its Intended Nationally Determined Contribution (INDC) to the United Nations Framework Convention on Climate Change (UNFCCC). Herein it is stated that the "implementation of this INDC will represent a major challenge to the government of Namibia. Moreover, the loss of soil fertility and increased soil erosion is explicitly mentioned in the INDC as an observed impact in Namibia that will be exacerbated in the future.[98]

5.2.10 The Regional Planning and Development Policy

This policy drafted in 1997 under the supervision of the National Planning Commission acknowledges trends of increasing degradation of pastures, rangelands and woodland and gives attention to soil, water and forest management as development tools. It promotes strategies such as soil conservation and controlled grazing cycles.[99]

5.3 *Statutory Legislation*

A wide number of enactments are pertinent—directly or indirectly—to issues relating to the protection of soils. Environmental framework legislation of cross-sectoral nature such as the Environmental Management Act No. 7 of 2007 are rather broad in scope, while sectoral legislation such as the Soil Conservation Act No. 76 of 1969 and the Agricultural Land Reform Act No. 6 of 1995 are more specific in nature.

5.3.1 The Environmental Management Act No. 7 of 2007

The Environmental Management Act (EMA) is an important tool in terms of environmental protection in general.[100] The Act requires adherence to the principle

[96] See Government of the Republic of Namibia (2011), p. iv.

[97] Ruppel (2016c), p. 360.

[98] Available at http://www4.unfccc.int/submissions/INDC/Published%20Documents/Namibia/1/INDC%20of%20Namibia%20Final%20pdf.pdf.

[99] Government of the Republic of Namibia (1997).

[100] See Ruppel-Schlichting (2016) for further details on the EMA.

of optimal sustainable yield in the exploitation of all natural resources. The Act gives effect to Article 95(l) of the Constitution by establishing general principles for the management of the environment and natural resources. It promotes the coordinated and integrated management of the environment and sets out responsibilities in this regard. Furthermore, it intends to give statutory effect to Namibia's Environmental Assessment Policy; it enables the Minister responsible for the environment to give effect to Namibia's obligations under international environmental conventions; and provides for associated matters. In order to promote the sustainable management of the environment and the use of natural resources, the EMA states its objective in Section 2 and has established a bundle of principles for decision-making on matters affecting the environment in Section 3. The EMA promotes inter-generational equity in the utilisation of all natural resources. Environmental impact assessments and consultations with communities and relevant regional and local authorities are provided for to monitor the development of projects that potentially have an impact on the environment.

According to the EMA, Namibia's cultural and natural heritage is required to be protected and respected for the benefit of present and future generations. The Act provides for a Sustainable Development Advisory Council to be established[101] to advise the Minister on issues that promote cooperation and coordination between organs of state, non-governmental organisations, community based organisations, the private sector and funding agencies, on environmental issues relating to sustainable development. An Environmental Commissioner advises Government bodies on the preparation of environment plans, receives and records all applications for environmental clearance certificates, determines whether a particular listed activity requires an environmental assessment, reviews environmental assessment reports, issues environmental clearance certificates and conducts inspections to monitor compliance with the EMA.

The EMA does not include any provisions relating explicitly to soil protection and management. However, many provisions of the EMA aiming at environmental protection in general will also be relevant for the protection of soils. The definition of environment in Section 1 is very broad and includes references to "land" and "all organic and inorganic materials". Of particular importance is Part VII relating to environmental assessment, which lists activities requiring an environmental clearance certificate to be issued before such activities can be undertaken. Listed activities include land use and transformation; resource renewal; agricultural processes; waste and sewage disposal; as well as any other aspects which the minister considers necessary to list. These activities often relate to soil management and protection, and the requirement of environmental clearance certificates provides a valuable protection mechanism in this context.

[101] In February 2012, the Government of Namibia gazetted the Regulation for the implementation of Environmental Management Act No. 7 of 2007. Subsequently, the Ministry of Environment and Tourism invited nominations for appropriate persons from the public, organisations, associations or institutions to sit on the Sustainable Development Advisory Council. The first Sustainable Development Advisory Council has been inaugurated in January 2013.

Section 48 in Part IX additionally allows for the minister to introduce legislation or make such regulations which give effect to international agreements to which Namibia is a signatory and which are binding on Namibia. The provision lists areas which can be covered by such legislation and regulations. However, to date, no regulations relating specifically to soil and related activities or agreements have been promulgated.

5.3.2 The Agricultural (Commercial) Land Reform Act No. 6 of 1995

Preservation of arable land in Namibia is imperative, as roughly 78 % of the country is being used for farming purposes and supports more than 1.2 million people living on and off such land.[102] It must be kept in mind that, even though commercial farming, most often focusing on cattle farming, contributes 7.7 % to the GDP,[103] farming practices utilised on communal land have a severe impact on the quality of soil and other resources. In order to fully develop sustainable farming practices, communal farming methods must also be addressed sufficiently and sustainable farming practices need to be propagated by means of educational awareness programmes. Furthermore, appropriate and continuous support must be given to 'emerging' commercial farmers who have received land under the policy of land redistribution in Namibia.

Approximately 36.2 million hectares, which amount to 44 % of the total land area or 52 % of agriculturally utilisable land, is held under freehold title. This land is commonly referred to as the commercial farming sector and it is regulated mainly by the Agricultural (Commercial) Land Reform Act of 1995. This Act was promulgated to provide for the acquisition of agricultural land by the state for the purposes of land reform and for the allocation of such land to Namibian citizens who do not own or otherwise have the use of any or adequate agricultural land, and foremost to those Namibian citizens who have been socially, economically or educationally disadvantaged by past discriminatory laws or practices. The Act vests in the state a preferential right to purchase agricultural land and it empowers the state to compulsorily acquire certain agricultural land for the purposes of land reform. It also regulates the acquisition of agricultural land by foreign nationals and establishes a Lands Tribunal to adjudicate disputes that may arise in land matters. The Act is thus important for the issue of land distribution and the use of agricultural land in the country in general rather than for the protection of soils.

[102] Mendelsohn (2006), p. 10.

[103] See Index Mundi (2015a).

5.3.3 The Communal Land Reform Act No. 5 of 2002

The Communal Land Reform Act provides for the allotment and administration of all communal land in the areas described in the first schedule to this Act or in any other area declared to be communal land under Section 16(1)(a). An obligation is placed upon the minister to establish Communal Land Boards to execute the functions conferred on such a board by the Act within the area for which each board is established. The boards are intended to control the allocation and the cancellation of customary land rights by chiefs or traditional authorities. The boards consider applications for the right of leasehold, establish and maintain a register and a system of registration of customary land rights and leasehold rights, and give advice to the minister.

Provision is made for the prevention of land degradation as the Act provides that the minister may make regulations with regard to "the combating and prevention of soil erosion, the protection of the pastoral resources and the limitation and control of the grazing of stock".[104] Rights are afforded to communal farmers and traditional authorities and representation on Communal Land Boards is provided for in the Act. Of note is the provision of Communal Land Boards, with representation of officials from the Ministry of Environment and Tourism and the Ministry of Agriculture, Water and Forestry, as well as representatives from any of the conservancies.

Communal areas may be declared by the president of Namibia if they are non-alienated state land. Communal areas are vested in the state, in trust, for the benefit of the traditional communities residing in those areas and for the purpose of promoting the economic and social development of the people of Namibia, especially the landless and those with insufficient access to land. Customary land rights are to be allocated upon application for a limited period. Only specific customary land rights may be allocated in respect of communal land, and size limits are imposed.

Recognition of customary land rights is further provided for, in addition to the granting of a right of leasehold for agricultural purposes or a right of grazing on communal land. The Act makes provision for the prevention of land degradation and, therefore, indirectly contributes to the conservation of biological diversity. Fundamental environmental provisions of the Act refer to the allocation of customary land rights. If a land right is being used predominantly for a purpose not recognised under customary law, customary land rights may be cancelled according to Section 27 of the Act.

Special provisions are made with regard to grazing rights. A chief or traditional authority is vested with the power to prescribe conditions relating to the kind and number of stock that may be grazed on communal land, as well as to the section or sections of the commonage where stock may be grazed, and the grazing in rotation

[104] See Section 45(1)(i).

on different sections. The latter provision, in particular, ensures the sustainable use of grasses and herbs.

Section 45 of the Act addresses issues pertinent to the conservation and sustainable management of certain natural resources. The minister may make regulations in relation to watercourses, woods and the use of water (Section 45(g)) and to the combating and prevention of soil erosion, the protection of the pastoral resources and the limitation and control of the grazing of stock.

In 2007, a process of reviewing, updating and consolidating the Agricultural (Commercial) Land Reform Act, and the Communal Land Reform Act, into one Land Act was initiated. Although this process had been finalised in cooperation with relevant stakeholders, a bill emanating from the process has not been tabled in parliament.

5.3.4 The Soil Conservation Act No. 76 of 1969

The Soil Conservation Act of 1969 remains applicable in Namibia, and is specifically referred to in the Communal Land Reform Act No. 5 of 2002. The Soil Conservation Act gives wide ranging powers to the minister, which include powers to issue directives relating to the cultivation of land,[105] the management of water and drainage,[106] as well as to the protection and stabilisation of soil surfaces.[107]

Provision is made by the Act for the prevention and control of soil erosion and the protection, improvement and conservation of soil, vegetation and water supply sources and resources. Although the jurisdiction of the original Act was limited to commercial land, the recent Communal Land Reform Act of 2002 specifically requires compliance in terms of conservation and prevention of soil erosion (Clause 31), implying that these measures apply to communal land areas too.

The Act furthermore provides the Minster with discretion regarding the construction and maintenance of soil conservation works.[108] The costs of such construction may be allocated to the state or to owners of such land.[109] Furthermore, the Act empowers the minister to carry out soil conservation for the purposes of research or demonstrations,[110] subject to the landowner's consent.

The Soil Conservation Committees as provided for in Part III of the Act are generally appointed for certain areas, and act in an advisory capacity to the minister.[111] However, no such committee is currently operational.

[105] Section 3(1)(a).

[106] Section 3(1)(c), (d), (f).

[107] Section 3(1)(e), (g) and (h).

[108] Part II of the Act.

[109] Section 7(2).

[110] Section 8.

[111] Section 9(1) and Section 10.

A wide power to expropriate land is given to the minster in terms of Section 18. It is stated that expropriation may be required for the prevention of soil erosion and stabilisation of land, as well as prevention of drift sand and protection of catchment areas.[112]

One of the enforcement mechanisms for compliance is set out in Section 21, which provides for penalties in the instance of non-compliance. This conduct is further criminalised as an offence which can be punishable by a fine or imprisonment.[113]

One of the biggest obstacles which hinder effective soil conservation in Namibia is the fragmentation of responsibilities relating to soil. Even though the Soil Conservation Act remains in force in Namibia, its implementation is haphazard, with Soil Boards not being operational, and duties having been delegated across directorates and ministries.[114]

5.3.5 The Minerals (Prospecting and Mining) Act No. 33 of 1992

The Minerals (Prospecting and Mining) Act No. 33 of 1992 Act makes provisions for various types of licences relating to mining operations such as exclusive and non-exclusive prospecting licences, mining claims, mineral licences, reconnaissance licences, mineral deposit retention licences, and mining licences.[115]

The Act contains some relevant provisions for pollution control related to mining activities in the country. The Act not only provides that a holder of a licence claim must take all reasonable steps necessary to prevent or minimize any pollution of the environment,[116] it also requires that the holder of a mineral licence must prepare an environmental impact assessment indicating the extent of any pollution of the environment before any mining activities can be carried out as well as an estimate of any pollution likely to be caused by the respective mining activity.[117] The application for a mining licence must inter alia contain particulars "of the manner in which it is intended to prevent pollution, to deal with any waste, to safe guard the mineral resources, to reclaim and rehabilitate land disturbed by way of the prospecting operations and mining operations and to minimize the effect of such operations on land adjoining the mining area."[118] Under specific circumstances, the Minister has the power to reserve land from mining operations. Should the Minister for instance consider it necessary in terms of the protection of the environment or natural resources, he or she may declare that a mining activity may only be carried

[112]Section 18(1).

[113]Section 21(1).

[114]See Legal Assistance Centre (2009).

[115]For more details see van den Berg and Koep (2016).

[116]Section 41(1)(e).

[117]Section 50(f)(i).

[118]Section 91(f)(iii).

on with the special permission of the Minister.[119] The Act furthermore provides that mineral licence holders are liable for any damage to land, water, plant or animal life caused by spilling or pollution and must take all such steps as may be necessary to remedy such spilling, pollution, loss or damage at his or her own costs.[120]

5.3.6 The Forest Act No. 12 of 2001

The Forest Act No. 12 of 2001 establishes a regime for authorisation of the harvesting of trees[121] in order to combat deforestation and thereby prevent the exacerbation of related natural hazards. The Act is formulated around the tenets of sustainable management of forests, and the purpose for which forest resources are managed and developed. This also includes the planting of trees where necessary, as well as soil conservation, the safekeeping of water resources, the maintenance of biological diversity, and the use of "forest produce in a way which is compatible with the forest's primary role as the protector and enhancer of the natural environment".[122]

Since well managed forest ecosystems have the potential to contribute to the protection of soil, the Forest Act No. 12 of 2001 states in its Section 10 that "[T]he purpose for which forest resources are managed and developed, including the planting of trees where necessary, in Namibia is to conserve soil and water resources, maintain biological diversity and to use forest produce in a way which is compatible with the forest's primary role as the protector and enhancer of the natural environment." If it is necessary to protect the soil, water resources, protected plants and other elements of biological diversity, protected areas can be established according to Section 21 and regulations can be made relating to the protection of soil (Section 48).

5.3.7 The Agricultural Pests Act No. 3 of 1973

The Agricultural Pests Act No. 3 of 1973 pertains to the registration of nurseries; the control and eradication of plants, insects and diseases at nurseries; the control and eradication of exotic (vertebrate) animals (excluding farm animals) and plants infected by insects or plant diseases; control of plant, insect and plant disease imports; honey bees, honey and exotic animals; and the eradication of plant diseases, insects and locusts. The Act also defines the powers of inspectors. It is principally intended to prevent the introduction and spreading of plants, insects, non-farming exotic vertebrates and diseases that may prove harmful to the

[119] Section 122(2)(b).

[120] Section 130.

[121] Section 27 of the Forest Act No. 12 of 2001.

[122] Section 10 of the Forest Act No. 12 of 2001.

agricultural sector. Provisions regarding the destruction of exotic animals as well as any plants infected by insects or diseases are provided in Section 9. Section 11 outlines the regulations concerning plant and exotic animal imports, and the prohibition on the importation of plants, insects, plant diseases, honey bees, honey, beeswax and exotic vertebrates without permits, while Section 12 contains the regulations allowing the importation of biological control agents needed for the control or eradication of weeds and pests. This Act may require amendment to ensure that it remains effective by, for instance, incorporating a wider spectrum of alien invasive species and ensuring that fuller use is made of the existing measures of inspection and enforcement administered jointly by Customs and Excise and the Phytosanitary Section in the Ministry of Agriculture. This Act will has been repealed by the new Plant Quarantine Act,[123] although any permits issued under Section 11(1) that are in force at the commencement of the new Act will remain valid and be deemed to be permits as specified in Section 4(1).

5.3.8 The Plant Quarantine Act No. 7 of 2008

In the first place, the Pland Quarantine Act No. 7 of 2008 aims to prevent, monitor, control and eradicate plant pests and is concerned with the facilitation of the movement of plants, plant products and other regulated articles within and into or out of Namibia. The Act furthermore provides for the certification of the phytosanitary standards of plants and plant products exported from Namibia. The Act, which came into force in 2012 contains several provisions relating to soil. Interestingly, it contains a definition on soil stating that this term "means material wholly or partly derived from the upper layer of the earth's crust which is capable of sustaining plant life and which contains solid organic substances such as parts of plants, humus, peat, or bark, but excluding any medium which is sterile, composed entirely of unused peat, or is otherwise incapable of harbouring or transmitting plant pests."[124] According to Section 4 of the Act, soil may not be imported into the country unless a specific permit is issued or exemption granted.

6 The Customary Law of Traditional Communities

Article 66 of the Namibian Constitution lays the foundation for the constitutional recognition of customary law. It states that both the customary law and the common law of Namibia in force on the date of Independence shall remain valid to the extent that such customary or common law does not conflict with the Constitution or any

[123] Act No. 7 of 2008. The Act came into force on 1 July 2012 as per Government Notice No. 157 in Government Gazette No. 4975 of 2012.

[124] See Section 1.

other statutory law. Section 3 of the Traditional Authorities Act[125] gives certain powers, duties and functions to traditional authorities and members thereof. It is the overall responsibility of traditional authorities to supervise and ensure the observance of the customary law of that community by its members. As to nature conservation it is one of the duties of a traditional authority to ensure that members of the traditional community use the natural resources at their disposal on a sustainable basis and in a manner that keeps the environment and maintains the ecosystem for the benefit of all Namibians. Customary law plays an important role in the sustainable development of natural resources and the protection of biological diversity as it incorporates a broad knowledge of ecosystems relationships.[126] Many of the customary laws contain specific provisions for environmental protection. With a particular focus on land degradation, the laws of the Mashi for example recognise that "the increase in the cattle population has led to overgrazing and to erosion, which also increases where trees are chopped down for firewood and winter grass is burnt off to provide new grass, which is then overgrazed."[127] The Chief or Traditional Authority can thus prescribe conditions under which communal land may be used, such as the kind and number of stock that are permitted to be grazed.

7 Selected Hotspots

Namibia has promulgated numerous statutes and policies addressing the management of land resources and designed to prevent soil degradation. These statutes are deemed to become more and more subject in law suits, filed by affected industry, state and local governments, indigenous groups, conservation groups and private citizens. Although soil degradation is one of the major environmental concerns in Namibia, so far no court case from Namibia's superior courts has been reported addressing legal provisions on soil degradation or the judicial review of administrative decisions dealing with questions of soil conservation.

However, many projects and programmes by governmental and non-governmental organisations are concerned with issues circulating around the preservation and conservation of soil with different approaches. A small selection of these programmes will be examplarily sketched in the following.

[125] Act No. 25 of 2000.

[126] Hinz and Ruppel (2008), p. 57.

[127] See the Laws of the Mashi in Hinz (2010), p. 463.

7.1 *Mining and Soil*

Mining accounts for roughly 11.5 % of Namibia's gross domestic product (GDP), but provides in excess of 50 % of foreign exchange earnings.[128] Furthermore, the products of mining and quarrying account for 47 % of Namibia's exports. The importance of mining as an industry in Namibia is evidenced by the extensive reach of mining permits across the Namibian landscape, in addition to the large contribution to employment figures.

As a consequence, negative environmental impacts have become more evident recently. This is especially true from the viewpoint of soil protection, since uranium production, one of the largest mining sectors, has an extremely high adverse impact upon the environment. Namibia is the fourth largest uranium producer in the world.[129] Furthermore, Namibia's exports in land-mined diamonds and other gemstones, in addition to an increase in offshore diamond mining, have further contributed to increased environmental concern.[130]

Most of the policies concerning soil relate to the prevention of contamination and the rehabilitation of mined areas after the mining activities have come to an end. The provisions relating to the prevention of soil contamination are generally aimed at both soil and water contamination prevention measures. However, the protection of water resources has a more pronounced place within the policies.

The Annual Report of the Ministry of Mines and Energy for the 2013–2014 period illustrates programmes of the Ministry relating to a sharp increase in geological data mapping. These programmes were intended to improve the quality and availability of geological maps for the benefit of research institutions, mineral exploration companies and professionals.[131] Furthermore, these mapping programmes were a joint venture between the Geological Survey of Namibia, the Council for Geosciences of South Africa and the Ministry.

Additionally, in 2013 the Ministry, in cooperation with the Department of Mining and Metallurgical Engineering of the Polytechnic of Namibia, undertook a soil contamination mapping survey.[132] This survey was limited to the area of Berg Aukas, where an abandoned mine had led to disturbing levels of, among others, arsenic, copper, lead, zinc in the ground. An additional field assessment survey of the central Namib area was also completed in order to determine the impact of abandoned mines in the Erongo area. Overall these samples are intended to form the basis of determination of mining environmental liabilities, classification of risk levels and future action.[133]

---

[128] See Index Mundi (2015b).

[129] Index Mundi (2015b).

[130] Legal Assistance Centre (2009), p. vi.

[131] Government of the Republic of Namibia (2014b), p. 15.

[132] Government of the Republic of Namibia (2014b), p. 25.

[133] Government of the Republic of Namibia (2014b), p. 25.

The Environmental Management Act No. 7 of 2007 and the Minerals (Prospecting and Mining) Act No. 33 of 1992 are the most important statutes dealing with mining in Namibia. The provisions relating to environmental impact assessments, certificates and plans are most relevant in the mining context, as a mineral licence may not be issued unless an environmental clearance certificate has been furnished. In order to obtain such a clearance certificate, an application has to be lodged with an environmental commissioner and it is at the discretion of the environmental commissioner that a certificate is issued or that an environmental impact assessment is requested.

Mineral licences may require an applicant to provide information regarding company ownership and location, as well as location and geological information of the desired claim area in addition to information regarding existing environmental conditions.[134] The applicant may also be required to provide information regarding the potential environmental degradation resulting from the prospecting or potential mining activities. The Environmental Assessment Policy of 1994 provides the guidelines for such information. One of the challenges in this regard is that the policy is non-binding and the establishment of institutional infrastructure is still in its fledgling phase. This infrastructure and respective enforcement mechanisms are provided for in the Environmental Management Act and relevant regulations.[135] An environmental impact assessment may need to be completed in order to obtain a mineral licence. Furthermore, if the mining operations are intended to be larger, it is common for the minister to require guarantees that could be used to remedy damage caused by these mining activities. These would be additional to the requirement that the holder of a mining licence must take all steps to remedy damage caused by the mining operations, to the satisfaction of the minister.[136]

Even though many policies concerning the mitigation and monitoring of environmental impacts of mining exist, the focus is generally not on soil protection. Rather, other resources potentially impacted, such as water and air, are emphasised and regulated in the current framework. It is recommended that new and existing policies be devised and updated to include soil protection and management, in line with a growing international trend. Even though soil may have less popular appeal than other environmental resources, it is one of the largest and most important areas of the environment, and its contamination will have severe and long-term impacts upon biodiversity and food security.

Abandoned mine sites occupy significant tracts of land in Namibia. The subsequent soil contamination, and its influence on continued use of the land for residential and agricultural purposes, is clearly illustrated by a study on soil surrounding the abandoned ore-processing site at Berg Aukas.[137] The site at

(continued)

[134]Legal Assistance Centre (2009), p. 6.

[135]See Ruppel-Schlichting (2016) and van den Berg and Koep (2016).

[136]Section 130 of the Minerals (Prospecting and Mining) Act of 1992.

[137]The case study will form the basis for the discussion in this section. See Mapani et al. (2009).

Berg Aukas served as a mine and roasting site until 1979.[138] At the time of the study, the National Youth Service was using the area for residential purposes as part of an agricultural vocational school, and some of the area was being used for agricultural purposes.[139]

Analysis of the water and soil was undertaken during the course of the study. The results of the water analysis overall showed no material health risk, except at three boreholes.[140] Overall, the groundwater quality had not been impacted significantly and the only contamination detected was in water which was pumped directly from the old mine shafts.

However, the results of the soil analysis showed severe adverse impacts. The soil sampling was undertaken in a comprehensive manner,[141] which ensured the reliability of the results. In the areas of former mining and processing, high levels of arsenic were detected.[142] Elevated concentrations of cadmium encircled the whole area of Berg Aukas, and it was concluded that these had been caused by emissions from the roasting of ores. Extreme contamination of soil by lead could be found in selected areas, such as in an area of 800 m × 600 m located in the central and southwestern part of Berg Aukas.[143] This contaminated soil is where the central part of the National Youth Service was located. According to German Guideline Values used in the study, these areas are unsuitable for residential purposes as such lead contamination is hazardous to human health. This contamination was possibly caused by tailing material having been spilled and such material, in addition to slag, having been used for road construction.[144] The agricultural produce grown at the National Youth Service site was also tested, and the values found exceeded the World Health Organization (WHO) limits for food.[145] Slightly elevated levels of mercury were also found.[146] However, the most severe contamination is attributed to zinc contamination originating from the former smelting area.[147]

The study overall found that severe contamination occurred in the sites previously used for processing of the ores, including smelting and roasting

(continued)

[138] Mapani et al. (2009), p. 25. The ore contained lead, vanadium and zinc.

[139] Mapani et al. (2009), p. 26. The agricultural land is used as an experimental crop farm.

[140] Mapani et al. (2009), p. 27.

[141] Mapani et al. (2009), p. 28.

[142] 29. >2.6 ppm with a median content of 4.99 ppm in the topsoil.

[143] Mapani et al. (2009), p. 30. The concentrations varied from 10,000 ppm, with a further 1 km halo surrounding this area in which concentrations between 400 and 10,000 ppm were measured.

[144] Mapani et al. (2009), p. 32.

[145] Mapani et al. (2009), p. 33. The values of arsenic, lead and zinc were primarily evaluated.

[146] With a maximum concentration of 9.6 ppm.

[147] Mapani et al. (2009), p. 32. The values vary between 107,000 and 377,000 ppm.

processes. The contaminants were spread further by wind erosion and other natural processes, which led to a larger contaminated area around Berg Aukas.[148] It was concluded that persons living and working in this area face health risks from being continuously exposed to the contaminated area.[149] The study classifies the severely contaminated area as a high hazardous risk zone, and recommends that the site should not be used for further industrial, residential or agricultural developments.

Remediation possibilities spelt out include rehabilitation of the soils by means of covering top soils with organic matter, the removal and proper disposal of contaminated soils, and reprocessing of soil where the contamination penetrated deeply into the soil horizon.[150] In light of this study, the area was also evacuated by the Ministry of Mines and Energy and people were relocated to a safer area. The Ministry is further tasked with ensuring remedial action is monitored.

In 2014, a preliminary feasibility study was published by China Africa Resources, for the Berg Aukas Project.[151] Apart from extensive research into the financial and geological viability of reopening the mine, the study—as discussed above—was also addressed briefly.[152] It was observed that no remediation or re-arrangement for land usage was undertaken after the National Youth Service had been evacuated. However, in compliance with the legal framework regarding mines, environmental impact assessments and environmental management plans will have to be formulated if this mine is to become operational.[153] Currently, the final feasibility study is in the process of being initiated.[154]

7.2 Agriculture and Soil

An example of the impact of agricultural activities on land is contained in the research outputs of a project referred to as The Future Okavango Project (TFO). The TFO is a trans-disciplinary research project, focusing upon the issue of sustainable land management as a whole.[155] The project is aimed at providing

[148] Mapani et al. (2009), p. 34.

[149] The health risks arise from inhalation and ingestion of dust and crops grown in the area.

[150] Mapani et al. (2009), p. 35.

[151] China Africa Resources (2014).

[152] China Africa Resources (2014), p. 79.

[153] China Africa Resources (2014), p. 80.

[154] Unsted (2015).

[155] Pröpper et al. (2015), p. 5.

scientific knowledge and data upon which recommendations and future policy can be formulated, in the context of sustainable development. While the TFO spans a wide variety of data, this section will focus upon the impacts of agricultural activity on soils, more specifically small scale subsistence farming.

Irrigated industrial agriculture is implemented in approximately 2 200 ha of the Namibian Kavangos,[156] whereas semi-permanent smallholder agriculture is prominent in the more densely populated areas of the Kavangos.[157] The core site of Mashare was analysed and provided data for TFO and this case study. The smallholder agriculture, which is the focus of this case study, is mainly conducted on communal land, which is regulated by the Communal Land Reform Act of 2002. Even though this Act is intended to decelerate deforestation and protect arable land by means of Land Boards who supervise such communal lands, the implementation has left much to be desired.[158]

National projects, such as "green schemes", are also being implemented in the Kavangos. However, subsistence farming on communal land provides the livelihood of a large part of the population in Mashare. Unsustainable land practices are common, which have affected the quality of soil negatively. These unsustainable land practices include slash and burn agriculture, deforestation, overgrazing and irregular ploughing and harvesting.

Many subsistence farmers do not own oxen and other instruments necessary to plough fields, and are thus forced to lease these when the time to service the fields has come.[159] However, the time period after the rains, in which the ploughing and servicing of the fields is necessary, is very short. It is quite common for subsistence farmers to be unable to arrange for the instruments at such times, which means that many fields go un-serviced.[160]

Irregular servicing of arable land, can lead to soil degradation, especially if such land was previously forest area, which was converted into crop-land. The loss of biodiversity and subsequent change in the chemical composition of the soil are such negative consequences, which are aggravated if the land is not serviced and if no soil conservation measures are taken.

Due to socio-economic and environmental pressures, such as poverty, low levels of education, erratic rainfall and seasonal flooding, small-scale subsistence farming often lacks any input-based soil conservation measures, which further increases soil degradation and negatively impacts the productivity and availability of the arable communal land.[161] The naturally occurring soil in this area has a sandy and humus-poor character, which adversely affects natural soil fertility and is exacerbated by the above factors.[162]

---

[156]The Kavangos collectively refers to East and West Kavango regions of northern Namibia.

[157]Pröpper et al. (2015), p. 32.

[158]Pröpper et al. (2015), p. 80.

[159]Pröpper et al. (2015), p. 43.

[160]Pröpper et al. (2015), p. 43.

[161]Pröpper et al. (2015), pp. 80 and 92.

[162]Pröpper et al. (2015), p. 85.

Slow implementation of national policies relating to soil conservation is but one of the factors affecting soil protection in Namibia. Unsustainable agricultural practices, exacerbated by a lack of knowledge and conservation efforts are further impediments to achieve soil conservation and soil degradation neutrality. Impoverished communities, reliant upon subsistence farming on communal lands very often do not have the skill set, nor knowledge to practise sustainable land use and implement soil conservation measures, as evidenced by the Mashare case study.[163] TFO recommends better implementation of direct and indirect support through empowerment, education and access to small-scale financial credits. Even though certain measures taken within the framework of TFO have aided in promoting sustainable land use, a continued support structure is necessary to ensure long term and sustainable results.[164]

7.3 *Climate Change and Soil*

Namibia is considered to be highly vulnerable to the predicted effects of climate change.[165] It is one of the biggest and driest countries in sub-Saharan Africa with characteristic high climatic variability in the form of persistent droughts, unpredictable and variable rainfall patterns, high temperature variability and scarcity of water.[166]

The agricultural sector is particularly vulnerable and climate change effects on agriculture are already being observed.[167] Water scarcity related to the effects of climate change poses negative consequences for agriculture, and is likely to also increase the rate of soil erosion, affecting crop production and livestock. "Livestock and crop production under rainfed conditions has been declining by about 33 % on average every year in the last few farming seasons. The year 2013 recorded the worst drought in 30 years and the GDP contribution by agriculture and its sub-sectors fell significantly with livestock farming recording the highest decline of 37.6 %."[168]

Based on the predictions of the Fifth Assessment Report (AR5) of the Intergovernmental Panel on Climate Change (IPCC),[169] that with high confidence, climate change will amplify existing stress on water availability in Africa and that climate change will interact with non-climate drivers and stressors to exacerbate

[163]Pröpper et al. (2015), p. 95.

[164]Some TFO findings with legal relevance are captured in Hinz et al. (2012).

[165]Mapaure (2016), p. 339.

[166]Available at http://www4.unfccc.int/submissions/INDC/Published%20Documents/Namibia/1/INDC%20of%20Namibia%20Final%20pdf.pdf.

[167]Government of the Republic of Namibia (2015b), p. 22.

[168]Government of the Republic of Namibia (2015b), p. 15.

[169]Reports online available at https://www.ipcc.ch/report/ar5/.

vulnerability of agricultural systems,[170] land degradation, soil erosion, bush encroachment and deforestation will continue to be serious challenges for Namibia and extreme weather events, especially severe droughts and floods will worsen soil erosion and destroy crops.

"Conservation agriculture provides a viable means for strengthening resilience in agroecosystems and livelihoods that also advance adaptation goals."[171] One example for projects on climate change and soils out of many more is a project on adaptation of agriculture to climate change in Northern Namibia, which has recently been launched to assist small-scale farmers in Northern Namibia to apply successfully farming practices that are adapted to the impacts of climate change.[172] Based on the basic principles of conservation agriculture that soil is not tilled; crop rotation is practised using a variety of crops; and that the soil is always covered with vegetation or plant residues, the MAWF formulated a comprehensive programme for widespread introduction of conservation agriculture in 2015. This Comprehensive Conservation Agriculture Programme for Namibia aims to counter and reverse land degradation and to adapt to climate change through the adoption of conservation agriculture as a basis for sustainable crop production and improved food security.

Options for indigenous peoples to enjoy land rights through an innovative approach of integrating conservation and development, known as community based natural resource management (CBNRM) has been investigated through the lens of climate change.[173] In this light it is important to acknowledge that indigenous peoples are among the groups that are most vulnerable to actual and potential detrimental impacts of climate change. They live in the most vulnerable ecosystems so are often the first groups to be impacted by climate change.[174]

In its Intended Nationally Determined Contribution (INDC) to the United Nations Framework Convention on Climate Change (UNFCCC) Namibia is well aware that "climate change is a reality in the everyday life of all Namibians.[175] Being complacent will only aggravate the situation, as climate change is here to stay and all will have to live with it in the longer term."[176] Broad avenues for adaptation will (inter alia) need to come from agricultural adaptation strategies including coordinating the timing of ploughing and crop planting with rainfall events; using drought-resistant crop varieties and livestock breeds; shifting livestock to alternative grazing areas and; improving ecosystem management, protection and conservation.[177]

---

[170]Niang and Ruppel (2014), p. 1202.

[171]Niang and Ruppel (2014), p. 1203.

[172]For further details see Deutsche Gesellschaft für Internationale Zusammenarbeit (GIZ) (2016).

[173]LAC (2013).

[174]Ruppel (2011), p. 198.

[175]Available at http://www4.unfccc.int/submissions/INDC/Published%20Documents/Namibia/1/INDC%20of%20Namibia%20Final%20pdf.pdf.

[176]Ibid.

[177]Available at http://www4.unfccc.int/submissions/INDC/Published%20Documents/Namibia/1/INDC%20of%20Namibia%20Final%20pdf.pdf.

In order to address the aforementioned more effectively in future, more coherent legal and policy measures concerning soil conservation may need to be implemented. This, however, does not only apply to Namibia but more or less to the entire African continent.[178]

8 Concluding Remarks

Soil protection in Namibia is imperative in light of the far-reaching and heavily adverse social and economic effects that result from its neglect. Soil management in its current state is not sufficient to address and prevent adverse impacts of land degradation effectively owing to the fragmented nature of policies and the wide array of affected actors. Although current laws, policies and regulations cover many aspects of land degradation control in Namibia, it seems that as of yet, there is no cohesive policy to coordinate the effectiveness of existing laws and regulations with regard to soil protection in Namibia.

Most of the laws and policies outlined above impose obligations relating to soil protection on specific ministries and directorates. So far, cross-sectoral collaboration, and vertical and horizontal cooperation between ministries has left much to be desired. Even though elements relating to soil protection are evident in many pieces of primary and subordinate environmental legislation, the consequential and effective implementation and enforcement of such elements can be subject to further development and improvement. The Ministry of Agriculture, Water and Forestry and the Ministry of Mines and Energy both address soil concerns in their annual reports, and together with the Ministry of Environment and Tourism efforts need to be coordinated and concerted in order to increase the efficacy of current legal and policy frameworks. The SDGs and especially Goal 15 have been grappled with in the context of NAP3, while emphasising land degradation neutrality.[179]

Many of the laws and policies pertinent to soil protection do emphasise community-based natural resources management (CBNRM) and, in effect, intend to limit the function of the government and its entities to regulation and technical support. This is especially applicable in the context of agricultural activity on communal lands, as well as in newly redistributed farmland, where the intention is to enable farmers to develop their capacity to manage resources.

Efforts to promote sustainable farming practices are already being made by farmers.[180] Soil conservation is also supported by urban dwellers moving towards more environmentally friendly lifestyles, in which less waste is produced, less water wasted, and concerted efforts to protect the Namibian pristine landscapes are made.

[178]Cf. Ruppel (2015).

[179]Government of the Republic of Namibia (2015a).

[180]Ministry of Agriculture, Water and Forestry (2014).

In order to ensure that soil protection is achieved, an integrated approach with a concerted effort by all guardians of the Namibian soil, namely government entities, non-state actors and the broader civil society is necessary to allow for a coordinated and comprehensive implementation of existing policies and the formulation of a new and collaborative approach to soil protection. This seems particularly important in light of Namibia's extreme vulnerability to the impacts of climate change.

References

Alori E, Nwapi C (2015) The international legal regime for sustainable soil. In: Ako R, Olawuyi D (eds) Food and agricultural law: readings on sustainable agriculture and the law in Nigeria. Afe Babalola University Press, Ado-Ekiti, pp 98–114, doi:10.2139/ssrn.2641138. Accessed 16 Nov 2015

Amoo SK (2014) Property law in Namibia. Pretoria University Law Press, Pretoria

Amoo SK, Skeffers I (2008) The rule of law in Namibia. In: Horn N, Bösl A (eds) Human rights and the rule of law in Namibia. Macmillan Education, Windhoek, pp 17–38

Bethune S (2003) Review of legislation and policies pertinent to combating desertification – a case study from Namibia. Revi Eur Commun Int Environ Law 12(2):178–182

China Africa Resources (2014) Berg Aukas project preliminary feasibility study. http://www.chinaafricares.com/downloads/pdf/Berg-Aukas-PFS-Final-April-2014.pdf. Accessed 21 Nov 2015

CIA (2015) The world fact book on Namibia. https://www.cia.gov/library/publications/the-world-factbook/geos/wa.html. Accessed 18 Aug 2015

Deutsche Gesellschaft für Internationale Zusammenarbeit (GIZ) (2016) Adaptation of agriculture to climate change. https://www.giz.de/en/worldwide/34175.html. Accessed 26 Mar 2016

Du Plessis L (1999) An introduction to law, 3rd edn. Juta, Kenwyn

FAO (2015) After the international year of soils, we won't take soils for granted anymore. http://www.fao.org/soils-2015/news/news-detail/en/c/353737/. Accessed 20 Jan 2016

FAO (2016) International Year of Soils 2015 – IYS 2015 http://www.fao.org/globalsoilpartnership/iys-2015/en/. Accessed 20 Jan 2016

Government of the Republic of Namibia (1995) The national agricultural policy. Ministry of Agriculture, Water and Rural Development, Windhoek

Government of the Republic of Namibia (1997) Regional planning and development policy. National Planning Commission, Windhoek

Government of the Republic of Namibia (2001) National resettlement policy. Ministry of Lands, Resettlement and Rehabilitation, Windhoek

Government of the Republic of Namibia (2004) Namibia vision 2030: Policy framework for long-term national development. AIM Publications, Windhoek

Government of the Republic of Namibia (2005) Policy review on issues pertinent to the improvement of land management and biodiversity conservation in Namibia. Ministry of Environment and Tourism, Windhoek, http://www.met.gov.na/Documents/Desertification%20%20and%20biodiversity%20policy%20review.pdf. Accessed 16 Nov 2015

Government of the Republic of Namibia (2007) Land use planning and environmental sustainability – contributions to an analytical framework for sustainable land use

Government of the Republic of Namibia (2011) National policy on climate change for Namibia. Ministry of Environment and Tourism, Windhoek

Government of the Republic of Namibia (2012a) Fourth national development plan 2012/13 to 2016/17. Office of the President, National Planning Commission, Windhoek, http://www.gov.na/documents/10181/14226/NDP4_Main_Document.pdf/6a57bd4f-a1ff-4aa5-9bd1-264e4a3e1e1a. Accessed 25 Mar 2016

Government of the Republic of Namibia (2012b) Namibia 2011 population and housing census. National Planning Commission, Windhoek

Government of the Republic of Namibia (2014a) Third national action programme for Namibia to Implement the United Nations Convention to Combat Desertification 2014-2024. Ministry of Environment and Tourism, Windhoek, http://www.unccd.int/ActionProgrammes/Namibia-2014-2024-eng.pdf. Accessed 22 Nov 2015

Government of the Republic of Namibia (2014b) Annual Report 2013-2014. Ministry of Mines and Energy, Windhoek, http://www.mme.gov.na/files/publications/264_MME_Annual%20Report%202013-2014_(3mm%20bleed).pdf. Accessed 26 Mar 2016

Government of the Republic of Namibia (2015a) Land degradation neutrality – national report. Ministry of Environment and Tourism, Windhoek, http://www.unccd.int/en/programmes/RioConventions/RioPlus20/Documents/LDN%20Project%20Country%20Reports/namibia-ldn-country-report-updated-version2.pdf. Accessed 27 Mar 2016

Government of the Republic of Namibia (2015b) Third national communication to the United Nations Framework Convention on Climate Change. Ministry of Environment and Tourism, Windhoek, http://unfccc.int/resource/docs/natc/namnc3.pdf. Accessed 26 Mar 2016

Groenewaldt AL (2008) Legal implications of grazing policies on the carrying capacity of rangeland in Namibia. Unpublished dissertation submitted in partial fulfilment of the degree of Bachelor of Laws (LLB), University of Namibia

Hanns Seidel Foundation Namibia (2012) Facts and figures about Namibia. July 2012. Hanns Seidel Foundation, Windhoek. www.hsf.org.na. Accessed 9 Sep 2012

Hinz MO (ed) (2010) Customary law ascertained. Volume 1: The customary laws of the Owambo, Kavango and Caprivi communities. Namibia Scientific Society, Windhoek

Hinz MO, Ruppel OC (2008) Legal protection of biodiversity in Namibia. In: Hinz MO, Ruppel OC (eds) Biodiversity and the ancestors: challenges to customary and environmental law. Case Studies from Namibia. Namibia Scientific Society, Windhoek, pp 3–62

Hinz MO, Ruppel OC, Mapaure C (eds) (2012) Knowledge lives in the lake. Case studies in environmental and customary law from Southern Africa. Namibia Scientific Society, Windhoek

Index Mundi (2015a). Namibia GDP - composition by sector. http://www.indexmundi.com/namibia/gdp_composition_by_sector.html. Accessed 28 Mar 2016

Index Mundi (2015b) Namibia economy profile 2014. http://www.indexmundi.com/namibia/economy_profile.html. Accessed 28 Mar 2016

Klintenberg O, Seely M (2004) Land degradation monitoring in Namibia: a first approximation. Environ Monit Assess 99:5–21

Legal Assistance Centre (2009) Striking a better balance – an investigation of mining practices in Namibia's protected areas. Legal Assistance Centre, Windhoek

Legal Assistance Centre (2013) Indigenous peoples and climate change in Africa report on case studies of Namibia's Topnaar and Haillom Communities. Legal Assistance Centre, Windhoek

Mapani B, Ellmies R, Kříbek B et al (2009) Human health risks associated with historic ore processing at Berg Aukas, Grootfontein area, Namibia. Communs Geol Surv Namibia 14:25–40. http://www.mme.gov.na/files/publications/717_Mapani%20et%20al_Berg%20Aukas.pdf. Accessed 26 Mar 2016

Mapaure I (2016) Review of the climate change situation in Namibia: projected trends, vulnerability and effects. In: Ruppel OC, Ruppel-Schlichting K (eds) Environmental law and policy in Namibia – towards making Africa the tree of life. Hanns Seidel Foundation, Windhoek, pp 339–358

Melber H (2011) From anti-colonial struggles to post-colonial realities: reflections on the limits to liberalisation in Southern Africa. In: Ruppel OC, Winter G (eds) Justice from within: legal pluralism in Africa and beyond. Liber amicorum Manfred O. Hinz in celebration of his 75th birthday. Verlag Dr. Kovac, Hamburg, p 27–4.

Mendelsohn J (2006) Farming systems in Namibia. RAISON – Research & Information Services of Namibia, Windhoek

Mendelsohn J, Jarvis A, Roberts C et al (2009) Atlas of Namibia: a portrait of the land and its people, 3rd edn. Sunbird Publishers, Cape Town

Ministry of Agriculture, Water and Forestry (2014) Annual Report 2013/2014. Ministry of Agriculture, Water and Forestry, Windhoek

Ministry of Environment and Tourism (2006) Vital signs of Namibia 2004. An integrated state of the environment report. Directorate of Environmental Affairs, Ministry of Environment and Tourism, Windhoek

Montanarella L (2015) Agricultural policy: govern our soils. Nature 528:32–33, http://www.nature.com/news/agricultural-policy-govern-our-soils-1.18854. Accessed 21 Jan 2016

Namibia Statistics Agency (2012) Namibia household income and expenditure survey 2009/2010. Namibia Statistics Agency, Windhoek

Niang I, Ruppel OC (2014) Africa. In: Field CB, Barros VR, Dokken DJ et al (eds) Climate change 2014: impacts, adaptation, and vulnerability. Contribution of Working Group II to the fifth assessment report of the Intergovernmental Panel on Climate Change. Cambridge University Press, Cambridge, pp 1199–1265

Pröpper M, Gröngröft A, Finckh M et al (eds) (2015) The future Okavango – findings, scenarios, and recommendations for action. University of Hamburg, Hamburg. http://www.future-okavango.org/downloads/TFO_Report_engl_compiled_small_version.pdf. Accessed 26 Mar 2016

Ruppel OC (2011) The International, African and Regional Institutional, Legal and Policy Framework of Climate Change. In: AU and AMCEN. Addressing Climate Change Challenges in Africa. African Union, AMCEN and United Nations Environment Programme, Nairobi, pp 170–202

Ruppel OC (2013) Intersections of law and cooperative global climate governance – challenges in the anthopocene. In: Ruppel OC, Roschmann C, Ruppel-Schlichting K (eds) Climate change: international law and global governance. Volume I: Legal responses and global responsibility. NOMOS Law Publishers, Baden-Baden, pp 29–93

Ruppel OC (2015) Climate change, law and development in Africa: a reflection on selected aspects, relations and responses. In: König D, Koch HJ, Sanden J, Verheyen R (eds) Legal regimes for environmental protection, governance for climate change and ocean resources. Brill & Nijhoff Publishers, Leiden

Ruppel OC (2016a) Environmental law in Namibia: an overview. In: Ruppel OC, Ruppel-Schlichting K (eds) Environmental law and policy in Namibia – towards making Africa the tree of life. Hanns Seidel Foundation, Windhoek, pp 29–54

Ruppel OC (2016b) Introduction to international environmental law. In: Ruppel OC, Ruppel-Schlichting K (eds) Environmental law and policy in Namibia – towards making Africa the tree of life. Hanns Seidel Foundation, Windhoek, pp 55–70

Ruppel OC (2016c) Legal and regulatory aspects of climate change in Namibia and SADC. In: Ruppel OC, Ruppel-Schlichting K (eds) Environmental law and policy in Namibia – towards making Africa the tree of life. Hanns Seidel Foundation, Windhoek, pp 359–370

Ruppel OC, Ruppel-Schlichting K (2012) Legal and judicial pluralism in Namibia and beyond. Journal of Legal Pluralism and Unofficial Law. No. 64/11, 2012, pp 33–64

Ruppel OC, Ruppel-Schlichting K (2016) Namibia and its legal setup. In: Ruppel OC, Ruppel-Schlichting K (eds) Environmental law and policy in Namibia – towards making Africa the tree of life. Hanns Seidel Foundation, Windhoek, pp 1–8

Ruppel OC, Bethune S, von Finckenstein A (2016) Land and agricultural laws and policies relevant for environmental protection in Namibia. In: Ruppel OC, Ruppel-Schlichting K (eds) Environmental law and policy in Namibia – towards making Africa the tree of life. Hanns Seidel Foundation, Windhoek, pp 177–189

Ruppel-Schlichting K (2016) General principles of environmental management in Namibia. In: Ruppel OC, Ruppel-Schlichting K (eds) Environmental law and policy in Namibia – towards making Africa the tree of life. Hanns Seidel Foundation, Windhoek, pp 113–121

Seely M, Montgomery S (2009) Proud of our deserts: combating desertification. DRFN, Windhoek. http://www.drfn.info/docs/napcod/napcod_book_small.pdf. Accessed 16 Nov 2015

Sweet J, Burke A (2006) Country pasture/forage resource profiles – Namibia. FAO, Rome, http://www.fao.org/ag/agp/agpc/doc/counprof/PDF%20files/Namibia-English.pdf. Accessed 28 Aug 2015

Unsted S (2015) Update: China Africa Resources loss narrows, focus on Berg Aukas mine. Alliance News 3 Sep 2015. http://www.lse.co.uk/AllNews.asp?code=mjh0foc8&headline=UPDATE_China_Africa_Resources_Loss_Narrows_Focus_On_Berg_Aukas_Mine. Accessed 26 Mar 2016

USAid (2010) Country profile: property rights and resource governance. http://www.usaidlandtenure.net/sites/default/files/country-profiles/full-reports/USAID_Land_Tenure_Namibia_Profile.pdf. Accessed 20 Nov 2015

USAid (2014) USAID country profile property rights and resource governance: Namibia. http://www.usaidlandtenure.net/sites/default/files/country-profiles/fullreports/USAID_Land_Tenure_Namibia_Profile.pdf. Accessed 20 Nov 2015

Van den Berg HM, Koep P (2016) Mining and energy in Namibia. In: Ruppel OC, Ruppel-Schlichting K (eds) Environmental law and policy in Namibia – towards making Africa the tree of life. Hanns Seidel Foundation, Windhoek, pp 211–232

Vigne P, Whiteside M (1997) Encouraging sustainable smallholder agriculture in Namibia. Environment and Development Consultancy Limited, Windhoek

Zaire DU (2014) Namibia and the United Nations until 1990. In: Bösl A, Zaire DU, Du Pisani A (eds) Namibia's foreign relations – historic contexts, current dimensions and perspectives for the 21st Century. Macmillan Education Namibia, Windhoek, pp 37–50

Brazilian Federal Court of Accounts (TCU), Department of Agriculture and Environmental Audit: Soil Governance Audit

Vinícius Neves dos Santos, Tiago Modesto Carneiro da Costa, and Junnius Marques Arifa

1 Introduction

The performance audits carried out by the Brazilian Federal Court of Accounts (TCU) are intended to make an independent and objective examination of the savings, efficiency, efficacy and effectiveness of government organizations, programs and activities with the aim of improving public management.

TCU has never before undertaken a project focused on analyzing public policies related to Brazilian soils. With the advent of the Soil Governance Conference in 2015, the Court carried out a Non-Urban Soil Governance Performance Audit with the aim of assessing soil governance in rural and forest areas in Brazil in terms of "Institutionalization", "Plans and Goals", "Coordination and Consistency" and "Monitoring and Assessment" as defined in the TCU "Framework for Assessing the Governance of Public Policies".[1]

According to the mentioned TCU Framework, "Governance in public policies refers to the institutional arrangements that shape the way policies are made, implemented and assessed for the benefit of society."

As a result of this study, TCU Plenary Ruling 1,942/2015[2] was drafted with Minister Walton Alencar Rodrigues as the rapporteur based on deliberations intended to improve the governance of both Brazilian soils and of several related initiatives and to make measures to fight soil degradation more efficient.

[1]http://portal2.tcu.gov.br/portal/pls/portal/docs/2686056.PDF.

[2]http://portal.tcu.gov.br/fiscalizacao/agricultura-desenvolvimento-agrario-e-meio-ambiente/destaques/fiscalizacoes-concluidas.htm.

V.N. dos Santos (✉) • T.M.C. da Costa • J.M. Arifa
Brazilian Court of Accounts, Brasila, Brazil
e-mail: Viniciusnds@tcu.gov.br; TIAGOMC@tcu.gov.br; JUNNIUSMA@tcu.gov.br

H. Ginzky et al. (eds.), *International Yearbook of Soil Law and Policy 2016*,
International Yearbook of Soil Law and Policy,
DOI 10.1007/978-3-319-42508-5_21

The TCU audit adopted methodologies in accordance with the Brazilian Federal Court of Accounts Auditing Standards (NAT) and with TCU's Performance Auditing Standards. The auditing procedures were applied in the states of Paraná, São Paulo, Rio de Janeiro, Mato Grosso do Sul, Pará, Ceará, Paraíba and Pernambuco, as well as at the Ministerial headquarters in Brasília. The main players involved in the execution of policies and initiatives relating to soil organization, use, conservation and recovery at these federal agencies were consulted.

In this context, the aim of this paper is to present the situations identified in the TCU study and describe the situation of Brazilian public policies that deal with the issues of land use and occupation and soil and water sustainability in an integrated fashion in connection with the current system of governance.

This paper describes the conclusions and findings identified by the Court of Accounts in this study and is divided into three sections: detailing of strategic areas, covering aspects linked to standardisation, institutional arrangements and competencies; information on plans and goals, as well as the Federal Government's level of knowledge of Brazilian soil, coordination of the players involved and the linking of their actions; aspects related to systems for monitoring and assessing the selected public policies and the use of performance indicators and feedback, in addition to the integration of Information Technology (IT) systems.

2 Overview

Soil is the basis for the production of food, fuels and fibers and it is essential for ecosystem functions. The United Nations Conference on Environment and Development (Rio + 20) recognized the economic and social importance of sound land management, with emphasis on soil, especially its contribution to economic growth, biodiversity, sustainable agriculture and food security, poverty eradication, the fight against climate change and improving water supply. The official document that resulted from "The Future We Want" Conference deals specifically with the issues of desertification, soil degradation and drought, emphasizing that these are global challenges and still pose serious obstacles for the sustainable development of all countries.

The "Global Soil Alliance" Conference held in Rome by the Food and Agriculture Organization of the United Nations (FAO) in 2014, highlighted the fact that 33 % of the soils in the world suffered from moderate to high degradation. Degradation of soil is a consequence of the loss of its biological, physical and chemical (fertilizing) capacity to remain productive, which prevents it from retaining carbon dioxide (CO_2). Environmental degradation imposes high costs on society, apart from pushing rural producers into poverty.

Brazilian Ministry of the Environment (MMA) data shows that 140 million hectares of Brazilian land are degraded, which corresponds to 16.5 % of the country. Of this land, around 30 million hectares of pasture land have degraded to some extent, rendering their productivity for producing animal food which is very low. On top of this, in recent years Brazil has been confronted with the phenomenon

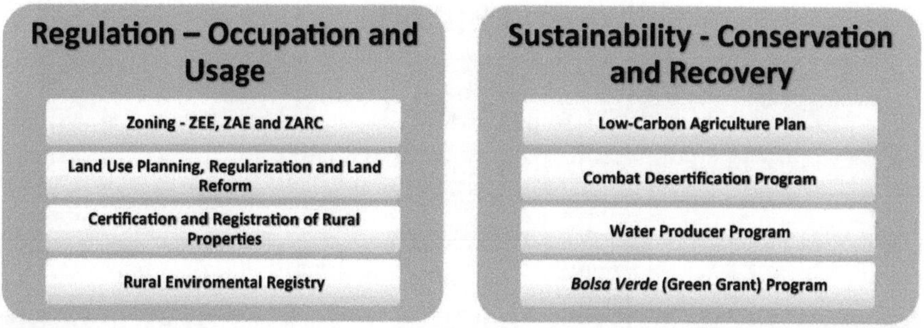

Fig. 1 Audit scope into non-urban soil and land governance. Source: Prepared by TCU Plenary Ruling 1,942/2015

of desertification, which has reached several regions such as the Semi-Arid region in the Northeast, the Tocantins Savannah area and the north of Mato Grosso and Minas Gerais states, as well as the phenomenon of, silting up which has struck the plains of Rio Grande do Sul state. As these arid areas form, the temperature increases and air humidity falls, making the soil infertile, reducing food production and increasing hunger and poverty.

Brazil does not have a specific public policy for soil protection, conservation and recovery and coverage of this subject is found scattered about several laws, making it difficult to understand. Due to the fact that there is no all-embracing public policy specifically dealing with the management of soil resources in Brazil, the audit's assessment focused on Federal Government initiatives relating to land use and occupation and to soil sustainability (conservation and recovery) (Fig. 1).

3 Governance from the Point of View of Institutionalisation in Laws and Standards

In the performance audit, it was not possible to identify an institutional definition of the Brazilian State's priorities for either the regulation of land occupation and use or the promotion of soil and water sustainability in non-urban areas. There are several government initiatives designed to deal with these issues, but these are not interconnected or aligned with each other.

An analysis of the legal framework that sustains the federal government's territorial organization and regulation actions showed that there are several scattered rulings that govern soil occupation and use from the points of view of territorial organization, legalisation, and ownership, tenure and access to land.

This failure to consolidate provisions for this purpose has created a complex legal system that generates conflicts and disputes between the different players interested in occupying non-urban soil, such as landless rural workers, family farmers, agrarian reform settlers, indigenous people, large and medium-sized

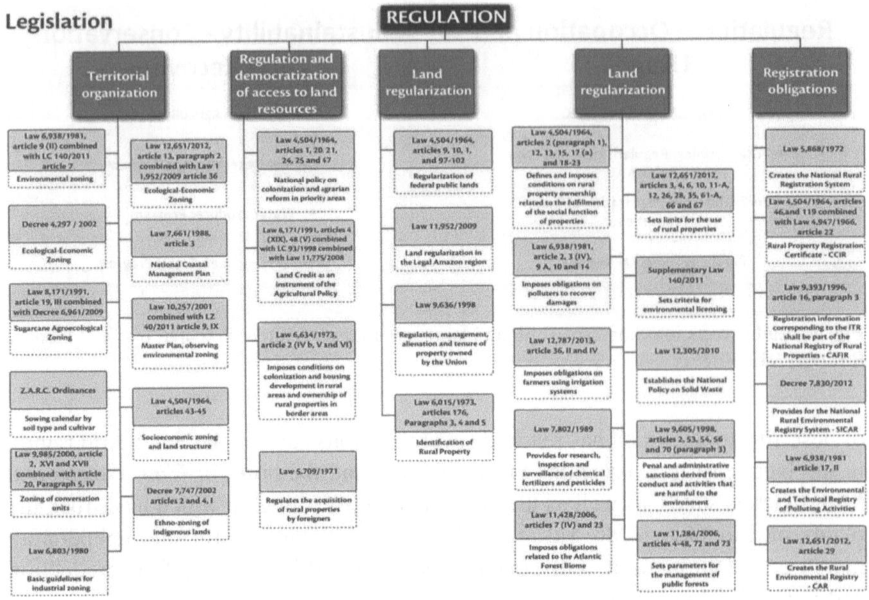

Fig. 2 Systematized legislation of territorial organisation and regulation. Source: Prepared by TCU Plenary Ruling 1,942/2015

food producers and the government, with consequences for public security. Figure 2 shows some of the laws which provide for regulation of land use and occupation.

This situation shows that no priority has been given to reviewing and consolidating normative provisions dealing with land organisation and access to land resources with the aim of reducing the legal uncertainty that the different players involved in non-urban soil occupation and use are exposed to in Brazil.

As an example, Law 4,504/1964, the Land Statute, regulates the rights and responsibilities of rural properties with the aim of promoting a better distribution of land with a focus on social justice and increased productivity. According to this law land ownership depends on the performance of its social function, namely on promoting the well-being of workers and owners, productivity, the conservation of natural resources and fair labor relations. Law 11,952/2009 was passed to legalise lands belonging to the Brazilian State in the Amazon Region to be sold or donate to small farmers under rights of use.

With regard to the sustainability of soil resources, the normative provisions are just as fragmented and usually linked to government programs, as shown in Fig. 3.

In spite of their being inseparable, soil and water rights are dealt with in different laws in Brazil. While there is a Water Act (Law 9,433/1997) at federal level, there are only general considerations of sustainable use of land, scattered throughout several laws.

Law 4,504/1964, which provides for the Land Statute, states that anyone who ensures the conservation of natural resources is fulfilling the social function of

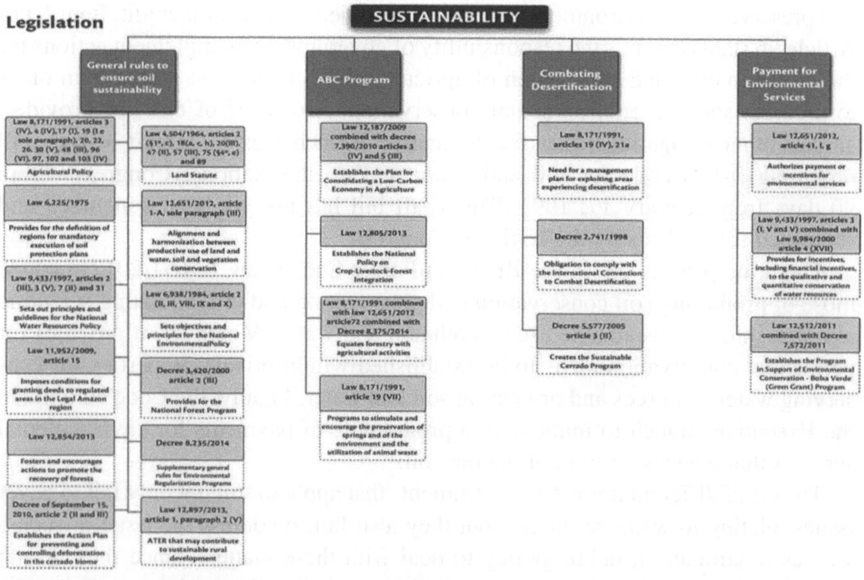

Fig. 3 Systematised legislation on soil sustainability. Source: Prepared by TCU Plenary Ruling 1,942/2015

ownership and, in the event of non-compliance with this provision, the property is subject to expropriation. Article 75, paragraph 4(e) defines help and assistance for the rational use of soil and combatting and preservation of natural resources as technical assistance goals in Agrarian Reform areas.

Law 6,225/1975 creates legal obligations that require the Ministry of Agriculture (MAPA) to identify regions where land can only be farmed or exploited economically against the compulsory execution of plans to protect the soil and combat erosion, inhibiting finance for investing in land that fails to meet this requirement. The Ministry of Agriculture has not yet identified these regions, thus rendering the practical application of this law impossible.

Law 6,938/1981, which provides for the Brazilian Environmental Policy, defines as a principle the rationalisation of soil, subsoil, water and air use, the planning and inspection of the use of environmental resources, the recovery of degraded areas and the protection of areas threatened with degradation.

Article 102 of Law 8,171/1991 states that soil must be respected as a natural asset of the country and that erosion must be tackled by government and rural landowners. Furthermore, several of its articles contain provisions on soil, such as item V of Paragraph V, which imposes on the Ministry of Agriculture the obligation to keep an integrated information system with the states, the Federal District and municipalities with data on the registration, cartography and type of soil of rural properties. In addition, item III of Article 48 defines "stimulating the adoption of rational methods in production systems with the aim of increasing productivity, improving the living standards of rural populations and properly conserving the soil and preserving the environment" as one of the objectives of rural credit. Item VI of Article 96 states that it is the responsibility of governments to implement actions for the dissemination and stimulation of agricultural mechanisation practices in order to promote soil and environmental conservation. Article 97 of this law provides, among other obligations, that the Executive Branch was to submit a draft bill providing for "a code for soil and water use" to the National Congress within 90 days from January 17, 1991. This draft bill has not yet been concluded and submitted for analysis by the National Congress.

The principles set out in the Brazilian Forest Code (Law 12,651/2012) include those of promoting soil conservation and preservation and of stimulating scientific research into the sustainable use of both soil and water. Moreover, it provides for permanent conservation areas to be established within private properties for conserving water resources and protecting soil and water. Finally, the Code authorized the Executive Branch to implement a programme of payments for environmental services that conserve and improve the soil.

The set of different normative instruments that apply to soil not only fail to cover issues relating to water resources, but they also fail to address the issue in such a way as to structure a public policy to deal with these matters, since they do not indicate the agencies or bodies that will be responsible for applying policies, for making decisions and settling disputes, and for ensuring the necessary funds for implementing them.

The weaknesses resulting from the absence of a legal framework for integrated soil and water management are the determining factor in the increased risk of losing a large quantity of soil resources every year, reducing the fertility of the country's arable land, and causing the silting up of rivers and other bodies of water. Consequently, the availability of water resources in urban and rural areas is reduced, resulting in huge losses in water supply and food production.

As a basis of good practice for establishing an integrated legal framework for soil and water, mention should be made of Law 9,343/1997, which provides for the National Water Resources Policy and includes most of the necessary items for establishing goals, actors, roles, monitoring systems and punishments for non-compliance with the law, which today constitute gaps in the domestic law on soils.

Similarly, the TCU audit did not identify a consolidated, formally established legislative instrument defining the roles of the different government bodies, institutions and spheres that operate in these areas or setting operational limitations between them to avoid overlapping, conflicts or duplication of efforts.

No appropriate legislative instrument establishing a hierarchy or official coordinating body for land-use planning covering all non-urban areas in their different uses (farming, protected and indigenous lands, among others) was identified. No limitations have been imposed on the competencies of bodies with competing land-use planning competencies. A similar situation exists with regard to the analysis of organisational competencies for promoting land and water sustainability.

As an example of this situation, several institutions with competencies to recover degraded areas were identified. However, there is no distinction of competencies between agencies and institutions, resulting in conflicts and overlapping competencies. The task of recovering degraded areas is split among the Ministry of Agriculture, the Ministry of Agrarian Development (MDA) and the Ministry of the Environment, as well as among other government bodies, such as the Chico Mendes Institute for Biodiversity Conservation (ICMBio), the National Institute for Colonisation and Agrarian Reform (Incra), the Brazilian Forestry Service (SFB) and the National Water Agency (ANA).

The lack of a consolidated legal framework defining the competences and operational limits of the different agencies operating in this area has led to overlapping competencies, with the consequence of a waste of funds and human resources. In addition, the current situation makes it difficult for the Executive Branch to establish rules defining how public policies are to be coordinated at federal, state and municipal level. Furthermore, it undermines the transparency, accountability and social control of the policy, as there is no clear definition of who should be held accountable for the results of the actions proposed by the government.

Another issue addressed in Ruling 1,942/2015 concerning legal provisions relating to the regulation of land use and occupation refers to existing obligations and restrictions on owners of non-urban properties and on such properties. Once again, the rulings that stipulate these obligations and restrictions have not been consolidated in such a way as to avoid hindering the implementation of public policy as a result of excess red tape and detail.

Among the rules that regulate the use of non-urban soils, especially farmland and forest soils, rules have been identified that define obligations such as: making rural ownership conditional on compliance with its social function; establishing environmental restrictions and obligations; defining criminal and administrative sanctions relating to land use; and imposing standards for the economic exploitation of forests and protected areas. Furthermore, other provisions define registries where rural real estate owners must register their properties and are obliged to periodically submit information and which operate as requirements for registration at notaries, inheritance, environmental licensing, and payment of fees or access to agricultural credit.

The government has created several obligations for rural landowners and a lot of records that need to be filled in, imposing financial and non-financial burdens on private individuals and government institutions to comply with these obligations without assessing the possibility of simplifying and consolidating these requirements according to their purpose. Furthermore, it can be seen that federal, state and municipal public agents use the collected information inefficiently, due to difficulties in consolidating it as a result of the contradictory nature of its different objectives, thus limiting its use in public policies.

The ideal situation would be to have a single register where owners would fill in all the data about their properties and land use, thus reducing the red tape involved in maintaining different registries and preventing the problem of contradictory information.

4 Soil Governance Assessed from the Viewpoint of Coordination and Consistency, and Plans and Goals

The audit carried out by TCU could not identify a general formulation defining an intervention logic through plans that allow for the implementation of necessary actions outlined on the basis of guidelines, objectives and goals. In addition, no integrated planning instrument for public organizations in charge of operationalising the initiatives that fell within the scope of the audit has been found.

It was not possible to identify any strategic planning by the Federal Government to coordinate the efforts of the different government initiatives relating to the regulation of soil occupation and use and to the promotion of soil and water sustainability aligning inputs, activities, products, effects and impacts according to the problems to be tackled with the necessary consistency between these different initiatives in all spheres. The Ministry of Agriculture (MAPA) which is the agency in charge of a number of soil-related issues in rural areas, does not even address the issue of soils in its Strategic Planning for 2006–2015.[3]

Similarly, no consolidated federal or interfederal plans contemplating guidelines, priorities and general objectives for integrated soil and water management were found in Brazil. No plans revealing the existence of a long-term strategic vision contemplating the government's role in matters relating to the regulation of non-urban soil use and occupation or of soil and water conservation and recovery actions at the national level could be found either.

Due to the lack of such strategic long-term plans, several federal government initiatives are taken here and there without appropriate links among them and they lack, as a whole, a logical sequencing required to fulfill the purpose of ensuring the sustainable occupation and use of Brazilian soil. Examples of these initiatives include the following ones: Environmental Economic Zoning (ZEE), Agro-ecological Zoning (ZAE), Agricultural Climate Risk Zoning (ZARC), National Agrarian Reform Plan (PNRA), Land Regularisation, Rural Certification and Registration, Rural Environmental Registry (CAR), Low-Carbon Agriculture Plan (ABC), Water Producer Program, Green Grant Program and the Program to Combat Desertification.

By comparing the objectives and goals of these initiatives, the audit concluded that the actions and specific objectives of the interventions carried out by the different entities are not in alignment and that the common outcomes as expected through implementation fail to reinforce one another mutually. Moreover, both overlapping and duplication of efforts have been observed, as well as the existence of parallel actions among the initiatives studied.

For the planning of any public policy, information must be available about the current situation of the object of the policy, according to the good practice defined by the TCU Framework for Assessing the Governance of Public Policies, namely,

[3]http://www.agricultura.gov.br/arq_editor/file/2720_13.pdf.

"Clear definition of the initial reference stage, i.e. the baseline that will be used as input to support the analysis of the outcome of a public intervention" (Moura 2013). How this reference stage should be defined needs to be analysed. In the audit, knowledge of the soil was assessed from the standpoint of two aspects: in relation to public or private land occupation and with regard to the chemical and physical properties of the different types of soils available in Brazil.

For evaluating the occupation of the Brazilian territory, the audit relied on the main georeferencing systems that identify land areas in the territory cartographically, which are available in the land archives and are fed with data provided by several public agencies and entities. When the data was analysed, several examples of overlapping of federal areas were identified, revealing situations in which the same area had been registered as set apart for two or more different purposes in different registries. These overlapping areas indicate that the Federal Government is either not very aware of the situation of its own territory or has yet to decide what its land should be set apart for.

As an example of these overlapping areas, TCU saw in the federal georeferencing information systems that settlement projects overlap with 18,144,538.9 ha of protected areas and with 690,515.75 ha of indigenous areas. Furthermore, indigenous areas overlap with 6,882,025.56 ha of protected areas.

When data from the Brazilian Institute for Geography and Statistics (IBGE) was also aggregated and analysed together with the other data available in the land archives, the number of inconsistencies found in the federal government's official data increased, especially in the Amazon Region. Figure 4 resulted from the plotting of the different federal, privately-owned and overlapping areas on the map. The gray areas on the right-hand side of the map are areas not yet included in government land registries and the ones in red are overlapping areas, showing that the same area was set apart for two or more purposes.

The unavailability of sufficient, reliable, relevant data on the destination and occupation of public and private land makes it more difficult to plan public policies and for the other players interested in topics related to land, soil and water to play a part in such effort, besides making it harder to measure public policy progress and achievements, since a baseline as defined as good practice by the TCU Framework is lacking.

As far as the knowledge available in Brazil with respect to the physical and chemical aspects of its soils is concerned, the audit identified a similar situation to that found regarding the information available for planning public policies.

No spatial information is available in Brazil on soil classes and properties for most of the country. According to Perez et al. (2014), all of the soils in the country have been mapped at the 1:1,000,000 scale (schematic survey) and about 8.6 % of Brazilian land has been surveyed with published soil maps at scales between 1:100,000 and 1:250,000 (reconnaissance survey), while only 0.6 % of the territory has been mapped out at more detailed scales than 1:50.000 (semi-detailed and detailed survey). Such level of information about Brazilian soil is insufficient for planning public policies for soil and water use, conservation and recovery at micro-watershed level.

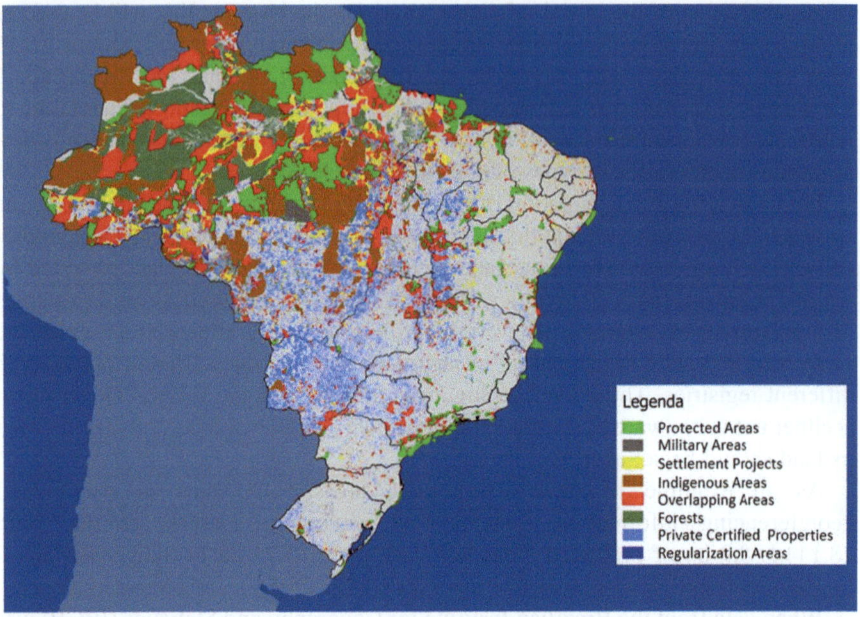

Fig. 4 Map of overlaps between federal and private areas in Brazil. Source: IBGE, Land Database of the MMA, Funai, Incra, SNIF

For assessing the sufficiency of the information on soils available in Brazil, a comparison can be made to the information on soil available in the U.S., which is a country of a similar continental size than Brazil and one of its main competitors in the international agricultural market. The U.S. has digital soil maps of its territory at scales ranging from 1:12,000 to 1:31,680, which gives them a large competitive advantage over Brazil with respect to the planning of soil use, conservation of soil and water resources, production sustainability, mitigation of the effects of global warming and costs for fertilizing and recovering its production areas.

As far as the breakdown of the information on Brazilian soil is concerned, the question of the quantity of already-known soil profiles in the country must be taken into consideration. The audit identified that there are 8958 soil profiles and samples registered for consultation in Brazil's Soil Information System, corresponding to a sample density of 1 profile for roughly every 100,000 ha. The Americans, in turn, have about 62,000 soil profiles registered in their database, corresponding to a sample density of around 1 profile for each 15,000 ha. It is worth mentioning that the greater the scale of a map, the more soil profiles are necessary for correctly attributing characteristics to them and the more fieldwork is required to check the outlines of the soil polygons.

In comparison to other countries in Latin America, with respect to the mapping of most of their territory (over 50 %), Brazil's scale is greater than only that of Argentina, Costa Rica, Honduras, Guatemala and Nicaragua. The soils of the other countries have been mapped at scales ranging from 1:250,000 to 1:500,000 (exploratory survey, enabling soils to be analyzed at the regional level). It is worth noting Latin American countries such as Bolivia (soil maps for all the country at a scale of 1:250,000), Colombia (soil maps for all the country at scales between 1:500,000 and 1:100,000), Cuba (national map at the 1:25,000 scale), Mexico (national map at 1:250,000) and Venezuela (95 % of the territory at 1:250,000), among other countries of a similar or lower socio-economic status than Brazil, have greater knowledge of their soil.

The Brazilian Agricultural Research Agency (Embrapa) is beginning to develop a soil information system that will enable each and every piece of information about soil produced in Brazil to be brought together.

The most recent federal public systematic analysis of soil information in Brazil took place between 1970 and 1985 with the RADAM-BRASIL project. Since then, no government initiatives have been taken to increase the scales used at that time or to break down the existing information, in spite of its great potential use for the country for economic and ecological purposes. Therefore, in addition to the need for studies in new areas, the soil information gathered under public policies of the past needs to be updated due to the different transformation processes that these soils have been through.

The literature contains several studies that demonstrate the favorable cost-benefit ratio of mapping the soil, confirming that the benefits obtained from soil mapping exceed the costs of carrying it out. The cost-benefit ratio obtained by the authors above ranges from 6:1 to 123:1. This variation derives from the intensity of the surveys, as more detailed scales result in a much greater cost-benefit ratio. For instance, based on the study carried out by Giasson et al. (2006), it can be estimated that the total cost of mapping the soil for all the Brazilian territory (851 million hectares) at the 1:50,000 scale would amount to approximately US$240 million. In interviews with researchers specializing in the area, estimates ranging from US$240 million to US$800 million were mentioned, depending on the priority areas, while for potential economic benefits, considering only the agricultural area of the country (282 million hectares), an estimate of US$10.5 billion was mentioned.

The audit concluded that the current level of soil knowledge is not sufficient either for planning land use and forest and agricultural activities in Brazil or soil and water conservation and recovery actions at micro-watershed level, as provided for in Law 8,171/1991, Article 20 and Decree 7,127/2010, Article 21, Paragraph III(d).

5 Governance from the Perspective of Monitoring and Assessment

The TCU audit revealed that no comprehensive routines have been formalized and materialized at federal level to ensure integrated monitoring and follow-up on the management of soil and water occupation, use, conservation and recovery, which is a situation that affects the quality of decisions made and the transparency of such initiatives.

Decree 7,127/2010 provides that it is up to MAPA's Agricultural and Livestock Development Department (SDC) to plan, promote, coordinate, supervise and assess government activities designed to protect, conserve and manage soil and water in rural areas using micro-watersheds as planning units. Through to 2011, MAPA had a medium-term plan (PPA)[4] for a specific action designed to ensure compliance with that regulatory mission, but since 2012 these actions have been focused on reducing greenhouse gas emissions in agriculture. The Strategic Plan for 2006–2015 of the Ministry of Agriculture (MAPA), published in 2009, also fails to mention issues related to soil conservation and recovery.

So no specific integrated soil and water conversation objectives were identified among either the MAPA strategic objectives or follow-up reports. The above-mentioned plan featured only initiatives and projected results that do not address soil-and-water-related problems in a comprehensive manner.

Furthermore, the TCU audit did not detect any performance indicators designed to measure the actual situation of soils in the country. The set of performance indicators for federal government initiatives related to land-use planning and to promoting soil and water sustainability that were analyzed (Water Producer Program, Land-Use Framework Planning, Environmental Economic Zoning (ZEE), Green Grant Program, Rural Environmental Registry (CAR), Low-Carbon Agriculture Plan (ABC) and the Program to Combat Desertification) are not sufficient for identifying and measuring the main aspects of public intervention.

When analysed jointly with their main objectives, the set of indicators for the initiatives that were studied should be capable of providing information allowing for the actual situation of policies designed to regulate soil occupation and use and to promote soil and water sustainability to be described, classified, ordered, compared or quantified. However, in general, these indicators, when they are actually available, are focused on their implementation procedures rather than on their performance and fulfillment of stated goals, making it difficult to build scenarios for these public policies and to monitor the progress of government interventions, target-setting, the assessment of actions and the definition of strategies.

The lack or insufficiency of quality indicators militates against interested parties being able to take part in initiatives in an informed manner, apart from affecting the transparency of funds disbursed and results achieved. In addition, it hampers the assessment and monitoring of inputs, processes, products, results and impacts of the government initiatives that were assessed, in addition to undermining their planning, execution, supervision and review.

The Federal Government is not provided with relevant information systematically to ensure the actual implementation of public policies designed to promote the sustainability of soil and water use in non-urban settings. Furthermore, it ends up not checking whether the funds and activities carried out by it and by other

[4]http://www.planejamento.gov.br/assuntos/planejamento-e-investimentos/plano-plurianual.

federated entities for managing soil use, conservation and recovery are being implemented as planned and whether their intended results are being actually achieved.

As an example of a good practice, the monitoring and follow-up on water and soil conservation carried out by USDA in the U.S., called RCA (Soil and Water Resources Conservation Act)[5] Appraisal, should be mentioned. This study was developed by ten U.S. agencies as part of the obligations provided for in the Soil and Water Resources Conservation Act and covers, among other things, land use, agrarian and agricultural structure, soil and water health, conservation actions, climate change, etc. Another U.S. initiative that can be mentioned as an example is the progress report of its strategic objective to conserve soil and water resources. This information is stored on a website (www.performance.gov) where progress reports for each fiscal year are available, along with the challenges to be faced in the following years and any external risk factors that may prevent goals from being achieved.

The audit also assessed the integration between different information systems for rural properties and activities. The main information systems identified were the National Rural Registration System (SNCR), which contains a register of all rural real estate; the Rural Properties Registry (Cafir), which is used for taxing rural properties; the Information System for Agrarian Reform Projects (SIPRA), which controls agrarian reform information; and the Rural Environmental Registry System (Sicar), which controls environmental reserves within rural properties.

It was seen that the information contained in those systems is not integrated between them, in breach of Law 5,868/1972 and Decree 4,449/2002, which provide for the existence of a common information database on the country's real estate. Moreover, a large part of these systems contain the same and often conflicting information on the ownership, location, size, and geographical or georeferencing data for non-urban properties, which limits its use. Figure 5 presents an overview of the current situation.

It should be stressed that Law 10,267/2001 created the National Registry of Rural Properties (CNIR) to comprise the databases of several federal and state public institutions that produce and use information on Brazilian rural areas that was supposed to be jointly managed by Incra and the Brazilian Internal Revenue Service. However, up to the present time, the CNIR has not yet been implemented.

Besides the CNIR managing institutions, Decree 4,449/2002 determines that "along with Incra and the Internal Revenue Service, all other agencies of the Federal Public Administration are to act as producers, feeders and users of the CNIR information base". However, no contribution or participation by other federal administration agencies in developing the CNIR, which will contain only SNCR and Cafir data initially, was found.

On the other hand, Decree 6,666/2008 established the National Spatial Data Infrastructure (INDE) system within the federal executive branch with the aim of

[5]http://www.nrcs.usda.gov/wps/portal/nrcs/main/national/technical/nra/rca/.

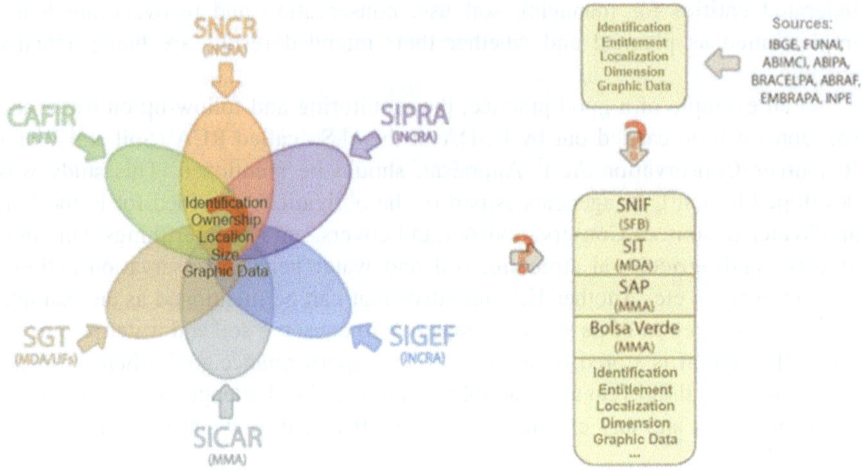

Fig. 5 Insufficient integration between information management and control systems for rural properties. Source: Prepared by TCU Plenary Ruling 1,942/2015

integrating and standardizing spatial data produced by different federal institutions. INDE's ambition is to bring together all of the systems that contain geospatial information, including rural information systems, for the purpose of avoiding overlapping and waste of funds. Nevertheless, the integration and sharing of spatial data by agencies and entities of the federal administration in the INDE system is still incipient.

Figure 6 shows the ideal situation for the integration and interaction of soil occupation and use information control systems, which would rely on a common database, the CNIR, that would be fed and shared by all users, would be a part of INDE and would be available to be accessed by public policy managers.

The following may occur as a result of these systems and their lack of integration: increased process time due to limited integrated information; increased costs and increased risk of exposure to errors due to the human labor involved in taking the data found in one database or system and putting it into another; delays on the part of the institution to identify and deal with events that could have been observed if events registered in other information systems were communicated; and inconsistencies in information generated by the different federal databases involved. Furthermore, this could lead to legal uncertainty as a result of difficulties for government to locate specific areas, illegally occupied areas and public lands.

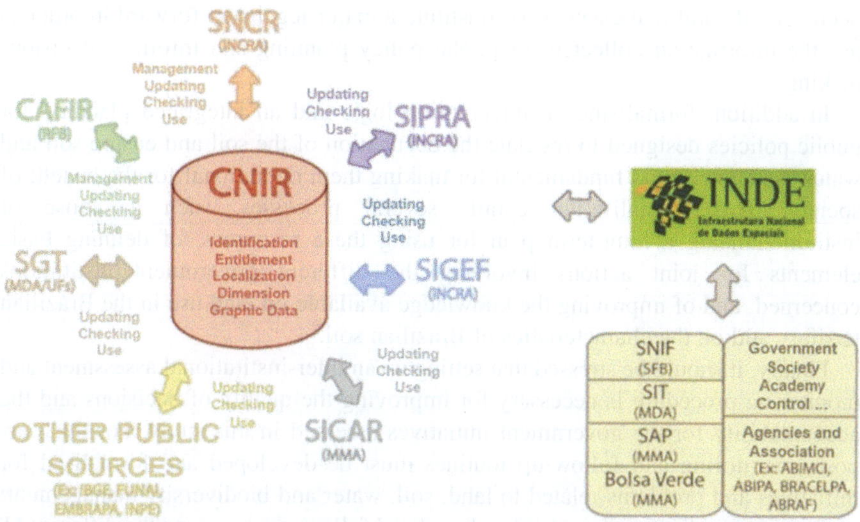

Fig. 6 Ideal situation of integration among the systems, according to Law 10,267/2001 and Decree 6,666/2008. Source: Prepared by TCU Plenary Ruling 1,942/2015

6 Conclusion

Soil is an essential resource for a series of activities that are fundamental for human life, such as food, fiber and energy production, the provision of environmental services, sustaining biodiversity and maintaining water resources. The United Nations Conference on Environment and Development (Rio + 20) recognised the economic and social importance of sound land management, stressing the importance of the soil and especially its contribution to economic growth, biodiversity, sustainable farming and food security, poverty eradication, the fight against climate change, and to improving water availability.

The analysis of the current status of soil governance carried out in this study, which covered the main government land, soil, water and biodiversity initiatives, unveiled a series of opportunities for improvement in different areas such as: in the legal framework for soil management; in institutional arrangements; in registration obligations related to non-urban properties; in inter-institutional plans, in objectives and coordination; in knowledge; and in public policy monitoring and follow-up.

This study's first conclusion is that the institutional definition of the Brazilian State's priorities for regulating soil occupation and promoting soil and water sustainability is fundamental for the success of the public policies involved. This definition involves the consolidation, harmonization and integration of different laws, normative instruments and institutional competencies related to the management of the territory and to promoting soil and water sustainability, since the Brazilian legislative framework deals with these laws in a fragmented and isolated way. The consolidation and simplification of registry requirements for owners or occupiers of rural real estate also constitute a major legal step forward in order to use the information collected for public policy planning and informed decision-making.

In addition, formalising an intervention logic and an integrated planning for public policies designed to regulate the occupation of the soil and ensure soil and water sustainability is fundamental for making them operational for the benefit of society. This formalisation entails several processes, such as those of institutionalising a long-term plan for using these resources, of defining basic elements for joint actions involving the different government institutions concerned, and of improving the knowledge available on land use in the Brazilian territory and on the characteristics of Brazilian soil.

Finally, it should be stressed that setting up an inter-institutional assessment and monitoring procedure is necessary for improving the quality of decisions and the accountability for the government initiatives assessed in this study. For this purpose, monitoring and follow-up routines must be developed and formalised for initiatives and problems related to land, soil, water and biodiversity management; performance indicators must be developed and followed up on; and the different IT systems that store information on rural properties in Brazil must be integrated.

References

Baca JFM (2015) Proposta de um Programa Nacional de Mapeamento de Solos do Brasil. In: Soil governance conference, 1st

Bazaglia Filho O (2012) Comparação entre mapas de solos obtidos pelos métodos convencional e digital numa área complexa. Dissertation (Masters degree in soils and plant nutrition)

Brazilian Federal Court of Accounts (TCU) (2015) TCU plenary ruling 1,942/2015

Brazilian Federal Court of Audit - TCU (2014) Referencial para Avaliação de Governança em Políticas Públicas, 1st ed

Carrick S, Vesely ET, Hewitt A (2010) Economic value of improved soil natural capital assessment: a case study on nitrogen leaching. In: 19th World Congress of soil science, soil solution for a changing world, Brisbane, Australia, 2010

Carvalho CCN, Nunes FC, Antunes MAH (2013) Histórico do levantamento de solos no Brasil: da industrialização brasileira à era da informação. Revista Brasileira de Cartografia, No. 65/5

Craemer R, Barber M (2007) Building a business case for investment in soil information: market dynamics and institutional strategy. National Land & Water Resource Audit, Canberra

de Moura AMM (2013) Governança das polítcas ambientais no Brasil: desafio à construção de um sistema integrado de avaliação. Texto para discussão. Instituto de Pesquisa EconÔmica Aplicada – IPEA

EMBRAPA (2006) National Center for Soil Research. Sistema brasileiro de classificação de solos, 2nd ed

Gardi C et al (2014) Atlas de suelos de América Latina y el Caribe. Comissión Europea – Oficina de Publicaciones de la Unión Europea, Luxemburgo

Giasson E, Inda Jr AV, Nascimento PC (2006) Estimativa do benefício econômico potencial de dois levantamentos de solos no Estado do Rio Grande do Sul. Ciência Rural, Santa Maria, V. 36, No 2

Klingebiel A (1966) Costs and returns of soil surveys. Soil Conservation, No. 32

Manderson A, Palmer A (2006) Soil information for agricultural decision-making: a New Zealand perspective. Soil Use Manag 22

Ministry of Agriculture, Livestock and Supply - MAPA (2009) Plano Estratégico, 2nd ed

Ministry of Planning, Budget and Management - MPOG. (2011) Planejamento e Investimentos Estratégicos. Plano plurianual 2012-2015, 1st ed

Odeh I, Mcbratney A (1996) Estimating the value of soil information: an example. ACLEP Newsl 5(4)

Oliveira VA (2014) Mapeamentos de solos no Brasil: situação e perspectivas. Boletim Informativo da Sociedade Brasileira de Ciência do Solo, V. 39, No. 1

Oliveira SRM, Zurmely HR (2014) Sistema de Informação de Solos Brasileiros. Consultas: orientações passo a passo

Paixão S (2010) Design of a Conceptual Land Information Management model for the Rural Cadastre in Brazil. PhD Dissertation. – Department of Geodesy and Geomatics Engineering, Technical Report No. 270, p. 404, University of New Brunswick, Fredericton, New Brunswick, Canada

Perez DV, Coelho MR, Fontana A, Santos HG (2014) O solo e a sustentabilidade agrícola no Brasil: um enfoque pedológico. Boletim Informativo da Sociedade Brasileira de Ciência do Solo, V. 39, No. 3

Santos E (2015) O poder do solo. Revista XXI Ciência para Vida – Embrapa, No. 9

United States. United States Department of Agriculture - USDA. (2011) RCA Appraisal: Soil and Water Resources Conservation Act. Washington

United States. United States Department of Agriculture - USDA. (2015) Strategic goal: "ensure our national forests and private working lands are conserved, restored, and made more resilient to climate change, while enhancing our water resources". Available at: www.performance.gov. Accessed 7 Feb 2015

United States. US Code, Title 16, Chapter 40, Soil and water resources conservation. Available at: http://uscode.house.gov/browse.xhtml. Accessed 7 Oct 2015

USDA (1999) Soil Survey Staff. Soil taxonomy: a basic system of soil classification for making and interpreting soil survey. Agriculture Handbook, vol 436, 2nd edn

Legislative Progress on Soil Contamination Prevention and Control in China

Tianbao Qin and Fang Dong

1 The Status Quo of Soil Contamination

The Status of the World's Soil Resources produced by the Food and Agriculture Organization of the United Nations concludes that the majority of the world's soil resources are in fair, poor or very poor condition and that conditions are getting worse. In particular, 33 % of land is moderately to highly degraded, due to soil erosion, salinization, compaction, acidification, and chemical pollution of soils soils (United Nations 2015).

The outlook for China's soil contamination is also not optimistic. From April 2005 to December 2013, the Ministry of Environmental Protection and Ministry of Land and Resources carried out the first nationwide investigation on soil contamination, covering all farmland and a portion of forest land, grassland, unused land and construction land totaling 6.3 million square kilometers. On April 17, 2014, the Ministry published the *Report on National Soil Contamination Investigation*,[1] which showed that 16.1 % of China's soil was polluted, among which the slightly polluted, mildly polluted, moderately polluted and heavily polluted accounted for 11.2 %, 2.3 %, 1.5 % and 1.1 %, respectively. Inorganic pollution contributed the most, accounting for 82.8 % of all pollution; organic pollution was second; and

[1]"Ministry of Environmental Protection and Ministry of Land and Resources issued a national survey report of soil pollution" (2014). http://www.zhb.gov.cn/gkml/hbb/qt/201404/t20140417_270670.htm. Accessed 15 May 2016.

T. Qin (✉) • F. Dong
Research Institute of Environmental Law, Wuhan University, Wuhan, China
e-mail: tianbaoq@hotmail.com

H. Ginzky et al. (eds.), *International Yearbook of Soil Law and Policy 2016*,
International Yearbook of Soil Law and Policy,
DOI 10.1007/978-3-319-42508-5_22

compound pollution only constituted a relatively small share.[2] The soil condition across China is rather disturbing in general. The Yangtze River Delta, the Pearl River Delta and the traditional Northeast Industrial Bases are the most seriously polluted regions. Farmland quality is deteriorating and pollution from mining and industrial waste is high.

From 1999 to 2014, the China Geological Survey, together with the Provincial People's Government and its respective Department of Land and Resources, with the support of the Ministry of Land and Resources and Ministry of Finance, carried out a nationwide geochemical investigation of farmland. On June 25, 2015, the Geological Survey published the *Report on the Geochemical Survey on China's Farmland*.[3] According to the results, 2.5 % of the land was moderately or heavily polluted or beyond the threshold for heavy metals, and 5.7 % of the land was slightly or mildly polluted or beyond acceptable levels.[4] It revealed that 29.3 % of China's farmland is becoming more alkalized, compacted and less fertile, making it difficult for crops to absorb nutrients. Additionally, 21.6 % is highly acidified, which not only reduces soil quality but also adds to the activeness of heavy metals, thereby making the ecological status of the soil and groundwater more prone to deterioration.[5]

When soil becomes polluted, hazardous substances accumulated in the crops find their way into the human body through the food chain, thereby causing various diseases and damaging human health. It is estimated that as much as 12 million tons of food are contaminated by heavy metals every year in China, resulting in 20 billion CNY (about 3.64 billion USD) of direct economic loss. Thus, soil contamination not only directly affects the structure and function of the soil ecosystem and eventually threatens ecological security, but it also poses broader consequences that necessitate its elimination or reduction (Ministry of Environmental Protection of the PRC 2006).

---

[2]Inorganic pollution contains a variety of harmful metals, salts, acids, alkaline substances, inorganic suspended solids and so on.

Organic pollution is composed of natural organic matter, such as carbohydrates, proteins, amino acids and fats, as well as some other biodegradable synthetic organic compounds.

Compound pollution is the pollution that many elements or a variety of chemicals to the same medium (soil, water, air, biological).

[3]The investigation covered 1.507 million square kilometers of land in total, among which the area of farmland was 0.924 million square kilometers, accounting for 68 % of the total farmland nationwide (1.354 million square kilometers). This investigation assessed the geochemical condition of farmland according to the amount of harmful elements it contains.

[4]The investigation is based on the standards of the *National Assessment of Soil Pollution (2008)* (as described in Sect. 2.1 of this article) and the *Soil Environmental Quality Standards (GB15618-1995)*.

[5]In reference to another soil threat, the survey also indicated that soil organic carbon levels are dropping markedly in some regions like the northeast, northwest, Qinghai and Tibet, which may have severe impacts on fertility.

2 The Current Legislation on Soil Contamination Prevention and Control in China and the Existing Problems

2.1 Relevant Regulatory Documents

In recent years, a series of regulatory documents have been released at the national level on preventing and controlling soil contamination. As is clearly stated in the *State Council's Decision on Implementing the Scientific Concept of Development and Strengthening Environmental Protection,* released in December 2005, "We should strengthen environmental protection in rural areas with soil contamination prevention and control as our focus, carry out a national investigation on soil contamination and manage farmland that is beyond standards in a comprehensive way. We should improve environmental legislation. There must be no delay in drawing up laws and regulations on soil contamination." The fourth session of the 10th National People's Congress in March 2006 ratified the *Outline of the Eleventh Five-Year Plan for National Economic and Social Development*, which pointed out that China should "carry out a national investigation on soil contamination and manage it in a comprehensive way." In the same year, the State Environmental Protection Administration released the *Notification of a National Survey on Soil Contamination* and a nationwide investigation was conducted.[6]

In June 2008, the Ministry of Environmental Protection released the *Opinions on Strengthening Soil Contamination Prevention*, further putting forward some opinions on strengthening soil contamination prevention and control.[7] After that, the *Outline of the Twelfth Five-Year Plan for National Economic and Social Development (2011)* clearly stated that China should strengthen governance of soil contamination and carry out pilot restorations of contaminated sites. *The Report to the Eighteenth National Congress of the Communist Party of China (2012)* devoted a chapter to promoting an ecological approach, stressing that the nation should "insist on prevention-centered management in a comprehensive way, give priority to prominent environmental problems that threaten harm to human health and strengthen the prevention and control of water, air and soil pollution."

On January 23, 2013, the General Office of the State Council released the document *On Printing and Distributing the Notice on the Recent Soil Environment Protection and Comprehensive Governance*, which lays out detailed requirements

[6]Based on the national investigation, the *Report on National Soil Contamination Survey* was published on April 17, 2014, which provides important data support for China's legislation on soil contamination prevention and control.

[7]The opinions included four parts, such as "To fully understand the importance and urgency of strengthening soil pollution prevention and control", "To provide clear guiding ideology, basic principles and main objectives of the prevention and control of soil pollution", "To highlight the key areas of soil pollution control" and "Identify relevant measures to strengthen soil pollution prevention and control".

for the purposes, objectives and safeguard measures for soil protection. Among them, one of the purposes is to build a national soil environment protection system so the national soil environmental quality will be significantly improved by 2020.[8] In 2014, *Document No. 1* of the Central Government proposed to "push forward the subsidy program on increasing the organic matter of soil", to "work harder on preventing and controlling agricultural non-point source pollution", and to "launch a pilot project for the restoration of farmland polluted by heavy metals" in order to enhance soil contamination prevention and control. Finally, in 2014 the State Council released the *Opinions on the Division of Labor Between Departments for Implementing the Major Tasks of the Report on the Work of Government*, which requests cooperation between the Ministry of Environmental Protection, National Development and Reform Commission and the Ministry of Science and Technology in "carrying out soil restoration". This sequence of national policy statements and directives highlights that soil contamination prevention has gained national prominence.

2.2 Existing Laws and Regulations

The proposal for legislation on soil contamination prevention and control started in the late 1970s. China first legislated on soil protection and soil contamination prevention and control in Article 10 of the 1979 Environmental Protection Law (trial[9]), stating that "attention shall be paid to using the land rationally according to local conditions, improving the soil and increasing the vegetation, so as to prevent erosion, hardening, alkalinization and desertification of the soil." Further, Article 21 states that "efforts shall be made to develop pesticides which are of high effect, low toxicity and low residue, comprehensive and biological methods of pest control shall be promoted, and rational use of wastewater for irritation shall be practiced so as to prevent pollution of the soil and crops." Later, Article 10 of the 1982 Constitution of People's Republic of China and Article 3 of the Land Management Law (revised in 2004) both put forward the principle of "giving rational use to the land". Article 32 of the Environmental Protection Law (revised in 2014) stipulates that "the State shall strengthen the protection of air, water and soil, and establish and improve the corresponding investigation, monitoring, assessment and

[8]The objectives are to strictly control new soil pollution, to determine soil and environmental protection priority areas, to lessen the environmental risk from soil pollution, to carry out remediation of soil pollution, to improve the environmental supervision capacity of the soil, and to accelerate the construction of soil environmental protection engineering. The safeguard measures are to strengthen the organization and leadership, to improve the investment mechanism, to improve regulatory policies, to strengthen scientific and technological support, and to guide public participation in strict target assessment of soil contamination prevention and control.

[9]Trial text is a special type of normative document, providing for the trial implementation of the covered subject. It only gives legal effect to the provisions during the trial period.

restoration system". Articles 33, 42, 49, 50 stipulate protection of the agricultural environment and prevention of agricultural non-point source pollution, and these stipulations also appear in the 2002 Agriculture Law.

Other special enactments, including the Law on the Prevention and Control of Air Pollution, the Law on the Prevention and Control of Water Pollution and the Law on the Prevention and Control of Environmental Pollution by Solid Wastes provide for the prevention of soil contamination by controlling different sources of pollution. The Law on the Prevention and Control of Air Pollution (revised in 2014) provides that areas affected by acid rain and sulfur dioxide pollution are delimited in light of the meteorological, topographical, soil and other natural conditions, and this Law stipulates the prevention and control of acid rain pollution and the precautions necessary to prevent soil acidification. Article 51 of the Law on the Prevention and Control of Water Pollution (revised in 2008) focuses on preventing sewage from polluting the soil, groundwater and agricultural produce due to use of industrial and urban wastewater for irrigation. Also for the purpose of preventing soil contamination, the Law on the Prevention and Control of Environmental Pollution by Solid Wastes (revised in 2013) controls the treatment of various solid wastes from production to disposal, forbids their discharge into the environment, encourages the reduction of their production and harmfulness as much as possible, and promotes the comprehensive utilization and non-hazardous treatment of solid wastes.[10] In addition, various articles on preventing and controlling soil contamination appear in a range of administrative laws and regulations, including the Regulations on the Protection of Basic Farm (1998), Rules for Implementing the Law of the People's Republic of China on the Prevention and Control of Water Pollution (2000), Regulation on Land Reclamation (2011), and Regulations on the Control of Agricultural Chemicals (2001).

At the local level, the prevention and control of soil contamination is mostly mentioned in the regulations concerning the protection of the agricultural ecosystem, farmland protection and the maintenance of farmland productivity. In recent years, a small number of local legislative practices have established the embryonic form of the legislation on the prevention and control of soil contamination. This included the risk assessment and restoration system of soil in the Regulations of the Zhejiang Province on the Prevention and Control of Environmental Pollution by Solid Wastes (revised in 2013), the joint supervision system in the Measures of the Shenyang City for Environmental Governance and Restoration of Polluted Places (trial 2007), and the ecological risk control of polluted places in the Regulations of the Guangdong Province on the Prevention and Control of Environmental Pollution by Solid Wastes (2004). Some regions have issued regulatory documents on the governance and restoration of soil contamination. For example, the third chapter of the Regulations of the Sichuan Province on the Water Conservation of the Yangtze River (revised in 2004) lays out specific rules for soil protection. The Regulations of the Anhui Province on the Protection of the Agricultural Ecosystem (revised in

[10] Luo (2007), p. 100.

2006) protects soil as one element of the agricultural ecosystem and provides that whoever pollutes is held responsible for governance. Additionally, *The Outline of the Plan of the Hainan Province for Constructing an Ecological Province (revised in 2013)* stipulates that soil and environment quality shall be protected in unison.[11] *The Circular on Some Questions Concerning Carrying out Soil Environment Assessment on the Original Location of Relocated Industrial Enterprises (2007)* released by Beijing city was also deigned to safeguard the security of formerly polluted land being redeveloped.

2.3 Current Legal Gap

Generally speaking, the current legislation on soil contamination is supported by only a small number of provisions that are general principles, which lack practicality, and there is no specific legal regime. Although there is no such legislation at the national level, some provinces like the Fujian and Hubei provinces have legislated on the prevention and control of soil contamination and an embryonic form of a legal regime has been established in some local legislative regimes. But that is far from enough. Specifically speaking, China's legislation on the prevention and control of soil contamination suffers from the following problems.

2.3.1 The Standard System Is Unsound

China has always referred to the Environmental Quality Standard for Soil[12] ever since it was formulated in 1995. Although the standard was amended in 2008, many provisions are now outdated to effectively control the current level of soil contamination in China. The acceptable levels of heavy metals are also unsatisfactory and cannot be referred to in judging the quality of the soil environment and carrying out soil contamination prevention and control. In January 2015, the Ministry of Environmental Protection released the draft amendment of the Environmental Quality Standard for Agricultural Soils and the Soil Contamination Risk Screening Guidelines for Construction Land, and is soliciting public opinion.

[11] Cai and Li (2008), p. 132.

[12] The Soil Environmental Quality Standard is the highest amount of pollutants permissible in soil. The pollutants in soil are limited to prevent crop fertility disorders, excessive accumulation in food products (according to food hygiene standards), and negative impacts on soil, water and general environmental quality. Additionally, the standards implement the People's Republic of China Environmental Protection of the ecological environment, agricultural and forestry production, and human health. The standards provide the Index value of the maximum allowable concentration of pollutants in soil and the methods according to soil application function, protection target and soil properties. The standards are applicable to farmland, vegetables, tea gardens, orchards, pastureland, forests, nature reserves and other soils.

The Environmental Quality Standard for Agricultural Soil was directly derived from the Environmental Quality Standard for Soil (1995), applicable for the environmental quality standard assessment and management of agricultural soils, like farmland, orchards, tea plantations and pastureland. The draft amendment has removed the "one-size-fits-all" natural background value[13] (first-grade standard value) and high background value[14] (third-grade standard value) under current standards and, according to the pH value of the soil, further divides the limit of cadmium from either 0.3 or 0.6 mg/kg to 0.3, 0.4, 0.5 or 0.6 mg/kg. Although they are more comprehensive, scholars are skeptical of the regulations. The new Standard removes the regulations for "natural background value", which may result in soils of high background value becoming listed as polluted soils, thereby making restoration of contaminated soil more difficult (Chen 2015a). Additionally, with regards to the principle of "protecting the soil environment and safeguarding human health", the Standard is more focused on limiting the amount of heavy metals and neglects soil acidification, loss of organic matter and the close relationship with safety standards for agricultural products. Finally, there is a gap in the risk prevention and control standard for soil contamination and in the technical standards for restoration.

2.3.2 The Systems Are Incomplete

First, the investigation system for the soil environment is not clearly regulated. Investigation is the foundation and starting point for all work on the prevention and control of soil contamination. The legislation does not address the scope and frequency of investigation of the soil environment in different regions of China and its different purposes. The department in charge of investigation, the collection of data and how it corresponds with the current land investigation system are also not clear. Soil investigation is chaotic in China at the moment; the media and scholars have completely different perceptions about the status of soil contamination. Before the Ministry of Environmental Protection and the Ministry of Land and Resources released the *Report on National Soil Contamination Investigation* in April 2014, there was no official data. The current investigation is preliminary and of a general nature, attempting to establish the general status of China's soil contamination. However, it is constrained by its objective conditions and low density of sampling point locations. For example, one point location is set for every 64 km^2 of farmland, which can only demonstrate the big picture for soil environmental quality of farmland in China. Therefore, this level of investigation cannot accurately indicate the area of polluted soil simply by calculating the ratio of point locations that exceed the standards. Thus, the current general investigation does not have clear assessment

[13]Natural background value is the composition of soil components and the content when not affected by human social behavioral interference (pollution) and damage. It is "the first-class standard to preserve the natural ecology of the region, setting limits for soil quality" under the Environmental Quality Standard for Soil (1995).

[14]High background value is the critical value of soil quality to ensure agriculture and forestry production and the normal growth of plants.

standards and a sufficient level of accuracy as it cannot elaborate on the distribution of major pollutants and the severity of pollution.

In addition, there is no monitoring network for soil contamination in China at present. The provisions in current legislation are too abstract to offer specific regimes concerning the scope and criteria for soil monitoring, the collection and release of information, the everyday monitoring of severely polluted areas, the emergency monitoring of sudden environmental events and the post-restoration monitoring of polluted soil.

The emergency mechanism particularly needs urgent improvement. For example, on August 12, 2015, a blast ripped through a warehouse storing hazardous products at Binhai New Area in Tianjin. The resulting environmental pollution attracted the attention of the public but the response was inadequate. Residue from the hazardous products remained on the land and percolated through the soil with rainwater infiltration. When pollution occurs, it is particularly important to undertake pollution control measures as soon as possible to prevent the spread of the pollution. *The National Plan for Environmental Emergencies*, revised in 2014, explicitly stipulated that emergency measures should be taken to cope with soil contamination. In some local areas, like Zhangye City and Minle County of the Gansu Province, created and distributed their own emergency plans for soil contamination accidents (The Emergency Management Office of Zhangye City 2014; People's Government of Mingle County, Gansu Province 2015), requiring systematic regulated monitoring, an early warning and information report, an emergency response, post-processing and emergency security. However, at the national level, there has yet to be an emergency system specially designed for addressing soil contamination accidents.

The classification management system also requires improvement. Agricultural land and construction land are two major categories of land use, but the pollutants that affect these lands differ substantially in their causes and characteristics. Soil contamination on agricultural land is due to non-point source pollution, which is dispersive and results from such agricultural activities as wastewater irrigation, overuse of chemical fertilizers, pesticides and plastic films, and manure from livestock and poultry production. Considering the need to maintain soil fertility and food security, soil contamination restoration on agricultural land is more complicated and usually resorts to use of biotechnology. Slow to take effect and involving farmers' fundamental asset, the restoration of agricultural land thereby relies on numerous capital and technology investments by the State. Soil contamination on construction land, on the other hand, is due to point source pollution and thus is typically more concentrated. It is mainly caused by the "three wastes" that remain after relocation—industry, mining and solid wastes. Construction sites also generally cover smaller areas of land than agricultural land and can be used for different purposes following restoration. Since this can be very lucrative, funds for restoration can be obtained through commercialized operations. In summary, due to their different causes and characteristics, the soil contamination prevention and control approaches for agricultural land and construction land should follow different paths and set different restoration goals, which are not embodied in current legislation at either the national or local level.

3 The History and Process of Legislation on the Prevention and Control of Soil Contamination

3.1 History of the Legislation

In China, the phrase "cadmium rice" first attracted public attention to the issue of soil contamination. The article "Cadmium Rice, a Potential Killer", published in the New Century Magazine in 2011, highlighted that residents of the Sidi Village in Guilin, Guangxi may have contracted itai-itai disease due to long-term consumption of rice containing excessive amounts of cadmium. In 2012, the Rice Product Quality Inspection and Supervision Center of the Ministry of Agriculture, undertook a security check for marketed rice across China. The results showed that 10.3 % of the rice contained excessive cadmium levels. On May 17 of the same year, the Guangzhou Food and Drug Administration announced that 18 groups of rice it had inspected also contained excessive amounts of cadmium (Caixin Century Weekly 2011).

As early as the second session of the 8th National People's Congress in 1994, representatives had proposed that land resource governance should be strengthened to concentrate on the conservation and rational use of land (The Website of National People's Congress 1994). Later, at the fourth session of the 9th National People's Congress, Bill No. 873 was tabled regarding establishing laws and regulations for environmental protection in rural areas and preventing and controlling agricultural chemical pollution. It suggested that relevant laws and regulations be developed to control chemical pollution in rural areas caused by irrational use of chemical fertilizers, pesticides, plastic films and wastewater irrigation (The Website of National People's Congress 2001). During the first session of the 12th National People's Congress in March 2013, 69 representatives, including Wu Qing and Du Guoling, proposed two Bills for laws on the prevention and control of soil contamination. They specified China's urgent need for a special law to prevent and control soil contamination through monitoring and assessment of soil contamination, pollution control, and post-pollution governance and liability (The Website of National People's Congress 2014).

The Ministry of Environmental Protection had established a draft legislation research group in 2006 to investigate legislation on the prevention and control of soil contamination. In 2012, soil contamination legislation was included in the third type of projects within the State Council's legislative work plan.[15] However, it was not until 2013 that the Soil Contamination Prevention and Control Law[16] was

[15]The legislative plan of the State Council is divided into three categories. The first category contains legislative projects which should be completed within the year. The second one indicates that the State Council needs to pay close attention to the subject and put forward legislation in a timely manner. The third category involves legislative projects which need to be actively researched and demonstrated.

[16]One of 69 legislative plans confirmed by the 12th NPC Standing Committee in 2013.

officially included in the first type of projects (drafts with conditions approaching and scheduled for submission for consideration). Its drafting was entrusted by the Environmental Protection and Resources Conservation Committee of the National People's Congress to the Ministry of Environmental Protection. In December 2014, the Ministry submitted the draft to the NPC Environmental Protection and Resources Conservation Committee. Then in 2015, the Committee set up a soil legislation expert group, a working group and a coordination group. Those groups prepared the legislative work plan, organized forums for the legislative departments of soil contamination prevention and finally developed the draft amendment of the law. At present, refinement of the law is still proceeding and the draft is expected to be submitted to NPC Standing Committee for review in 2017.

3.2 *The Legislative Process*

3.2.1 Legislation at the National Level

The Law of the People's Republic of China on the Prevention and Control of Soil Contamination (Draft Amendment) establishes the principles of "protection first, prevention as the focus, risk management and control, comprehensive governance, public participation and the polluter pays". It lays out regulations in detail concerning pollution investigation, monitoring, prevention and safeguard measures, control through supervision and inspection, and site restoration, with separate chapters devoted to construction land and agricultural land.

First, the Draft Amendment defines soil contamination as a phenomenon whereby harmful substances caused by human activities or natural events enter into the soil and, when accumulated to a high degree, deteriorate the soil environment, thereby threatening human health or reducing the effective utilization of soil. The general provisions clearly provide for government duties and create a target responsibility and appraisal system, which functions as the supervision and management system for the prevention and control of soil contamination. The Department of Environmental Protection manages the supervision system, with additional management duties for relevant sectors like agriculture. The Draft Amendment regulates the general obligations of citizens, legal persons and other organizations and their right to participation. Enterprises, institutions and other operators have the responsibility to prevent soil contamination, but if it occurs, they must provide compensation for damages. The State and government must therefore also adopt economic policies and measures in finance, taxation, credit and loan to encourage soil contamination prevention and control; establish a mechanism for immediate ecological compensation and remediation action; and strengthen technological support, publicity, education and commendation.

Second, the Draft Amendment provides for the setting of standards, undertaking investigations and monitoring. It explicitly requires that sound soil environmental quality standards be established as well as relevant standards and norms on investigation, monitoring, assessment, control, restoration and techniques. It provides for soil environment investigations to be carried out on a regular basis, setting up a monitoring network for the soil environment, compiling soil contamination control and restoration lists, sharing soil environment information and data, and formulating soil contamination prevention and control plans. It encourages third-parties agencies to carry out the soil contamination investigations, assessment, control and restoration.

Third, sufficient action must be taken to prevent soil contamination. The Draft Amendment mandates soil environmental impact assessments and prevents industrial activities, chemical storage and use, service businesses like gas stations, landfills and demolition from causing harm to the soil. It also gives prior protection to basic farmland, other lands for agricultural production and drinking water sources, provides guidance on agricultural technologies, encourages the development of eco-agriculture, enhances the protection of ecological land like national reserves and protects the soil environment during land reclamation.

Fourth, sufficient action shall be taken to control and restore polluted soil on land used for construction. The Draft Amendment regulates that investigations of soil contamination must be carried out, as well as soil contamination risk assessments for lands that contain pollutants beyond the soil environment quality standards and compilation of risk assessment reports. The land that, as those reports clearly indicate, needs to be controlled and restored should be included in the control and restoration list, and relevant measures should be taken by persons in charge or legitimate users, including the People's Government of the county where the land is situated. Restrictions should be imposed on the lands on the list and restoration plans developed where the soil needs to be restored. Secondary pollution during restoration shall be prevented. The environmental protection department of the People's Government of the county where the land is situated is then responsible for assessment and acceptance of the completed restoration project.

Fifth, action is required to manage and restore contaminated farmland soil. The Draft Amendment regulates the investigation of soil contamination, risk assessment, control, restoration list and corresponding measures. It especially provides that polluted farmland and other land for agricultural production shall be given priority governance through agronomic measures that prevent or reduce pollutants from contaminating edible food crops, thereby aiming to safeguard food security.

Sixth, the cost and economic safeguard measures for preventing soil contamination shall be specified. In principle, the polluter is held liable, but when the polluter is unidentifiable, the legitimate user of construction land and the People's Government of the county where the agricultural land is situated are to take charge. A fund for the prevention and control of soil contamination shall be set up by central and provincial governments, sourced from special capital, an additional tax on polluting products, pollution discharge fees from enterprises, an environmental tax, fines, a proportion of land transfer fees and interest income from the operation of the fund. The Draft Amendment also regulates the use and management of the fund and encourages finance, taxation and insurance systems.

Seventh, supervision and inspection of soil contamination prevention and control is required. The competent departments of environmental protection of the People's Government at or above the county level shall carry out on-the-spot inspections of enterprises, institutions and other businesses whose activities affect the soil environment. Based on their findings, the departments are entitled to take relevant compulsory administrative measures. They should supervise and inspect the control and restoration of polluted soils, oversee the implementation of relevant environmental protection measures and carry out on-the-spot inspection of lands on the control and restoration list. In terms of public participation, it also provides the right of public opinion supervision by the news media, including the right to report of the unit, individual and so on.

Lastly, the Draft Amendment provides for legal liability in the prevention and control of soil contamination, or the "polluter pays" principle.

In addition, after the "Ten Items of Air" and "Ten Items of Water" were released, the *Action Plan for Soil Environment Protection and Pollution Governance,*[17] referred to as the "Ten Items of Soil", was submitted by the Ministry of Environmental Protection to the State Council. The Minister of Environmental Protection, Chen Jining, suggested at the National Conference on the Environment in January 2016 that the official version would come out later in the year. Notably, according to the requirements of the State Council, the "Ten Items of Soil" will include areas defined as severely polluted by heavy metals, the amount of funding devoted to their management and the specific remediation measures to be taken. As a major plan for soil management and the prevention and control of soil contamination, it will lay out an explicit timetable for soil contamination governance in China. Generally, it classifies land into agricultural land and construction land and requires supervision and maintenance of each, respectively. According to grade, it assigns responsibilities and tasks for soil contamination governance to local governments and enterprises, intending to curb soil deterioration before 2020.

Meanwhile, significant progress has been made in formulating corresponding standards and regulations for the Draft Amendment legislation. In February 2014, the Ministry of Environmental Protection released five Technical Guides with standards for polluted land: On-the-Spot Environmental Investigation, On-the-Spot Environmental Monitoring, Risk Assessment of Polluted Land, Soil Restoration on Polluted Land and Terminologies for Polluted Land. In January 2015, the Ministry of Environmental Protection then released the draft recommendations: Soil Environment Quality Standard for Agricultural Land and Soil Contamination Risk Screening Guidelines for Construction Land, soliciting public commentary on both. As the revised version of the 1995 Soil Environment Quality Standard, the

[17] *Action Plan for Soil Environment Protection and Pollution Governance* is a normative document that is to enhance the prevention and control of soil pollution, and to gradually improve the quality of the soil environment. It is also referred to the "10-Chapter Of Soil", and is the third national action plan to curb pollution after the ones aimed at tackling air and water pollution. It was formulated from the latter half of 2013.

new Standard is the result of considerable revision due to the state of China's soil contamination and is a significant step in the construction of a technical standard system for China's soil contamination prevention and control.

3.2.2 Legislation at the Local Level

The Fujian and Hubei Provinces have promulgated their own legislation in the absence of national legislation for the prevention and control of soil contamination.

The Fujian Province Measures for the Prevention and Control of Soil Contamination

The Fujian Province Measures for the Prevention and Control of Soil Contamination, ratified on September 22, 2015, is the first local decree in China for the prevention and control of soil contamination. The decree is comprised of 49 items organized into six chapters titled General Provisions, Supervision and Management, Soil Contamination Prevention, Soil Contamination Control, Legal Responsibility and Appendix. It establishes the principle of "protection first, prevention as the focus, risk management and control, comprehensive governance, public participation and the polluter pays". It provides for a supervision system managed by the provincial department of environmental protection and separate management by relevant sectors to prevent and control soil contamination. The Measures also provide detailed rules on soil environment monitoring, investigation, assessment and restoration.

The General Provisions chapter differentiates between construction land, agricultural land and unused land, providing for governmental responsibility, a target assessment system for the prevention and control of soil contamination and strengthening of its technological support, publicity, education and commendation.

Under the Supervision and Management chapter, the government and departments concerned shall prepare a plan for the prevention and control of soil contamination. They must also formulate and release soil environment quality standards and technical norms for the investigation, monitoring, assessment and restoration of the soil environment. They must also coordinate the control, supervision and management of air, water and soil contamination. The government and departments concerned shall undertake monitoring of the soil environment on agricultural production land by establishing a monitoring network. They shall share monitoring data, strengthen the purchase of environmental monitoring services, carry out investigations on the soil environment on a regular basis, duly release investigation findings, establish records on polluted soil sites, and update the status and results of governance of soil contamination over time. They shall also carry out pilot governance and restoration of the soil environment. A credit archive shall be established for the soil environment protection actions by enterprises and institutions. It also strengthens instruction and supervision of third-party agencies in charge of soil environment monitoring, soil contamination assessment and restoration, and on-the-spot inspections on lands that are likely to experience soil contamination.

The Soil Contamination Prevention chapter requires that the government and departments concerned plan industrial layout, control the admission of industry, prevent newly admitted construction projects from causing new soil contamination, and carry out soil environment impact assessments and the "three simultaneous" system. There should be various grades of management over the soil environment on agricultural land, and certain agricultural activities are prevented along with manufacturing for leather production, mining, oil refining, operating a gas station, collecting radioactive substances or dangerous wastes, disposing medical wastes, burying household garbage, and dismantling equipment and facilities due to their potential to cause pollution and harm the soil.

The Soil Contamination Control chapter mandates that the People's Government make a list of major enterprises to be monitored due to their high levels of potential soil pollutants and carry out risk assessments on them. Those enterprises that are likely to damage human health and the environment must be controlled and prepare soil restoration goals and plans. The government shall prevent secondary pollution from occurring during restoration and when the restoration is completed, it shall assess the project and prepare a report. For those not listed as polluted lands, the government shall prepare a soil contamination control plan and carry out control activities. In terms of control and restoration, the Measures regulate that the polluter is held accountable in principle, with an obligation for the owner, legitimate user and actual user to prevent the pollution from spreading. When the polluter is unidentifiable, the People's Government at or above the county level where the land is located will take charge of control and restoration. The government will also establish a market-based mechanism for soil contamination control and restoration through a joint partnership between government, society and industry and create economic incentive measures to promote third-party governance of soil contamination. The government and departments concerned shall work out contingency plans for emergency soil contamination accidents.

According to the chapter on Legal Liability, entities will be held liable for evading, hindering or refusing on-the-spot inspections by the departments concerned, committing fraud in the assessment and monitoring of the soil environment, failing to entrust monitoring of the soil environment to a third party (specific to enterprises monitored as heavily polluting businesses), and failing to respond to soil contamination accidents in a timely manner by launching contingency plans. This chapter also regulates criminal liability in the prevention and control of soil contamination and civil liability to stop an infringement, eliminate obstruction, remove risks, restore to the original state and provide compensation under tort law. The Measures also provide for environmental public interest litigation.

The Hubei Province Decree on the Prevention and Control of Soil Contamination

The Hubei Province launched its legislation on soil contamination in March 2015. After four reviews by the Standing Committee of the Provincial People's Congress, the Hubei Province Decree on the Prevention and Control of Soil Contamination was approved by the fourth plenary meeting of the 12th People's Congress of the Hubei Province on February 1, 2016. The Hubei Decree differs from the Fujian Province Measures on the Prevention and Control of Soil Contamination due to its higher legal status, and it devotes two additional chapters to Environmental Protection for Soil with Specific Purposes and Information Disclosure and Public Participation. Further, it provides different regulations for specific regimes.

In the General Provisions chapter, the Decree defines "soil contamination" as a phenomenon whereby certain substances enter the soil and result in changes to the soil's chemical, physical and biological properties, thereby affecting the effective use of the soil, threatening human health, damaging the eco-environment and giving rise to the deterioration of the soil environment. It establishes the principle of "protection first, prevention as the focus, risk management and control, comprehensive governance, and the polluter pays", and it also creates the mechanism led by the government but also reliant upon coordination amongst the departments and public participation. The Decree implements a government liability system for soil contamination prevention and control, establishes special funds, improves a pluralistic fund investment and safeguard mechanism backed by financial and social funds, and strengthens technological support, publicity, education and commendation.

The Supervision and Management chapter establishes a comprehensive coordination mechanism for the prevention and control of soil contamination, as well as specific monitoring responsibilities for the environmental protection department and the departments of agriculture, housing and rural–urban development and land and resources. It also establishes a local supervision and law enforcement system for the soil environment, and provides the chief executive with both a responsibility system and target assessment system for soil contamination prevention and control. There are also provisions for soil environment audits on the departure, lifelong liability for damage to the soil environment, and establishment and improvement of soil environment quality standards and soil contamination control and restoration standards. The chapter improves the soil environment monitoring system, establishes a unified monitoring network and information-sharing platform, and requires census on soil environment quality across the province. It also establishes archives for soil environment quality data and requires preparation of soil environment functional zoning and soil contamination prevention and control plans. Finally, it regulates supervision and questioning by superior People's Governments and environmental protection departments.

The Soil Contamination Prevention chapter mandates that the government and departments concerned establish economic policies for soil contamination prevention and control; publish a list of polluting production projects that are forbidden from new construction, reconstruction or expansion; compile a list of obsolete techniques and equipment; and take measures to eliminate polluting processing equipment and to shut down polluting enterprises. The government and departments concerned shall devise a list of industries posing a high risk of causing soil contamination and require such enterprises to carry compulsory liability insurance for soil contamination and investigate its land and the surrounding soil at least once a year. The government shall carry out clean production assessments on major industries as well as environmental impact assessments and "three simultaneous" systems for planning projects and construction projects. Construction shall not commence before the environmental impact assessment document is approved. Measures shall be taken to prevent soil contamination from industrial production; mining; disposal of household garbage, sludge, radioactive substances or other toxic substances; refueling, washing and painting, or repair of vehicles and ships; maintenance, cleaning, storage and use of chemicals; and recycling of waste electronic products, batteries, vehicles and ships, tires and plastic products. Contingency plans shall be made for soil contamination accidents.

The Soil Contamination Control chapter provides for land to be listed as polluted according to permissible limits and the pollution shall be recorded in the land archive. The competent environmental departments shall organize and carry out soil environment risk assessments for polluted land and compile a list of land to be controlled and restored. Those persons in charge of controlling and restoration formulate plans and when the restoration is completed, the competent environmental departments are responsible for the review and acceptance. Soil contamination control zones shall be delimited and declared and measures must be taken within those zones to reduce pollution or prevent it from spreading. As for controlling and restoration, the Decree provides that the polluters shall be held accountable in principle, but the People's Government at or above the county level will take charge when the polluters are unidentifiable.

Under the Environmental Protection for Soils with Specific Purposes chapter, the Decree distinguishes land for agricultural production from that for dwelling, public administration and services, and commercial use. The agricultural land is administered by grade and different measures are taken for the three respective grades: clean, slightly or moderately polluted and heavily polluted. The government and its competent agricultural departments shall take measures to encourage agricultural producers to use agricultural investments properly, protect and improve the soil environment, develop organic, ecological and circulatory agriculture, examine the quality of irrigation water regularly and enhance supervision and management of pollution by livestock and fisheries. They shall clarify the different uses of construction land and undertake classification management. A soil environment quality assessment is required for construction land for residential, public administration and services, and commercial purposes, as well as that which has been newly added, changed uses, or the right of use has been transferred.

The Information Disclosure and Public Participation chapter provides for the government and departments concerned to establish regimes for the disclosure and release of information about the soil environment, optimize public participation procedures, and regularly publish the state of soil environment quality in their own administrative regions. It also provides that any organizations and individuals have the right to voice opinions, table proposals and file complaints about soil contamination, and qualified social organizations are eligible to initiate public interest litigation. The government shall take measures to encourage third parties to carry out investigations on soil environment quality, monitoring, risk assessment, pollution control and restoration, and they must establish a market-based mechanism for soil contamination prevention and control.

The chapter on Legal Liability provides for administrative liabilities for a variety of situations, including where highly polluting businesses do not monitor their land or the surrounding soil or fail to disclose information in accordance with the regulations. It also includes situations in which the people in charge of control and restoration do not carry out their duties as regulated, polluting equipment or processes are put into operation or use before the prevention and control facilities against soil contamination are completed, or facilities are dismantled or left idle without approval after completed. It also covers instances where third parties resort to deception and fraud in the investigations of soil environmental quality, monitoring, risk assessment, and pollution control or restoration. It also regulates criminal liability in the prevention and control of soil contamination and civil liability for stopping infringement, eliminating obstruction, removing risks, restoring to the original state and providing compensation under tort law.

4 Debates in the Legislation on Soil Contamination Prevention and Control

4.1 *Debate Over legislative Pattern*

Soil contamination prevention and control mainly consists of prevention before pollution occurs and post-pollution control. There is a debate over whether China should adopt a single legislative pattern or an integrated system that includes both the prevention and governance of soil contamination. For example, some scholars hold that it is unnecessary and impossible to regulate prevention and governance in detail in one single law.[18] Separate laws should be enacted according to this viewpoint. Since there have been respective laws on air pollution, water pollution and solid waste pollution, which are sources of soil contamination, there is no need to include overlapping provisions in soil contamination legislation. A special Law on the Prevention and Control of Soil Contamination enacted in accordance with the legislation covering the sources of soil contamination to fill the void in current legislation is urgently required, however, to specifically address soil contamination governance.[19]

[18]Wang (2012a, b), p. 31.

[19]Qiu (2008), p. 85.

First, other countries and regions such as Japan, Korea, the United States of America, and Germany have all adopted single legislative patterns for soil contamination control.[20] Second, soil contamination prevention and control exhibit more differences than similarities. It is hard to coordinate these actions through one single law and such an approach does not fit scientifically. Third, the inclusion of soil contamination prevention into the Law on Soil Environment Protection might awkwardly reduce it to being more of a supplement to the special law on pollution prevention and control, thereby deviating from the original intent behind its enactment. Soil contamination might stem from the improper discharge and disposal of wastewater, gas, solid wastes or other toxic substances. Given their different natures, they are governed by quite different institutions and in different ways. Some scholars think that including all sources in one law is nothing less than merging the special laws on pollution prevention and control and enriching them through regulations to fill the void and deficiency in current legislation. They argue that such a design is too simplistic and will not function as expected.[21] Of course, there are also scholars who believe that soil contamination prevention should not be interpreted as merely prohibiting or restricting discharges that might cause soil contamination. The functions that control soil contamination, such as formulating plans for soil contamination prevention and control, monitoring soil environmental quality, carrying out soil contamination investigations and environmental risk assessments, also contribute to the prevention of soil contamination.[22] These regimes are indispensible to the control of soil contamination.

4.2 Debate Over Legislative Focus

Whether advocating a single legislative system or an integrated one, scholars are also divided by their positions on legislative focus. Confronting serious soil contamination, should China focus on restoring and controlling current polluted soil or adhere to the "prevention first" principle of the Environmental Protection Law to prevent pollution from spreading? In answer to this question, some think that since addressing soil and underground pollution is difficult, expensive and spans a long period of time, the most fundamental thing is to cut off the pollution sources and then build soil health and address existing soil contamination. If the pollution sources are not cut off, the expected results cannot be achieved no matter how much attention is given to restoring polluted soils. Therefore, prevention must be prioritized. As some others point out, however, the regulations concerning soil contamination prevention scattered throughout relevant legislation mainly focus on preventing industrial pollution with little concern for preventing soil contamination

20Liang (2008a, b, c, d), p. 87.

21*Ibid*. p. 19.

22Wang (2008), p. 78.

from agricultural activities. As for soil contamination governance, the current legislation provides almost no specific regimes but rather several general principles. These deficiencies require the Law on Soil Environment Protection to place extra emphasis on control with due consideration for prevention: control includes soils polluted by industrial and agricultural activities and prevention mainly refers to agricultural land pollution from agricultural activities, such as improper use of pesticides, chemical fertilizers and wastewater irrigation.[23]

4.3 *Debate Over Local Environmental Standards*

Taking a panoramic view of the laws on environmental protection in China, we find that there are different regulations for local soil environment quality standards. The second item of Article 15 of the Environmental Protection Law provides that "the People's Government of provinces, autonomous regions and municipalities can make local environmental quality standards where they are no such regulations at the national level." The same provision is also found in the Law on the Prevention and Control of Water Pollution. Other special environmental laws have slightly different regulations and they do not require local environmental quality standards to be stricter than the national ones. The Law on the Prevention and Control of Air Pollution provides that "the competent environmental authorities of the State Council or the People's Government of provinces, autonomous regions and municipalities shall enact air quality standards that scientifically and rationally safeguard public health, protect the environment, and conform to the principle of social and economic development". Also, the Law on the Prevention and Control of Pollution from Environmental Noises provides that the "local People's Government at or above the county level shall, in accordance with national standards for acoustic environmental quality, divide their respective administrative regions into different zones for the application of different standards for acoustic environmental quality and exercise control, accordingly".

In terms of the legislation in China on the prevention and control of soil contamination, the draft national Law on the Prevention and Control of Soil Contamination (Draft Amendment), the Hubei Province Decree on the Prevention and Control of Soil Contamination and the Fujian Province Measures on the Prevention and Control of Soil Contamination all provide that when there are national standards for soil environment quality, the local authorities may formulate local standards that are stricter than the national ones. Based on reasonable consideration, soil contamination is complicated and distinctive, especially when the heavy metals contained in the soil are influenced by its natural background value. In many regions of southern China, the soil has always contained a high level of heavy metals, which will not cause harm to crops as long as the pH value remains

[23]Hu (2012a, b), p. 83.

stable.[24] If local soil environment quality standards must always be higher than the national level without considering specific local or regional circumstances, pollution levels will not reduce and the governance of soil contamination may thus be even harder. Therefore, some scholars hold that higher-level law should not be considered superior to lower-level law in all cases. For soils and regions with high background values, their soil environment quality standards should not have to be stricter than the national ones.[25]

4.4 Debate Over Legislative Structure

Scholars have seldom discussed the legislative structure of soil contamination prevention and control statutes. The controversies are mainly found in the legislative practices. The draft national Law on the Prevention and Control of Soil Contamination devotes chapters to discussing the investigation and monitoring of the soil environment and the prevention of soil contamination. The control and restoration of soil contamination are discussed in two chapters—Control and Restoration of the Soil Contamination of Construction Land and Control and Restoration of the Soil Contamination of Agricultural Land. One chapter is devoted to Cost Undertaking and Economic Safeguard. The Hubei Province Decree on the Prevention and Control of Soil Contamination includes the investigation and monitoring of the soil environment in the chapter on supervision and management. One sole chapter is devoted to the prevention and control of soil contamination. Later, the chapter on Environmental Protection of Soil with Specific Purposes is added, divided into two parts: Land for agricultural production and Land for residential, public administration and services and commercial use. Another chapter is devoted to Information Sharing and Public Participation. Similarly, the Fujian Province Measures on the Prevention and Control of Soil Contamination include soil environment investigation and monitoring in the chapter Supervision and Management and devote separate chapters to soil contamination prevention and control. Differently, the Measures do not distinguish "construction land" and "agricultural land" structurally but scatter them throughout the "prevention" and "control" parts. These three pieces of legislation are similar in content, but they differ in structure. Should they divide chapters in strict conformity with the sequence of soil environment investigation, monitoring, pollution prevention and control (restoration), integrate prevention and control systems for soils of different

[24]Chen (2015b), p. 52.

[25]*Ibid.*

uses, or distinguish construction land and agricultural land in terms of the different sources of their pollution? Should the applicable provisions be stipulated upon investigation, monitoring, prevention, assessment and control and restoration? Such decisions regarding the structure of the legislation not only affect the integrity of the legislation but also the focus of the legislation.

4.5 Debate Over the Establishment of Funds

Control of soil contamination spans a long period of time and is costly. Although China now abides by the polluter pays principle, the government pays for soil restoration when the polluter is unidentifiable. For example, under both the Hubei Province Decree on Pollution Prevention and Control and the Fujian Province Measures on Pollution Prevention and Control, when the polluter cannot be identified, it is the People's Government at or above the county level to which the land belongs that is responsible for preventing the spread of pollution and restoring the polluted land. Some scholars hold that the taxation principle applies when the polluter pays principle does not for the reason that the responsible party is unknown or incapable of cleaning up and governing. If the polluted land being cleaned up or governed is such that no one has use rights and the polluter is unknown or incapable of cleaning up or governing, the government should take charge. If the responsible party is discovered after the cleanup or governance, the government can then order that party to pay the costs incurred.[26] Thus, it is the taxpayers who pay the bill unless the latter scenario occurs and the polluter is found, which is unfair (China Environment Network 2015). Therefore, some scholars suggest that this method hinders the prevention and control of soil contamination and does not bring about good results (Kaiqi 2015).[27] It has been proposed that China should follow in the United States' footsteps and set up funds for the prevention and control of soil contamination (e.g., Superfund). When liability cannot be attached to the polluter or it cannot afford the cost, the funds will be able to provide financial support, not only protecting the rights of those victimized by the polluter's actions, but also contributing to the effective governance of the polluted soil.[28]

[26]HU (2012a, b), p. 92.

[27]Chen (2015b), p. 665.

[28]Wang (2012a, b), p. 35.

4.6 Dispute Over Restoration Goal

Regarding the goal of pollution cleanup and soil governance, there are two different propositions: one requires the polluted soil to be restored to its original state or be absolutely free of pollution; the other requires it to be restored so that it can satisfy its future use.[29] The restoration is not just environment-oriented. It is primarily intended to pursue the "maximum benefit" as a result of governance and redevelopment by coordinating with the overall development of society.[30] Some scholars are of the view that it is impossible and unnecessary to restore the soil to its original state or to being absolutely free of pollution.[31] The original state may be unknown and even if it is known, it may be, technically speaking, extremely difficult or even impossible to accomplish this level of restoration. Additionally, even though techniques may permit it, that option may still not be feasible or acceptable economically.[32] Thus, China should combine soil contamination control with the redevelopment of land, and according to its future use, determine the goal of control and restoration, establish restoration standards for lands of different uses, and integrate flexibility into the system for restoration with regards to the actual governance and restoration.[33]

5 Outlook

In conclusion, China is confronted with serious soil contamination. Relevant regulatory documents have been promulgated at the national and local level to deal with this issue. Sporadic provisions concerning the prevention and control of soil contamination have also been made in existing laws and regulations on other specific environmental media. Some progress has also been made in the development of legislation specific to soil contamination prevent and control as well. In response to the existing problems and realizing the status and characteristics of China's soil contamination, both the central and local governments will strive toward legislative cohesion, improving the relevant systems, and prioritizing solutions to the controversies over legislative mode, legislative focus and appropriate local standards. In this way, the legal system for soil contamination prevention and control will be established in conformity with China's national conditions.

[29]Liang (2008a, b, c, d), pp. 85–91.

[30]Ye (1998).

[31]Hu (2012a, b), p. 93.

[32]Liang (2008a, b, c, d), p. 20.

[33]*Ibid.*

References

Cai S, Li J (2008) Outline of soil pollution prevention and control law. J Henan Adm Inst Polit Law (3):132

Caixin Century Weekly (2011) Cadmium rice murder. http://magazine.caing.com/2011/cwcs437/. Accessed 15 May 2016

Chen N (2015a) My view on the amendment of soil environment quality standard for agricultural land. http://www.thepaper.cn/newsDetail_forward_1297112. Accessed 15 May 2016

Chen N (2015b) Soil Legislation from the Perspective of Pedology, China Economic Report (6):52

China Environment Network (2015) How to adapt foreign legislation on soil pollution prevention and control to Chinese use? http://www.epuncn.com/nengyuan/dikuang/87237_2.htm. Accessed 16 May 2016

Hu J (2012a) Legislative conception of soil environment protection law in our country. J Shanghai Univ (Social Sciences) (11):83

Hu J (2012b) Legislative conception of soil environment protection law in our country. J Shanghai Univ (Social Sciences) (11):92

Kaiqi C (2015) The legislative on soil pollution prevention and control in China. Presented to the 2015 National seminar on environment and resources law, p 665

Liang J (2008a) Choose a scientific legislative pattern for China's soil pollution prevention and control. Environ Prot :20

Liang J (2008b) The legislative patterns of major countries and regions on soil pollution prevention and control. Law Rev :85–91

Liang J (2008c) Study on the legislation mode of soil pollution prevention and control in the main countries and regions of the world. Law Rev (3):87

Liang J (2008d) Choose a scientific legislative pattern for China's soil pollution prevention and control. Environ Prot (4):19

Luo J (2007) Study over China's legislation on soil pollution prevention and control. Mod Law Sci (6):100

Ministry of Environmental Protection of the PRC (2006) A summary of the survey of the status quo of the whole nation. http://sts.mep.gov.cn/trhjbh/qgtrxzdc/200612/t20061231_99195.htm. Accessed 15 May 2016

Ministry of Environmental Protection of the PRC (2014) Ministry of environmental protection and the Ministry of land and resources issued a national survey of soil pollution. http://www.zhb.gov.cn/gkml/hbb/qt/201404/t20140417_270670.htm. Accessed 15 May 2016

People's Government of Mingle County, Gansu Province (2015) Notice on the issuance of emergency plan for soil environmental pollution event in Minle County. http://www.gsml.gov.cn/Item/115864.aspx. Accessed 15 May 2016

Qiu Q (2008) Soil pollution prevention law in Japan and South Korea. J Ecol Rural Environ (24):85

The Emergency Management Office of Zhangye City (2014) Notice on printing and distributing the emergency plan for soil environmental pollution events in Zhangye City. http://yjb.zhangye.gov.cn/Item/48329.aspx. Accessed 15 May 2016

The Website of National People's Congress (1994) Report on behalf of the eighth session of the second National People's Congress on the environment and Resources Protection Committee to consider the report submitted by the Commission to consider the results of the motion. http://www.npc.gov.cn/wxzl/gongbao/2000-12/14/content_5003271.htm. Accessed 15 May 2016

The Website of National People's Congress (2001) NPC Environmental and Resources Protection Committee on behalf of the ninth session of the Fourth National People's Congress on behalf of the delegation delivered a report on the deliberations of the bill. http://www.npc.gov.cn/wxzl/gongbao/2002-01/28/content_5284122.htm. Accessed 15 May 2016

The Website of National People's Congress (2014) Representatives suggested the development of soil pollution control law. http://www.npc.gov.cn/npc/xinwen/dbgz/yajy/2014-01/14/content_1823522.htm. Accessed 15 May 2016

United Nations (2015) The FAO report to call for the world to protect soil resources. http://www.un.org/chinese/News/story.asp?NewsID=25249. Accessed 15 May 2016
Wang S (2008) Thoughts over the Enactment of the Law of the People's Republic of China on soil pollution prevention and control. Law Rev (3):78
Wang N (2012a) Research on the legislative issues of soil pollution control in our country. Chinese Selected Master's Thesis Full-Text Database (4):31
Wang N (2012b) Research on the legislative issues of soil pollution control in our country. Chinese Selected Master's Thesis Full-Text Database (4):35
Ye J (1998) Soil pollution and land utilization—the draft of soil pollution prevention and control law. Lawyer Magazine

The Protection of Soil: Does the European Union Live Up to Its Own Ambitions?

Markus Raffelsiefen and Thomas Strassburger

1 Introduction

The *Treaty on European Union*[1] (TEU) defines in Article 3(1) the Union's aim: "to promote peace, its values and the well-being of its peoples". The article goes further by stressing in its third paragraph that the Union shall work for the sustainable development of Europe aiming at a high level of protection and improvement of the quality of the environment.

Article 191 of the *Treaty on the Functioning of the European Union*[2] (TFEU) stipulates that the Union shall have a Union policy on the environment that contributes, among other things, to the preservation of the quality of the environment, to protecting human health, to a prudent use of natural resources and to dealing with worldwide environmental problems. As such, the idea of a dedicated and effective Union policy on the sustainable use of Europe's soils is inevitable.

[1]Official Journal C 326, 26/10/2012, p. 1.

[2]Official Journal C 326, 26/10/2012, p. 47.

M. Raffelsiefen
Division "Soil Conservation and Resource Efficiency Legislation, Mining Legislation", Federal Ministry for the Environment, Nature Conservation, Building and Nuclear Safety, Bonn, Germany
e-mail: markus.raffelsiefen@bmub.bund.de

T. Strassburger (✉)
Division "Soil Conservation and Management of Contaminated Sites", Federal Ministry for the Environment, Nature Conservation, Building and Nuclear Safety, Bonn, Germany
e-mail: thomas.strassburger@bmub.bund.de

H. Ginzky et al. (eds.), *International Yearbook of Soil Law and Policy 2016*,
International Yearbook of Soil Law and Policy,
DOI 10.1007/978-3-319-42508-5_23

This article intends to sketch the current European approach to address efficient and sustainable use of one of its most vital assets - soil.[3]

Unlike for water and air, soil is still not subject to a comprehensive and coherent set of rules in the Union. Moreover, only a few Member States have specific legislation on soil protection, including for example Austria, Germany and the Netherlands.

The European Commission submitted a proposal for a **Soil Framework Directive** in 2006[4] to establish a certain level of protection for soil as with water and air. It was accompanied by a Thematic Strategy for Soil Protection which mentioned soil erosion, soil organic matter decline, decline in soil biodiversity, salinisation, soil compaction and floods/landslides as well as soil contamination and soil sealing as main threats to soil.[5] The Commission withdrew this proposal after lengthy and inconclusive discussions with Member States in 2014. With the official withdrawal, efforts to establish targeted legislation for soil protection at the European level have come to a preliminary halt.

The need for action to care better for our soils did not disappear, though. Just the contrary: The European Environment Agency's 2015 State of the Environment Report[6] (SOER) gives a bleak outlook on European land resources and soil functions. According to its synthesis report, both for the 10 year trend and the 20+ year outlook, for soil 'deteriorating trends dominate' - to stick to the wording of the SOER. This is also the message of a scientific assessment 'The State of Soil in Europe'[7] which demonstrates that soil is under increasing environmental pressure across the Union. Does Europe fall short of its own requirements?

Soil degradation comes at a cost and affects our economies. For example a study[8] estimates the damage from 357 recorded floods in the EU over the 2002–2013 period at a cost of €150 billion in damage at least and suggests that investing in green infrastructure (e.g. restoring natural features to have less soil sealing and therefore better water retention) could help reduce such costs. For soil erosion damage is estimated to cost €53 million per year in the United Kingdom alone.[9] In Italy, landslides and flooding events are estimated to account for approximately 1 billion euros in damage per year in the period 1951–2009.[10] Also of greatest concern are the costs of desertification and the consequences for farming, the landscape and the whole economy. As an extreme form of land degradation,

---

[3]According to the Soil Thematic Strategy (COM(2006)231 final, p. 2), "soil is generally defined as the top layer of the earth's crust, formed by mineral particles, organic matter, water, air and living organisms. It is the interface between earth, air and water and hosts most of the biosphere."

[4]The contents and context of this proposal will be examined in detail in Sect. 3 of this article.

[5]COM(2006)231 final of 22.09.2006, p. 7.

[6]European Environment Agency (2015).

[7]Joint Research Center (2012).

[8]Fenn et al (2014).

[9]DEFRA (2009), p. 11.

[10]Trezzini et al (2013).

desertification results in a serious impairment of all soil functions. Most visible forms of desertification are to be witnessed in parts of the Mediterranean region. Despite the lack of a sound scientific assessment, it is to be expected that prolonged droughts and irregular weather patterns will also affect regions beyond the Southern part of Europe. Already now, states like Bulgaria, Romania, Slovakia and Slovenia consider themselves affected. They are included in Annex V of the United Nations Convention to Combat Desertification.[11]

Adding to the loss of soil quality due to poor management, which may result in declining productivity or water retention potential or impacts on the ability of a soil to act as a habitat,[12] ever ongoing urbanization is taking its toll, leading to substantial loss of fertile land because of soil sealing. Available data show that close to half of the land take has come at the expense of arable farmland and permanent crops, nearly a third at the expense of pastures and mosaic farmland, and over 10 % at the expense of forests and transitional woodland shrubs. In the EU an area of at least 1000 km^2 is converted annually for infrastructure development—this is equivalent to the size of the city of Berlin. Or, reflecting more on the loss of quality soil, this corresponds to the entire acreage of the Netherlands in just 10 years. Reality is probably even worse since some land take goes unseen by the monitoring scheme. This comes at a price - and in order to compensate for the inappropriate handling of soil resources and the associated loss of its ecosystem services the shrinking pool of remaining soils has to be put to intensified use. The continued unsustainable use of soils does not only compromise the Union's domestic and international biodiversity and climate change objectives, but is also adding more and more pressure on food security. From 1990 to 2006 alone, an agricultural production potential of over 6 million tons of wheat has been lost due to land take in the Union. This trend is also at the expense of third countries, in which we increasingly outsource the production of food or other commodities. To compensate for the loss of one hectare of fertile land in Europe, an area up to ten times larger in another part of the world will be put to use.

2 The Union Acquis Regarding Soil Protection

The withdrawal of the proposal for a Soil Framework Directive mentioned above does not mean that there are no EU policies affecting the management and the protection of soil in the EU. Quite the contrary: While provisions on soil

[11]UNCCD, 2001—The purpose of this Annex is to provide guidelines and arrangements for the effective implementation of the Convention in affected country Parties of the Central and Eastern European region, in the light of its particular conditions.

[12]As a very dynamic system soil performs multiple functions and delivers services vital to human activities and to the survival of ecosystems: biomass production, storing, filtering and transforming nutrients and water, hosting the biodiversity pool and acting as a habitat, serving as a platform for most human activities, providing raw materials, acting as a carbon pool and storing the geological and archaeological heritage.

protection - and environmental protection in general, for that matter - have long been absent from the law of the European Communities, some of its policies have had a strong influence on soil and land use in Europe from its beginnings.

Today, a series of soil protection aspects can be found in the *acquis*, with different Community policies addressing soil aspects at varying degrees. Besides agriculture, policies in areas such as water, waste, chemicals, and prevention of industrial pollution contribute to the protection of soils.

At the level of the European Union, two principal areas of instruments for the implementation of policy objectives can be distinguished, apart from political initiatives and awareness raising: financial incentives and the creation of a regulatory framework.[13] The most important financial incentives impacting on soil in Europe are the subsidies granted within the framework of the Common Agricultural Policy (see Sect. 2.1). A regulatory framework contributing to soil protection can be (and has been, to a certain extent) created within the scope of two competences conferred to the European Union: the internal market (see Sect. 2.2) and environment policy (see Sect. 2.3).

The policy areas concerned in this paper are all subject to the shared competence of the European Union (Article 4 (2)(a), (d) and (e) TFEU). Therefore, the principle of subsidiarity stipulated in Article 5 (3) TEU, applies. This central element of European legislation means that "the Union shall act only if and in so far as the objectives of the proposed action cannot be sufficiently achieved by the Member States, either at central level or at regional and local level, but can rather, by reason of the scale or effects of the proposed action, be better achieved at Union level."

2.1 Soil-Related Aspects of the Common Agricultural Policy (CAP)

2.1.1 The CAP as a Principal Determinant of Land Use

The *Treaty of Rome* establishing the European Economic Community,[14] signed in 1957, on the one hand established a Common Market, including for agricultural products (Article 38 (1) subpara. 2 TFEU[15]). On the other hand, some Member States, especially France, and the farming professional organisations wanted to maintain a high level of state intervention in agriculture. This intervention had to be transferred to the Community level.[16] Therefore, the Treaty of Rome also provided

[13]European Commission (2015a), ch. 1.

[14]Now the TFEU.

[15]For the sake of convenience, all references to the European treaties cite the provisions of the current version of the treaties after the entry into force of the Lisbon Treaty.

[16]European Parliament (2016).

for the establishment of a Common Agricultural Policy (CAP) (Article 38 (4) TFEU) which was realized in 1962.[17]

On the background of the food shortages in Europe after World War II, this policy focused on maximizing agricultural production by providing price guarantees to farmers and thus, promoted a short-term maximization of agricultural land use.[18] The objectives of the CAP, laid down in Article 39 TFEU, did not deal with the environmental aspects of food production. The first major reform of the CAP in 1992 ("MacSharry reform"[19]) aimed at limiting rising production. It replaced the price guarantee system by "direct payments" depending on the size of arable land cultivated by a farmer. It also created 'set-aside' payments to withdraw land from production, payments to limit stocking levels, and introduced measures to encourage retirement and afforestation. While these measures were for the most part not directly aimed at the protection of the environment, they helped to relieve some of the pressure linked to the intensification of land use with its negative effects such as soil compaction and pollution.[20]

2.1.2 The Legal Basis for the Integration of Soil Protection Concerns into the CAP

Meanwhile, the legal basis for the adoption of secondary law for the implementation of the CAP (Article 43 TFEU) offers also the basis for environment and soil related legislation, insofar as the provisions concerned are related to agricultural practices.[21] The most obvious example is the introduction of environmental criteria into the legislation concerning subsidies under the CAP, but this competence might also be invoked in order to regulate the use of fertilisers in agriculture.[22]

The need for the integration of soil protection requirements into the CAP legislation is expressed by the TFEU itself: Article 11 TFEU stipulates that "environmental protection requirements must be integrated into the definition and implementation of the Union's policies and activities, in particular with a view to promoting sustainable development."

[17]Tangermann and von Cramon-Taubadel (2013), p. 17; Zobbe (2001), pp. 4–5.

[18]Delayen (2007), p. 1.

[19]The basic document of this reform is Commission Communication COM(91) 100 final: "The development and future of the CAP—Reflections paper of the Commission".

[20]Lee (2006), p. 73.

[21]Epiney (2015), margin number 26.

[22]The placing on the market of fertilisers has been regulated on the basis of the internal market competence of the Union.

2.1.3 Agri-Environment Measures and Soil Protection

A major step was the CAP chapter of the Agenda 2000,[23] decided on the basis of the *Council Strategy on Environmental Integration and Sustainable Development in the Common Agricultural Policy* (the so-called "Cardiff process") adopted by the European Council in 1999. It introduced the EU Rural Development Policy, also known as the second pillar of the CAP. Its second axis concerns the improvement of the environment and the countryside through support for land management as well as helping to fight climate change ("agri-environment measures"). Such projects could for example concern preserving water quality, sustainable land management, planting trees to prevent erosion and floods. For 2007–2013 EU expenditure on agri-environment measures amounted to nearly 20 billion EUR or 22 % of the expenditure for rural development.[24] It is only fair to mention that soil related measures have been part of the Rural Development Policy, but at a decent level. Soil has never been a top priority.

Agri-environment measures provide payments to farmers who subscribe, on a voluntary basis, to environmental commitments related to the preservation of the environment and maintaining the countryside.[25] Farmers commit themselves, for a minimum period of at least 5 years, to adopt environmentally-friendly farming techniques that go beyond existing legal obligations. Agri-environment payments thus encourage farmers to adopt agricultural activities or levels of production intensity that deliver positive environmental outcomes, while not being necessarily the first choice from the point of view of profitability. In return, farmers receive payments that provide compensation for additional costs and income foregone resulting from applying those environmentally friendly farming practices in line with the stipulations of agri-environment contracts.

Agri-environment measures may be designed at the national, regional, or local level through rural development programming (RDP) to be adapted to particular farming systems and specific environmental conditions (e.g. intensive or extensive farming, local landscape features). Depending on the political will and the environmental situation as assessed before the programme period, Member States may incentivize agricultural management practices with direct or indirect positive effects on soil.

Voluntary measures with a potential to address soil management are for example favouring the build-up of soil organic matter, the enhancement of soil biodiversity, and the reduction of soil erosion, contamination and compaction.[26]

[23]Bulletin of the EU, Supplement 5/97, p. 26.

[24]European Commission (2015b).

[25]European Commission (2015b).

[26]European Commission (2015c).

2.1.4 Cross-Compliance and Soil Protection

Since 2003, all farmers receiving direct payments are subject to compulsory cross-compliance.[27] The current legal basis is Regulation (EU) 1306/2013 and Commission Implementing Regulation 809/2014.

Cross-compliance includes two elements, both of which have a certain potential to contribute to soil protection:

Statutory Management Requirements (SMR) are 13 legislative standards of the acquis in the field of the environment, food safety, animal and plant health and animal welfare listed in Annex II of Regulation (EU) 1306/2013. Due to the lack of dedicated soil legislation at the European level, there are only a few standards that have some relevance for soil protection: the Sewage sludge directive 86/278/EEC, the Nitrates Directive 91/676/EEC and the Plant Protection Products Regulation (EC) No. 1107/2009.[28]

Good agricultural and environmental condition (GAEC) is a new requirement, first introduced in 2009 on top of the respect of the SMR. These standards go beyond requirements that are already part of existing EU legislation and can play a positive role for soil protection. As of today three GAEC requirements relevant to soil and soil carbon are listed in Annex II of Regulation (EU) 1306/2013:

- Maintenance of a minimum soil cover (GAEC 4),
- minimum land management reflecting site-specific conditions to limit soil erosion (GAEC 5), and
- maintenance of soil organic matter through appropriate practices including a ban on burning arable stubble, except for plant health reasons (GAEC 6).

These standards repeat what was included in the previous 2009 CAP regulation and indeed go back to 2003 and the introduction of cross-compliance.

Member States shall define minimum requirements on the basis of these standards. Minimum GAEC requirements should be defined by taking into account the specific characteristics of the areas concerned, including soil and climatic condition, existing farming systems, land use, crop rotation, farming practices and farm structures. The implementation of the GAEC framework results in a high variety of minimum requirements throughout Europe and sometimes even within the Member States when GAEC standards are defined at regional level. This flexibility is, in principle, desirable to account for the heterogeneity of agricultural conditions across Europe. Farmers who decide to ignore the restriction of cross-compliance will be penalized by the loss of a certain percentage of their payments. However, the criticism is made that Member States in the past have interpreted these standards in a lax way and also that they have not really been enforced sufficiently.

It is to be stressed here that GAEC reflects minimum standards only, not ensuring *best* practices. As such, its aspiration level may be below the standards

[27] European Commission (2015d).

[28] See Sect. 2.2 and 2.3.

of national farming advisory services. As such, national services are sometimes perceived by farmers as too demanding when comparing to GAEC.

2.1.5 The "Greening" of the First Pillar of the CAP

A long-awaited reform - agreed upon in 2013 for the CAP framework 2014–2020 - introduced a new environmental aspect into the first pillar by creating "Greening payments" now making up for 30 % of direct payments.[29] Each holding will receive a payment per hectare for respecting certain agricultural practices beneficial for the climate and the environment.

The three basic measures in Article 43 (1) of Regulation (EU) No 1307/2013 are: maintaining permanent grassland, crop diversification and maintaining an "ecological focus area" of at least 5 % of the arable area of the holding. One GAEC standard under the previous CAP rules - before 2013 - had been the protection of wetland and carbon-rich soil (GAEC 7). Following the reform it has been moved into the basic text to become part of the permanent grassland eligibility condition.

Overall, there is a rather small contribution of the "greening" as to soil protection. It is therefore hardly convincing that the farming sector has pointed at regulatory soil protection within the Common Agricultural Policy as an argument against the proposal for a Soil Framework Directive - so much for the alleged 'duplication of legislation'.

2.2 Soil-Related Provisions Within the Internal Market Legislation

The Union competence for legislative measures for the approximation of laws in order to ensure the establishment and functioning of the internal market has been part of the Treaties since 1957. It is laid down in Article 114 TFEU. In the context of soil protection, it can be used to set European standards for products impacting on the environment (cf. Article 114 (3)),[30] and also on soil. It is a common feature of these rules that they concentrate on the properties and the marketing of products, and not on the circumstances of their application, e.g. the accumulation of the input of certain contaminants. The harmonization of product standards at the European level can also be problematic as Member States are, according to Article 114 (4) to (6) TFEU, normally barred from introducing more stringent national rules, for example in favour of soil protection, for the products concerned, until the Commission has approved these rules.[31]

[29]European Commission (2015e).

[30]Nettesheim (2015), margin number 2.

[31]Maletić (2013), p. 19.

Some product standards relevant for soil protection are included in the following legal acts.

Regulation (EC) No 2003/2003 relating to fertilisers lays down rules for the placing on the market of fertilisers. The topic is obviously very important for soil protection. While the use of fertilisers aims at increasing soil fertility, they can also contribute to soil degradation in various ways (e.g. excessive accumulation of nitrogen, contamination, salinisation). The scope of the current version of the Fertilisers Regulation (a recast of Council Directive 76/116/EEC) is nonetheless limited. It includes primarily requirements for the free circulation of fertilisers marked "EC fertilisers" and listed in Annex I of the Regulation. These requirements concern identification and labelling as well as tolerances for nutrient contents. As to environmental protection requirements, Article 14 (b) contains the vague requirement that a type of fertiliser may only be included in Annex I if, under normal conditions, it does not adversely affect the environment. In practice, the inclusion of a fertiliser depends on the decision of the Committee according to Article 31 (1).

The Commission has recently published a proposal for a revised Fertilisers Regulation.[32] One of its objectives is to reduce soil and water contamination by EC-fertilisers.[33] This includes harmonised cadmium limits for some fertilizers in the product requirements in Annex I of the Regulation. To limit cadmium accumulation in soil is a prerequisite for the production of food with contaminant levels that will not give reason to worry for consumers. Since most of the fertilisers are already harmonised, Member States could not achieve this objective unilaterally.[34]

The **Biocidal Products Regulation** (EU) No. 528/2012 and the **Plant Protection Products Regulation** (EC) No 1107/2009 similarly contain harmonized rules for the placing on the market of products that may lead to the entry of dangerous substances into the soil.

Last but not least, the **Renewables Directive 2009/28/EC** may have a role in the development of European soil protection legislation.[35] This Directive mandates levels of renewable energy use in the European Union and contains various rules that are supposed to promote their use. In the context of the promotion of biofuels and bioliquids, the Directive provides for a sustainability scheme that promotes the use of restored degraded land for their production. To this end, it is necessary to establish definitions for severely degraded or contaminated land. The Commission has recently published a roadmap for a Commission Regulation on detailed definitions for degraded and heavily contaminated land[36] that could also be taken as a basis in future soil protection legislation[37] and in order to establish rules for the attainment of the Sustainable Development Target 15.3 of a 'land-degradation

[32]COM (2016) 157.

[33]COM (2016) 157, p. 3.

[34]COM (2016) 157, p. 6.

[35]Sanden (2016).

[36]European Commission (2015f).

[37]Sanden (2016).

neutral world'[38] (see Sect. 4) - for which we first need to have a common definition of what degraded land actually means.

2.3 *Soil-Related Provisions Within the Environmental Legislation*

The European Union today has a comprehensive legislative competence for environmental protection that is included in Articles 191–193 TFEU. This competence has only been introduced by the Single European Act in 1987. Earlier legal acts concerning the environment could, to a certain extent, be based on what is now Article 114 TFEU (see above) and on the flexibility clause in Article 352 TFEU.[39]

It should be noted that Article 192 (2) TFEU provides for a special legislative procedure (the Council acting unanimously and after consulting the European Parliament) regarding environmental measures affecting spatial planning and land use. This could also become relevant if future soil protection legislation contains planning provisions.

The current Article 192 (1) TFEU has been used extensively to legislate in the field of environment protection. In the absence of dedicated legal provisions for soil, a series of legal acts exist that contribute partially to soil protection, in particular the following directives:

2.3.1 Industrial Emissions Directive 2010/75/EU

The Directive on industrial emissions (IED), a recent recast of several prior Directives relating to emissions and waste control, contains provisions that are relevant for soil protection.[40] Pursuant to its Article 1, "it lays down rules on integrated prevention and control of pollution arising from industrial activities. It also lays down rules designed to prevent or, where that is not practicable, to reduce emissions into air, water and *land*". The Directive applies to a wide range of industrial activities listed in the Directive and its Annexes, all in all some 52,000 installations in Europe.[41]

The principal instrument required by the Directive is the obligation for the operator of an industrial installation to obtain a permit for the establishment and operation of the installation (Article 4). This permit system enables authorities to exercise an effective preventive control over industrial activities that may lead, i.a.,

[38] United Nations (2015), p. 24.

[39] Nettesheim (2015), margin number 3.

[40] Altvater et al. (2015), p. 28.

[41] European Commission (2015g).

to soil contamination. The permit shall include numerous requirements that are also relevant to soil protection (Article 14):

a) "emission limit values for polluting substances listed in Annex II, and for other polluting substances, which are likely to be emitted from the installation concerned in significant quantities,(. . .)". While the polluting substances listed in Annex II are only categorized under "air" and "water", it is obvious that many of them are also relevant to soil quality,
b) appropriate requirements ensuring protection of the soil and groundwater (. . .),
c) suitable emission monitoring requirements (. . .).

A new instrument introduced by Article 22 para 2 of Directive 2010/75/EU is the *baseline report* that is supposed to strengthen the polluter-pays principle.[42] The baseline report is required before the start of the operation of an installation "where the activity involves the use, production or release of *relevant hazardous substances* and having regard to the possibility of soil and groundwater contamination at the site of the installation (. . .)". (Article 22 (2) subpara. 1) "The baseline report shall contain the information necessary to determine the state of soil and groundwater contamination so as to make a quantified comparison with the state upon definitive cessation of activities" (Article 22 (2) subpara. 2). The existence of this baseline report makes it possible to require that "where the installation has caused significant pollution of soil or groundwater by relevant hazardous substances compared to the state established in the baseline report (. . .), the operator shall take the necessary measures to address that pollution so as to return the site to that state" (Article 22 (3)).

This obligation "to return the site to that [original] state" goes beyond the obligation to "take the necessary actions against any significant risk to human health or the environment due to the contamination of soil and groundwater as a result of the permitted activities" required by Article 22 para. 4 in absence of a baseline report. While the latter obligation can be fulfilled by implementing safeguarding measures (risk-based approach), the former requires the active restitution of the site ("no degradation").

It should be noted that unlike the Commission's proposal for soil protection, the IED was not subject to endless discussions, opposing the idea of common action, despite explicitly including the protection of soil resources. As to the IED, the prevailing opinion had been that none of the EU Member States can tackle today's challenges solely on its own. This shows a certain level of consistency in policy making, necessary to ensure a sufficient level of effectiveness of action, also because of the need for a level playing field within the Internal Market.

---

[42] Altvater et al. (2015), p. 30.

2.3.2 Waste Management Legislation

Several European legal acts in the field of waste management contain provisions that are relevant for soil protection, in particular the Sewage Sludge Directive and the Landfill Directive.

Council Directive of 12 June 1986 "on the protection of the environment, and in particular of the soil, when sewage sludge is used in agriculture" (86/278/EEC) is one of the earliest European legal acts that explicitly addresses soil protection issues. It does so quite precisely with regard to the particular issue of the use of sewage sludge in agriculture. Sludge tends to concentrate heavy metals and poorly biodegradable trace organic compounds. Sludge is, however, rich in nutrients such as nitrogen and phosphorous and contains valuable organic matter that is useful when soils are depleted or subject to erosion.[43]

In order to limit the negative effects of the use of sewage sludge on the soil, Directive 86/278/EEC requires Member States to lay down limit values for (only) seven heavy metals according to Annex I A of the Directive. Member States then have to "prohibit the use of sludge where the concentration of one or more heavy metals in the soil exceeds the limit values (…) and shall take the necessary steps to ensure that those limit values are not exceeded as a result of the use of sludge" (Art. 5 para. 1). Member States are also required to "regulate the use of sludge in such a way that the accumulation of heavy metals in the soil does not lead to the limit values referred to in paragraph 1 being exceeded." (Art. 5 para 2) Article 8 determines that when it is used, sludge shall be used "in such a way that (…) the quality of the soil (…) is not impaired".

Council Directive 1999/31/EC of 26 April 1999 on the landfill of waste provides a comprehensive framework[44] for the very important topic of landfills which directly benefits soil protection. It provides for measures to prevent or reduce the negative effects on the environment, in particular the pollution of water and soil (Art. 1). The principal standards for landfills are laid down in the Annexes of the Directive. Annex I sets up general requirements for all classes of landfills. It also addresses the protection of soil and notes that "[a] landfill must be situated and designed so as to meet the necessary conditions for preventing pollution of the soil […]" (Annex I para. 3.1). The protection of the soil is to be achieved, i.a. "by the combination of a geological barrier and a bottom liner during the operational/active phase and by the combination of a geological barrier and a top liner during the passive phase/post closure." Annex I para 3.2 requires the installation of a leachate collection and sealing system.

---

[43] European Commission (2015h).

[44] Altvater et al. (2015), p. 36.

2.3.3 Environmental Assessment Legislation

Directive 2011/92/EU on the assessment of the effects of certain public and private projects on the environment (a recast of the original Directive 85/337/EEC), as amended by Directive 2014/52/EU, requires an environmental impact assessment (EIA) for a wide range of defined public and private projects listed in the Annexes of the Directive. The aim is to ensure that environmental implications of decisions are taken into account before the decisions are made.[45]

Its Article 3 notes that the assessment of environmental impacts "shall identify, describe and assess in an appropriate manner, in the light of each individual case, the direct and indirect significant effects of a project" on, inter alia, "(c) land, soil (…)."

The reference to "land" has been introduced by Directive 2014/52/EU amending the EIA Directive. Recital 9 of this Directive gives extensive reasons for this amendment and makes references to recent developments regarding soil protection: "The Commission Communication of 22 September 2006 entitled 'Thematic Strategy for Soil Protection' and the Roadmap to a Resource-Efficient Europe underline the importance of the sustainable use of soil and the need to address the unsustainable increase of settlement areas over time ('land take'). Furthermore, the final document of the United Nations Conference on Sustainable Development held in Rio de Janeiro on 20–22 June 2012[46] recognises the economic and social significance of good land management, including soil, and the need for urgent action to reverse land degradation. Public and private projects should therefore consider and limit their impact on land, particularly as regards land take, and on soil, including as regards organic matter, erosion, compaction and sealing; appropriate land use plans and policies at national, regional and local level are also relevant in this regard."

Information on likely significant effects on a variety of environmental concerns, including soil, also has to be provided for a number of planning decisions ahead of relevant projects according to Annex I (f) of the Strategic Environmental Assessment Directive 2001/42/EC.

2.3.4 Environmental Liability Directive

Directive 2004/35/CE of the European Parliament and of the Council of 21 April 2004 on environmental liability with regard to the prevention and remedying of environmental damage provides a common legal framework for remediation/offsetting and prevention of several types of environmental damages that have occurred after the entry into force of the Directive (2007) (Article 1), including "land damage". This is defined as "any land contamination that creates a significant

[45]European Commission (2015i).

[46]United Nations (2012), p. 40.

risk of human health being adversely affected as a result of the direct or indirect introduction, in, on or under land, of substances, preparations, organisms or micro-organisms" (Article 2 (1)(c)). Damage to land is thus only covered by the Directive in the case of a contamination that poses a risk for human health. Furthermore, the directive applies only to certain occupational activities listed in Annex III and to other occupational activities only when the operator has been at fault or negligent (Article 3 (1)).

When all these criteria are fulfilled, Articles 5–8 of the Directive contain a rather complete set of obligations for operators of occupational activities that cause or about to cause an environmental damage: When there is an imminent threat of such damage occurring, the operator shall, without delay, take the necessary preventive measures (Article 5). Where environmental damage has occurred, the operator shall take the necessary remedial measures in accordance with Annex II (Articles 6 and 7). The operator shall bear the costs for the preventive and remedial actions taken pursuant to this Directive (Article 8 (1)). If the competent authority has taken preventive or remedial actions, it shall recover those costs (Article 8 (2)), thus enforcing the polluter-pays principle.

2.3.5 Habitats Directive 92/43/EEC

Council Directive 92/43/EEC of 21 May 1992 on the conservation of natural habitats and of wild fauna and flora is interesting for soil protection in as much it provides for the setup of a network of special areas of conservation (Article 3). While it concentrates on natural habitat types and maintaining species in a favourable conservation status, the protection of special areas of conservation is supposed to cover all natural components of a natural habitat. It can therefore also provide a tool to address soil degradation, e.g. the reduction of soil contamination, soil erosion and compaction as well as soil sealing.[47]

3 The Struggle for a Targeted European Legal Act on Soil Protection

As shown in recent scientific articles and reports including those of the European Environment Agency (see Sect. 1), the current patchy approach, combined with national legislation which is mostly limited to contaminated sites, has not succeeded in preventing soil degradation across the Union. On the other side, it has resulted in different national rules and conditions for economic operators in the

[47] Altvater et al. (2015), p. 36. For further details on the establishment of an ecological network of special areas of conservation (NATURA 2000) and soil-relevant conservation measures see Heuser (2005), p. 198 ff.

internal market. Even the full implementation of existing provisions will only yield fragmented and incomplete protection to soil.

3.1 The Soil Thematic Strategy

For the reasons cited above, the Commission proposed a Thematic Strategy for Soil Protection[48] in 2006 with the objective to protect soils across the EU. Inspired by Germany, the Commission had started in 1998 to discuss soil protection with interested European countries. Consequently, soil entered the policy arena and became an important part of the Union's 6th Environmental Action Programme.[49]

The strategy consists of a Communication[50] from the Commission to the other European Institutions, a proposal for a Framework Directive for the protection of soil (a European law),[51] and an Impact Assessment.[52] While the Communication sets the frame, the proposal for a Soil Framework Directive establishes common principles for protecting soils across the EU. These documents were supported by an analysis of the economic, social and environmental impacts of the different options that were considered in the preparatory phase of the strategy and of the measures finally retained by the Commission.

3.2 The Commission Proposal for a Soil Framework Directive

In a nutshell, the Draft Soil Framework Directive's objective had been to ensure the protection of soil, to prevent its degradation and to restore soil that has been subject to degradation.[53] Its intentions had been to leave it to EU Member States to decide how best to protect soil and how to use it in a sustainable way on their own territory.

In line with its design as an EU framework directive,[54] the proposal allows for flexibility to encompass local variations and national differences in soil protection regimes. It balances subsidiarity and EU action by setting common aims and a common approach for the diagnosis of soil degradation, whilst allowing Member

[48] COM(2006) 231.

[49] OJ L 242, 10.9.2002.

[50] COM(2006) 231.

[51] COM(2006) 232.

[52] SEC (2006) 1165 and SEC(2006) 620.

[53] Soil degradation or soil improvements have a major impact on other areas of Community interest, such as surface water and groundwater protection, human health, climate change, protection of nature and biodiversity, and food safety.

[54] Framework directives, in the parlance of the EU, are directives that leave the implementation of detailed rules to subsequent European legal acts or to the Member States.

States to define their own level of acceptable risk, their own environmental objectives, their own measures, and their own timetable to achieve these objectives.

The proposal for a Soil Framework Directive (SFD) has been structured along three lines: (1) Preventive measures: Member States must ensure a sustainable use of soil. If soil is used in a way that hampers its functions, mitigating actions must be undertaken. Other policies' impacts on soil must be assessed; (2) Identification of the problem: Member States must identify the areas where there is a risk of erosion, decline in organic matter, salinisation, compaction, or landslides. As far as contamination is concerned, Member States must draw up an inventory of contaminated sites; (3) Operational measures: Member States will then have to act upon the risks identified by adopting programmes of measures for the risk areas, national remediation strategies for the contaminated sites, and measures to limit or mitigate sealing. However, they are free to decide upon the level of ambition of their soil policy, to set their own targets, and to decide how and by when to achieve them.

Given the large variability of soil conditions there is not much sense in establishing one-fits-all environmental quality targets at the European level, for example, the natural humus content differs widely depending on the climate and geographical conditions. That is why, after a lengthy consultation process, the SFD proposed a harmonisation of parameters and of the methodology to be used by Member States.

The scope of the soil degradation processes to be considered by the SFD covers erosion, loss of soil organic matter, compaction, salinization, acidification, landslides and contamination. Member States are asked to identify areas with eminent soil risks. By setting their own, national risk reduction targets for those areas and through establishing effective programmes of measures to achieve them, it is up to Member States to ensure sustainable land management on their own territory.

As to contamination, the directive foresees to prevent further contamination, to establish an inventory of contaminated sites on their territory and to draw up national remediation strategies. It has an exhaustive and harmonised approach to the identification of contaminated sites with a detailed procedure and deadlines for systematic identification and remediation of contaminated sites. While the proposal allows for up to 25 years for the identification of all contaminated sites, it is left to Member States to decide on the timing and the procedure of necessary remediation work.

To further limit the risk of hidden contamination and to support the development of an inventory of contaminated sites, a soil status report has to be provided by the seller or the buyer to the administration and the other party in the transaction. This is obligatory when a site is being sold, where a potentially contaminating activity has taken or is taking place. Finally, the Member States are also called upon to limit or mitigate the effects of sealing, for instance by rehabilitating brownfield sites.[55]

[55]Brownfields are often abandoned, closed or under-used industrial or commercial facilities; redevelopment or reuse may be complicated by real or perceived contamination problems. They can be found both in developed urban areas and in the countryside. It requires intervention to bring them back to beneficial use in order to make them part of the land use cycle—promoting land use efficiency.

Remediation of contaminated sites allows for exploitation of capital assets, reducing new land take, and hence the encroachment on agricultural land and green areas.

The SFD proposal required Member States to preserve soil functions, to identify where degradation is already occurring, and to set their own level of ambition and their own timetable to combat such degradation. Member States will need to develop appropriate responses to ensure sustainable soil use. Such responses can build on existing measures and programmes, to avoid any duplication of efforts.

The proposal was subject to the co-decision procedure in accordance with Article 294 TFEU. This means that both the European Parliament and the Council had to agree on a common text on the basis of a proposal from the Commission and taking into account the opinions of the Committee of the Regions and the European Economic and Social Committee. In 2007 the European Parliament adopted its first-reading opinion; the Committee of the Regions and the Economic and Social Committee delivered their opinions. As to the Council, despite very intensive discussions throughout a number of years led by many different Presidencies, it was not possible to gather a qualified majority of Member States in favour of a proposal for a Directive on soil protection. The proposal always ran into a blocking minority with Austria, Germany, The Netherlands and The UK against and with France not in support of the proposal. Despite the positive support of 22 Member States, it was not possible to find agreement.

A last attempt to reach an agreement on a binding soil policy was made under the Greek Presidency in March 2014. A possible way forward at EU level was discussed during the Environment Council meeting. The debate indicated that protecting soils remained an important objective for the Union, despite the fact that, in its present format, the proposal for binding legal framework could not be agreed by a qualified majority.

Since the proposal had been pending for almost 8 years without the necessary qualified majority support in Council, the Commission decided to open the way for a future alternative initiative. In withdrawing the proposal for a Soil Framework Directive, the Commission indicated that "The Commission remains committed to the objective of the protection of soil and will examine options on how to best achieve this. Any further initiative in this respect will however have to be considered by the next college".[56]

3.3 *Recent Activities of the Commission*

Notwithstanding the impasse on the SFD, some progress as to sustainable soil management has been made at EU level over the last years, setting necessary tools in place.

[56]OJ C 153 of 21 May 2014 and *corrigendum* in OJ C 163 of 28 May 2014.

In September 2011, the European Commission adopted a 'Roadmap to a resource efficient Europe'.[57] This roadmap highlights the fact that the current land take trend is unsustainable. It is proposed that, by 2020, 'EU policies take into account their direct and indirect impact on land use in the EU and globally, and the rate of land take is on track with the aim of achieving **no net land take** by 2050; soil erosion is reduced and the soil organic matter increased, with remedial work on contaminated sites well underway'. An ambitious milestone indeed.

In order to contribute to the no net land take target and abiding by the scope of the Soil Thematic Strategy, the Commission has published the 'Guidelines on best practice to limit, mitigate or compensate soil sealing',[58] which is available in all EU official languages. Sealing is one of the main causes of soil degradation in the EU, affecting fertile agricultural land, threatening biodiversity, increasing the risk of flooding and water scarcity, and contributing to global warming. The Guidelines collect examples of policies, legislation, funding schemes, local planning tools, information campaigns and many other best practices implemented throughout the EU. The document highlights the importance of smart spatial planning and the existing development potential inside urban areas.

In 2012, the European Commission published a policy report[59] on the implementation of the Strategy and ongoing activities. The report provides an overview of the actions undertaken by the European Commission to implement the four pillars of the Strategy, namely awareness raising, research, integration of soil aspects into other Union policies, and legislation. It also presented ongoing soil degradation trends both in Europe and globally, as well as future challenges to ensure protection.[60]

The final decision of the Commission to withdraw the proposed soil directive was made against the background of the European Seventh Environment Action Programme[61] (EAP), entering into force at the beginning of 2014. The EAP provides that by 2020 "land is managed sustainably in the Union, soil is adequately protected and the remediation of contaminated sites is well underway" and commits the EU and its Member States to "increasing efforts to reduce soil erosion and increase organic matter, to remediate contaminated sites and to enhance the integration of land use aspects into coordinated decision-making involving all relevant levels of government, supported by the adoption of targets on soil and on land as a resource, and land planning objectives".

[57] COM (2011) 571.

[58] SWD (2012) 101 final/2.

[59] COM(2012) 46.

[60] The policy report has been supplemented by a reference report 'The State of Soil in Europe' of the European Commission's Joint Research Centre providing a comprehensive overview of our present understanding of soil resources and degradation processes. It has been published in collaboration with the European Environment Agency and is a contribution to the Environment State and Outlook report—SOER 2010.

[61] European Commission (2015j).

Additionally, the EAP states that "The Union and its Member States should also reflect as soon as possible on how soil quality issues could be addressed using a targeted and proportionate risk-based approach **within a binding legal framework**".

Against that background, the Commission has established an Expert Group in 2015, to implement the soil protection provisions of the 7th EAP, allowing for a structured and formal dialogue. The group is composed by experts mandated by Member States to support the work of the Commission. It met for the first time in 2015 with two more meetings scheduled for 2016.

In addition to the work of the expert group, a project has been launched to take stock of existing soil policy instruments in the EU in order to identify gaps in protection and to support the decision-making process for further policy action. The specific objectives of this work are to

- develop an inventory of existing and upcoming legislation and policies as well as non-binding instruments, measures and targets in the EU 28 Member States and its regions;
- compile information on and assess the effectiveness of soil legislation and measures in addressing the soil threats and soil functions in the EU, as well as the relevance and coherence of policies

This analysis shall help to identify any gaps in EU policies and national legislation in addressing soil threats and soil functions. The work of the expert group and the outcome of the study is expected to fuel the discussion regarding current and future policy initiatives at Member State and EU level.

4 Perspectives for European Soil Governance and Implementation of the SDG

European policy making will also have to respond to another international process: the outcome document 'The Future We Want' of the United Nations Conference on Sustainable Development held in Rio de Janeiro in June 2012 (Rio + 20) underlines the economic and social significance of good land management.[62] It goes on to recognise the need for urgent action to reverse land degradation and to strive to achieve a land-degradation neutral world in the context of sustainable development.[63]

[62]United Nations (2012), p. 40.

[63]Paragraph 206: "We recognize the need for urgent action to reverse land degradation. In view of this, we will strive to achieve a land-degradation neutral world in the context of sustainable development. This should act to catalyse financial resources from a range of public and private sources."

It is to be hoped that the respective call for implementation of Sustainable Development Goal 15 (Life on Land; its sub-target 15.3 demands to fight soil degradation and to achieve land degradation neutrality)[64] together with the plea of the EAP and the very need to achieve a more sustainable use of soil resources will fuel the debate on effective soil governance in Europe.

The growth in world population, the rising consumption of food products in the emerging economies, and the increased use of biomass for energy and other industrial purposes, coupled with basic nature and biodiversity protection aims, are likely to lead to increasing competing pressures on natural resources and ecosystems through land use changes. Climate change and variability are likely to drive land degradation and related effects on land productivity even further.

European soils are a vital asset, and they will become an even stronger pillar for our well-being in the future - provided we allow for that. At the beginning of the 1960s nearly half a hectare of arable land was available per head, at the beginning of this century it was less than a quarter of a hectare, and - according to FAO - it will halve again by the middle of this century. Thus, to handle with care what is at our feet and on which we depend so much should become a challenge and an example for smart policy making in the nearest future - globally, at the level of the European Union and in its Member States.

Finally and referring to the title of this article, it may be concluded that for the European Union there is still a lot to be done to claim to be on target as to the sustainable use of its soil resources.

> We abuse land because we regard it as a commodity belonging to us. When we see land as a community to which we belong, we may begin to use it with love and respect. (Aldo Leopold)

References

Altvater S, Dooley E, Roberts E (2015) Legal instruments to implement the objective “land degradation neutral world” in International Law. http://www.umweltbundesamt.de/publikationen/legal-instruments-to-implement-the-objective-land. Accessed 10 Apr 2016

DEFRA (2009) Safeguarding our soils. A strategy for England. https://www.gov.uk/government/uploads/system/uploads/attachment_data/file/69261/pb13297-soil-strategy-090910.pdf. Accessed 13 Apr 2016

Delayen C (2007) The common agricultural policy: a brief introduction. http://www.iatp.org/files/451_2_100145_0.pdf. Accessed 10 Apr 2016

Epiney (2015) TFEU Article 192. In: Landmann, Rohmer (2015) Umweltrecht, 77. EL August 2015

European Commission (2015a) Tool #15: the choice of policy instruments. In: Better regulation guidelines. Available via http://ec.europa.eu/smart-regulation/guidelines/tool_15_en.htm. Accessed 10 Apr 2016

European Commission (2015b) Agri-environment measures. http://ec.europa.eu/agriculture/envir/measures/index_en.htm. Accessed 10 Apr 2016

[64] United Nations (2015), p. 24.

European Commission (2015c) Agriculture and soil protection. http://ec.europa.eu/agriculture/envir/soil/index_en.htm. Accessed 10 Apr 2016
European Commission (2015d) Cross-compliance. In: Direct support. http://ec.europa.eu/agriculture/direct-support/cross-compliance/index_en.htm. Accessed 10 Apr 2016
European Commission (2015e) Greening. In: Direct support. http://ec.europa.eu/agriculture/direct-support/greening/index_en.htm. Accessed 10 Apr 2016
European Commission (2015f) Roadmap for a Commission Regulation on detailed definitions for degraded and heavily contaminated land. http://ec.europa.eu/smart-regulation/roadmaps/docs/2015_ener_018_degraded_heavily_contaminated_land_en.pdf. Accessed 10 Apr 2016
European Commission (2015g) The industrial emissions directive. http://ec.europa.eu/environment/industry/stationary/ied/legislation.htm. Accessed 13 Apr 2016
European Commission (2015h) Sewage sludge. http://ec.europa.eu/environment/waste/sludge/. Accessed 13 Apr 2016
European Commission (2015i) Environmental assessment. http://ec.europa.eu/environment/eia/home.htm. Accessed 13 Apr 2016
European Commission (2015j) General Union Environment Action Programme to 2020. Available via http://ec.europa.eu/environment/action-programme/. Accessed 13 Apr 2016
European Environment Agency (2015) Soil. In: State of the Environment Report 2015. http://www.eea.europa.eu/soer/europe/soil. Accessed 13 Apr 2016
European Parliament (2016) Rationale behind the CAP. In: The common agricultural policy (CAP) and the Treaty. http://www.europarl.europa.eu/atyourservice/en/displayFtu.html?ftuId=FTU_5.2.1.html. Accessed 10 Apr 2016
Fenn T, Fleet D, Garrett L, Daly E, Elding C, Hartman M, Udo J (2014): Study on economic and social benefits of environmental protection and resource efficiency related to the European semester, Final Report to DG Environment. http://ec.europa.eu/environment/integration/green_semester/pdf/RPA%20Final%20Report-main%20report.pdf. Accessed 13 Apr 2016
Heuser I (2005) Europäisches Bodenschutzrecht. Erich Schmidt Verlag, Berlin
Joint Research Center (2012) The State of Soil in Europe - a contribution of the JRC to the European Environment Agency's Environment State and Outlook Report - SOER 2010. http://esdac.jrc.ec.europa.eu/ESDB_Archive/eusoils_docs/other/EUR25186.pdf. Accessed 13 Apr 2016
Lee YL (2006) Nachhaltiger Bodenschutz international, europäisch und national. Universitätsverlag der TU Berlin, Berlin
Maletić I (2013) The law and policy of harmonisation in Europe's internal market. Elgar, Cheltenham
Nettesheim (2015) TFEU Art. 191. In: Grabitz, Hilf, Nettesheim, Das Recht der Europäischen Union, 57. EL August 2015
Sanden (2016) Der Fitness-Check oder wie Europa letztlich doch noch zu einer einheitlichen Altlastendefinition gelangen kann! In: W+B Zeitschrift für Deutsches und Europäisches Wasser-, Abwasser- und Bodenschutzrecht 1(2016):1
Tangermann S, von Cramon-Taubadel S (2013) Agricultural policy in the European Union - an overview. https://www.uni-goettingen.de/de/document/download/468756dd26772ba40606fb7034c7995d.pdf/Diskussionsbeitrag-1302.pdf. Accessed 10 Apr 2016
Trezzini F, Giannella G, Guida T (2013) Landslide and flood: economic and social impacts in Italy. In: Margottini C, Canuti P, Sassa K (eds) Landslide science and practice, vol 7. Springer, Berlin/Heidelberg, pp 171–176
United Nations (2012) Report of the United Nations conference on sustainable development. http://www.uncsd2012.org/content/documents/814UNCSD%20REPORT%20final%20revs.pdf. Accessed 28 Apr 2016

United Nations (2015) Transforming our world: the 2030 agenda for sustainable development. Resolution adopted by the General Assembly on 25 September 2015. http://www.un.org/ga/search/view_doc.asp?symbol=A/RES/70/1&Lang=E. Accessed 28 Apr 2016
Zobbe H (2001) The economic and historical foundation of the common agricultural policy in Europe. http://ageconsearch.umn.edu/bitstream/24212/1/ew010012.pdf. Accessed 10 Apr 2016

United States Soil Degradation

Jesse J. Richardson Jr. and Elizabeth Dooley

1 Introduction

Soils are very important in the United States, as elsewhere. The total land area in the US is 2.3 billion acres (931 million hectares).[1] As a large land mass supporting a population of over 300 million people,[2] US soil is impacted by many different soil threats resulting from myriad land uses throughout the country. Agricultural use, including cropping, grazing, and farmsteads, accounts for over 50 % of the total land area.[3] Thus, soil erosion has been a continuous concern targeted by government programs and policies (e.g., the first US Farm Bill was passed in response to the Dust Bowl droughts of the early 1930s) (see the chapter by Rees 2016, for a more detailed description).[4] In 2007, 960 million tons of soil were lost to water erosion and 765 million tons to wind erosion despite decreasing soil erosion on non-federal US cropland from 3.98 to 2.72 tons per acre per year from 1982 to 2010.[5]

Urban areas also occupy 68 million acres (27.5 million hectares), where residential, commercial, industrial, among other uses occur.[6] Urban development has

[1]Nickerson et al. (2011).

[2]U.S. Census Bureau (2015).

[3]Borchers (2015).

[4]Worster (1982). For a historical look at the development of the Farm Bill, see McGranahan et al. (2013).

[5]Natural Resources Conservation Service (2007); United States Department of Agriculture (2013).

[6]Nickerson et al. (2015).

J.J. Richardson Jr. (✉)
West Virginia University, Morgantown, WV, USA
e-mail: Jesse.Richardson@mail.wvu.edu

E. Dooley
Iowa State University, Beginning Farmer Center, Ames, IA, USA
e-mail: iowa.beth@gmail.com

H. Ginzky et al. (eds.), *International Yearbook of Soil Law and Policy 2016*,
International Yearbook of Soil Law and Policy,
DOI 10.1007/978-3-319-42508-5_24

sealed around 4.5 million hectares of farmland as built-up land increased by almost 60 % over the last 30 years.[7] Soil organic matter declines have occurred to varying extents based on regional variables (e.g., temperature, rainfall, soil texture, etc.), with as high as 50 % soil organic carbon losses documented in the Central Plains Grasslands due to cultivation.[8] The Environmental Protection Agency is also responsible for investigating sites where releases of hazardous substances have occurred or are threatened to occur, referred to as National Priorities List or Superfund sites, to determine whether emergency response is necessary. There are currently 1166 non-federal and 157 federal National Priorities List (NPL) sites throughout the country, with an additional 49 non-federal and 4 federal proposed sites, which present potentially serious consequences in terms of soil contamination as well as human health impacts.[9]

This chapter begins by providing a brief background to the legal structure in the United States aimed at preventing or influencing actions against soil degradation. The chapter then focuses on three recent developments related to soil degradation in the United States: the 2014 Farm Bill, the struggle over defining "waters of the United States" under the Clean Water Act and its impact on wetlands,[10] and a recent lawsuit in Des Moines, Iowa over nitrate management on farmland. Although none of these developments directly focus on soil degradation, each could have profound impacts on soil degradation policy in the United States, albeit each in very different ways.

2 Background

The allocation of rights to own, use, convey, alter, exclude or destroy land and soil in the US is traditionally differentiated between public and private.[11] Private actors control around 1.35 billion acres (around 500 million hectares) or 60 % of the surface area, dominated by three major land uses: rangeland, forest land, and cropland.[12] Only 6 % of the land is classified as non-rural developed (e.g., cities)

[7]Emili and Greene (2014).

[8]Burke et al. (1989).

[9]Environmental Protection Agency (2016a). The Superfund program was established under the Comprehensive Environmental Response, Compensation and Liability Act (CERCLA) to provide federal funds and quick clean-up of such sites with the possibility of cost-recovery by the government from liable polluters afterward, e.g., $500 million for the Gowanus Canal in New York. The New York Times, Superfund: Chronology of Coverage, http://topics.nytimes.com/top/reference/timestopics/subjects/s/superfund/index.html. However, failures to allocate federal budget to the program in recent years have slowed the remedial process.

[10]Please see 21 for a detailed description of U.S. wetland types, losses and restoration.

[11]53.

[12]Nickerson et al. (2011).

and is controlled by state and local governments.[13] The US federal government owns approximately 20 % of the nation's land and soil (e.g., large protected areas of public land as national parks and forests), which constitutes 640 million acres (259 million hectares).[14] The US is therefore predominantly rural with a significant amount of its land-based natural resources under the primary control of private actors.

The property regime of the United States is also constructed according to the federalist system of government, offering a high level of protection for private property rights at both the federal and state levels. Enshrined in the Fifth Amendment of the Federal Constitution, property, along with life and liberty, may not be denied to a person without due process of law. Known as the takings clause, the government may not take a person's private property for public use without just compensation. Given the localized nature of land and soil, however, the states have the power to govern the resources located within their boundaries.[15] Therefore, the private property regime for land and soil throughout the US states may slightly vary according to each state's context and political prioritization.

Soil protection within the US is accomplished through federal and state legislation, regulation and policies which specifically target elimination, reduction or mitigation of actions that threaten soil quality and quantity. There are a number of real property common and statutory law constructs that may impact soil use and quality. Some examples include easements, covenants, adverse possession, landlord-tenant, eminent domain, nuisance and trespassing. However, there are also principles and legal mechanisms within the US framework that may effectively result in barriers to regulation or work against soil protection. The subsections below provide examples of various legal mechanisms at federal and state level affecting soils and their potential drawbacks for soil protection.

2.1 Legal Mechanisms Aimed at Preventing Soil Degradation

2.1.1 Federal Level

Privately owned land and soil in the United States is subject to public environmental regulation to protect the wider public from potentially harmful uses.[16] This includes federal laws and regulations, such as the Resource Conservation and Recovery Act (RCRA) that targets industrial waste disposal to avoid soil

---

[13]U.S. Department of Agriculture (USDA) (2013).

[14]66.

[15]The federal government is enumerated certain powers under Article I, section 8 of the Constitution, and the remaining powers are reserved to the jurisdiction of the 50 states. See also 14; Pryor (2001).

[16]37.

contamination.[17] The Environmental Protection Agency's Construction and Development Effluent Guidelines and Standards under the Clean Water Act were promulgated to control the stormwater discharges from construction sites.[18] Those sites covered under the National Pollutant Discharge Elimination System (NPDES) permitting scheme (i.e., disturbing one or more acres of land) must abide by specific requirements to prevent soil erosion and pollution runoff.[19] The cross-compliance provisions under the Farm Bill (described below) also provide an example of federal regulation of privately owned lands, which tie farmers' compliance with restrictions on agricultural management of wetlands and highly erodible soils to their ability to access public funds in an effort to limit erosion.

Additionally, publicly owned and managed soil exists in the form of national parks, forests, rangelands, and protected areas. The US Code of federal statutes defines the boundaries of national parks and protected areas, for example, Yellowstone National Park,[20] Sequoia and Yosemite National Parks,[21] and Steese National Conservation Area.[22] The purpose stated for establishing the Sequoia National Park is to retain the "timber, mineral deposits, natural curiosities or wonders" within the park in their natural condition, toward which the authorization of livestock grazing permits, for example, must not be detrimental.[23] Thus, the soil threat of compaction would need to be a consideration in the determination whether to grant a grazing permit. Also, within the King Range National Conservation Area, the Secretary of the Interior is authorized to "install such soil- and water-conserving works and practices to reduce erosion" and enter into cooperative agreements with California's agencies, local government, and nonprofit organizations as deemed necessary in order to implement soil and moisture conservation.[24] These public conservation areas therefore provide for soil conservation on significantly large parcels of land as opposed to fragmented coverage on multiple small parcels, which may involve the federal government partnering with private, nonprofit, or other government entities to manage the soil.

2.1.2 State and Local Level

State-level legislation and regulations may additionally impose restrictions on use for environmental and public health, safety and welfare benefits as long as they do not conflict with federal regulations. Minnesota's requirements to setback

[17] 42 U.S.C. §§ 6901–6992k.

[18] 40 C.F.R. part 450.

[19] Environmental Protection Agency (2016b).

[20] 16 U.S.C. §§ 21–40c.

[21] 16 U.S.C. §§ 41–79-1.

[22] 16 U.S.C. §§ 460mm–460mm-4.

[23] 16 U.S.C. § 45c.

[24] 16 U.S.C. § 460y-4(7) and (8).

production and maintain a 50-foot buffer strip on soils adjacent to public waterways provide a good example of public regulation of private agricultural land in order to reduce runoff and water quality impairment.[25] Soil management at the local level in Iowa also provides an example of private (and public county) lands being potentially subject to large amounts of administrative oversight, assistance and alteration of soil management practices from soil and water conservation districts, though impacts may vary based on the extent and nature of districts' implementation.[26] State environmental agencies also set and enforce rules and regulations for private landowners in order to protect the states' natural resources, e.g., restrictions against open dumping.[27]

Conservation easements provide an example of a legal mechanism available in all states and specifically promoted to different degrees (in addition to the federal incentive program provided under the 2014 Farm Bill, described below) in an attempt to reduce soil sealing, particularly on high quality soils used for agricultural production, as well as indirectly reduce loss of soil organic matter, loss of biodiversity, and potential erosion.[28] Conservation easements allow the private property owner to voluntarily sell or donate certain rights (e.g., to subdivide or develop a parcel) to a public entity or organization.[29] The entity then holds the rights in perpetuity, meaning the easement transfers with the land.[30] Thus, the right to develop or subdivide (which the public entity or organization purposely does not use) restricts others beyond the present owner from developing that parcel in the future. Governmental agencies and qualified nonprofits may hold conservation easements. Qualified nonprofits are often land trusts, which are non-profit organizations that either purchase land in fee or easements or assist owners in practicing conservation stewardship.[31] Land trusts at the local, state and national level conserved over 47 million acres (19 million hectares) of US soil in 2010.[32] States may provide special tax provisions for conservation easements as well, such as Virginia's exemption for easement holders and use-value assessment at a lower rate for landowners who grant easements.[33]

Note, however, that conservation easements do not prevent development in general, but merely on particular pieces of land, which affects the spatial layout of development. Most elements affecting the rate of development are exogenous to

[25]Minnesota Statutes 2015, section 103 F.48.

[26]Iowa Code Chapter 161A.

[27]Iowa Code § 455B.307.

[28]28.

[29]E.g., The Nature Conservancy is a non-profit organization which accepts donated conservation easements, which it must defend in perpetuity, or it purchases them in high priority cases in order to prevent urban sprawl. See The Nature Conservancy (2016).

[30]28.

[31]16.

[32]*Ibid.* at 5.

[33]Virginia Code § 10.1-1011.

government and land use controls—birth rate, death rate, immigration rate, and employment, for example. Therefore, conservation easements may positively or negatively impact soil sealing, loss of soil organic matter, loss of biodiversity, and potential erosion. For example, placement of a conservation easement in a location more suited for development next to an urban area may push the development farther from population centers, causing negative effects (e.g., extension of infrastructure, fragmentation), or fail to prioritize higher quality soils for application of the mechanism.[34]

Another common set of rights and regulations that affect private landowners' soil are the state-level landlord-tenancy laws. In the agricultural context, for example, the length of the notice period before termination of a farm tenancy can impact soil conservation.[35] If there is no tenure security for the following growing season, there is an economic disincentive for the current tenant to invest in conservation practices because the next tenant would reap the benefits from them.[36] Thus, the longer or more stable the tenancy is, the better it may be for the soil. The State of Iowa, for instance, requires that notice of termination of a farm tenancy must be served by September 1 to allow for termination by the subsequent March 1, otherwise the tenancy automatically renews for another year.[37] In addition, leases may include provisions which incentivize lessees to invest in protecting the soil, such as cost-sharing agreements to implement certain management practices between the lessor and the lessee.[38] These types of tenancy provisions are highly relevant to soil protection due to the large amount of farmland under leasing arrangements (over 38 % nationwide[39] and over 50 % in some states[40]).

2.2 *Legal Principles That May Work Against Soil Protection*

As demonstrated in the preceding subsection, there are multiple laws in the US which aim to or indirectly prevent soil degradation. The legal principle of governmental takings of property could be utilized toward that purpose, e.g., agricultural land may not be taken using eminent domain for infrastructure expansion without due process and just compensation. However, the concept of regulatory takings may in effect prevent or limit the government's ability to regulate to protect soils. Ineffective land use planning and zoning may also work against soil protection within and adjacent to urban areas.

[34]See 52, pp. 4–5; 35, pp. 467–468.

[35]10, 1361, 8 N.W.2d 481 (1943).

[36]19.

[37]Iowa Code 562.7.

[38]20.

[39]USDA (2012).

[40]Iowa State University 33.

2.2.1 Federal Level

Regulatory takings involve the plaintiff challenging a governmental regulation that effectively nullifies the owners' rights to use and enjoyment of the property.[41] For example, *Lucas v. South Carolina Coastal Council* involved a challenge to the state's setback requirements on beachfront vacant lots, which meant that the plaintiffs could not build on their lot.[42] The State of South Carolina argued the state beachfront management statute was intended to prevent coastal soil erosion.[43] However, the Supreme Court of the United States ruled it a regulatory taking due to the complete elimination of economic use of the property. This case demonstrates the (often-contested) balance between the right to rational use and enjoyment of private property in the US versus the public interest in maintaining a certain level of environmental quality and benefits from the land and soil. In effect, the possibility that governmental regulations may be found to operate as a takings and require full compensation to the private landowners may inhibit public environmental and soil protection regulation.

2.2.2 State and Local Level

Land use planning and local zoning demonstrates another public legal mechanism that has an impact on privately held property. As stated above, states control land within their borders, but planning and zoning is typically delegated to the local level as the stakeholders most directly impacted by the development of urban areas.[44] Local planning and zoning commissions create comprehensive plans for development of urban areas based on where different land uses should be located in relation to each other, e.g., industrial, residential, commercial, mixed-use. Zoning boards of adjustment, as quasi-judicial bodies, hear applications and appeals for, inter alia, special permits, variances, and conditional use permits by landowners who want to use their property for a use different from its zoned use, e.g., residentially zoned property for a small commercial business. Failure to develop a smart planning approach to land surrounding urban areas can lead to fragmented soil sealing as well as inefficient growth patterns, e.g., commercial corridors. In contrast, the Vermont Municipal and Regional Planning and Development statute (Act 200) provides a good example of "smart growth" legislation to enhance coordinated and comprehensive planning at the local, regional and state level.[45] Additionally, urban soils may be positively or negatively impacted by uses within different zones. For instance, an industrial use close to a residential neighborhood with discharges could

---

[41] See, e.g., Woodbury Cty. Soil Conservation Dist. v. Ortner, 279 N.W.2d 276 (1979).

[42] 36 (1992).

[43] 39, p. 7.

[44] See, e.g., California Government Code, Sections 65000–66037.

[45] 65.

cause soil contamination and put local residents at risk. Oregon boasts the oldest comprehensive planning statute in the US, supplemented over the years to promote quality development "inside the boundaries" of existing urban areas as well as smart expansion of urban development.[46]

3 2014 Farm Bill

The Agricultural Act of 2014 replaced the former 2008 Farm Bill as an omnibus package of agricultural laws adopted every 5–7 years by the federal government. The Farm Bill provides for, inter alia, farm commodity support, horticulture, livestock, food assistance, and conservation and rural development programs applicable to private, local, state, and federal actors. One of the major changes incorporated into the 2014 Farm Bill was the elimination of direct payments, the counter-cyclical price (CCP) program and the Average Crop Revenue Election (ACRE) program.[47] Much of the savings from eliminating those payments was reallocated towards crop insurance or price and revenue (price times yield) protection for producers.

The issue most relevant to soil within Title I regarding commodity support programs is the continuation of conservation compliance. Swampbuster and Sodbuster, two programs instituted in the 1985 Farm Bill, were aimed at protecting against conversion of wetlands and requiring production with an approved conservation plan on highly-erodible soils, respectively, by tying the former direct payments to compliance.[48] Under the 2014 Farm Bill, farmers with land covered by the Swampbuster and Sodbuster programs are now similarly not eligible for any premium subsidy paid by the Federal Crop Insurance Corporation (FCIC) if they fail to file Farm Service Agency (FSA) certification of compliance, are not complying with a Natural Resources Conservation Service (NRCS) approved conservation plan, or have converted or planted a commodity on a wetland after February 7, 2014.[49] These restrictions aim to protect vulnerable soils from erosion, loss of soil organic matter, loss of soil biodiversity, and potentially landslides.

The Conservation Title II in the 2014 Farm Bill consolidated the number of programs available to farmers, but it also reduced mandatory funding by almost 4 billion USD over the 10-year baseline.[50] The Conservation Reserve Program (CRP) pays farmers to take marginal land out of production under 10 to 15-year contracts and set aside that soil for conservation purposes (i.e., payments for ecosystem services).[51] The enrollment will include 2 million acres of grassland as

[46]Oregon Department of Land Conservation and Development (2000); 11.

[47]55.

[48]16 U.S.C. §§ 3811 et seq.

[49]43.

[50]18, p. 8.

[51]7 C.F.R. § 1410.7.

well,[52] reaching a total enrollment cap of 24 million acres by 2018.[53] The Environmental Quality Incentives Program (EQIP) was reinstated by the 2014 Farm Bill,[54] which offers assistance to producers with conservation measures on working lands, as does the Conservation Stewardship Program (CSP).[55] The CSP was authorized to enroll 10 million acres annually, which is a reduction from the 2008 Farm Bill levels at 12.8 million acres annually.[56] These programs demonstrate the federal government's allocation of voluntary incentives to private landowners in exchange for a contract guaranteeing soil and land conservation practices for a certain length of time.

The newly created Agricultural Conservation Easement Program (ACEP) under the 2014 Farm Bill, in contrast, provides an example of federally funded easements to permanently prevent conversion or specific uses of covered soils.[57] Specifically, wetland reserve easements may be purchased to protect and restore wetlands,[58] and agricultural land easements are intended to preserve productive farmland and grassland from non-agricultural uses.[59] These mechanisms provide a consolidated approach toward preventing loss of soil organic matter through conversion of wetlands as well as potential soil sealing, compaction, loss of soil biodiversity, or erosion and loss of soil organic matter from conversion to non-agricultural uses (e.g., residential, infrastructure, commercial use, etc.). It is worth noting, however, that agricultural uses include constructing farm buildings, such as livestock barns and grain bins, which utilize impervious surfaces for their foundations and may also cause soil sealing. Similar to conservation easements arranged outside of the Farm Bill, the acceptance of applications for land to be included in the ACEP falls under the administrative discretion of the NRCS, so the highest quality wetlands and agricultural lands may, in reality, not be protected with the program's funds.[60] The other new conservation program created by the 2014 Farm Bill is the Regional Conservation Partnership Program (RCPP), which provides for public-private partnerships to address conservation of soil and land within regions or

[52]The CRP incorporated similar provisions to those in the former Grassland Reserve Program, which was repealed by the 2014 Farm Bill. See Federal Register (2015).

[53]18, p. 8.

[54]The 2014 Farm Bill included provisions for wildlife habitat practices, which replace the repealed Wildlife Habitat Incentives Program (WHIP) from the 2008 Farm Bill. 44.

[55]16 U.S.C. § 3838d.

[56]18, p. 8.

[57]58, p. 10.

[58]The Wetland Reserve Program was repealed by the 2014 Farm Bill and the ACEP incorporated provisions for wetland reserve easements. 64, p. 528-A.1.

[59]The ACEP incorporation of provisions for agricultural land easements provides a similar substitute for the Farmland Protection Program and the Grassland Reserve Program which were repealed by the 2014 Farm Bill. 64, p. 528-A.1.

[60]64, p. 528-D.5.

watersheds, including agreements between state and local governments, conservation organizations, farmer cooperatives, and Indian tribes.[61]

The reauthorization of biofuels energy programs within the 2014 Farm Bill[62] may have an impact on soil in the US as well due to the subsidization of cornstarch-based ethanol production incentivizing higher levels of corn production. Increasing percentages of the US corn crop were allocated to ethanol production following the introduction of the renewable energy provisions in the 2002 Farm Bill, rising from 7 % during 2001 to 23 % in 2007, which contributed to marginal lands being converted from native grass and prairies to plant corn.[63] That trend was concerning from the perspective of potential soil erosion and declining soil organic matter levels.[64] Programs for development of non-corn feedstock biofuels were therefore included in the 2008 Farm Bill, including the Bioenergy Program for Advanced Biofuels, Biomass Crop Assistance Program (BCAP) and Renewable Energy for America Program (REAP), and have been continued in the 2014 Farm Bill.[65] Biomass feedstocks for non-corn-based biofuels may pose reduced risks of soil threats, such as reduced erosion from perennial crops; however, the removal of crop residues (e.g., corn stover) has been cautioned not to exceed conservation retention rates due to the loss of nutrients and soil cover.[66]

Finally, Title X regarding Horticulture contains the programs reauthorized to support specialty crops and certified organic production.[67] Organic production may encourage soil management practices that build soil structure and water and nutrient retention capacities, such as through cover crops, inter-cropping, using symbiotic relationships, crop rotations, minimum tillage and organic fertilizer.[68] Such practices can combat soil erosion, which may reduce soil organic matter and biodiversity losses as well.[69] The 2014 Farm Bill continued, inter alia, the National Organic Program (NOP) as well as the Organic Agriculture Research and Extension Initiative (OREI), which provide technical support and continuing research regarding improved organic management.[70]

[61] 58, p. 12.

[62] Agricultural Act of 2014, Title IX.

[63] 54, p. 3.

[64] *Ibid.*

[65] 18, p. 15.

[66] 59.

[67] Agricultural Act of 2014, Title X.

[68] 22.

[69] 23.

[70] 18, p. 15.

4 Waters of the United States

The Clean Water Act (CWA) mainly deals with water pollution. Section 404 of the CWA[71] provides the federal government with a permit system governing discharges of dredge and fill material into "navigable waters". This permit system includes jurisdiction over certain wetlands, which contribute to the prevention of soil degradation.[72] Jurisdictional coverage under the CWA has been both controversial and uncertain, however.

Section 404 of the CWA prohibits discharge of dredged or fill material into "navigable waters" without a permit, 33 U.S.C. §§ 1311(a), 1342(a), and defines "navigable waters" as "the waters of the United States, including the territorial seas," § 1362(7). The United States Army Corps of Engineers (USACE) is the administrative agency responsible for determining which water bodies are covered and issuing the permits for discharge of dredged or fill material into navigable waters. The United States Environmental Protection Agency (EPA) shares administrative responsibility for the program with the USACE in a complex arrangement that, among other things, gives the EPA veto authority over certain USACE actions. The EPA and USACE jointly promulgate regulations implementing the legislation.

The interpretation of "waters of the United States" (WOTUS) forms the core of the controversy, and proves important to prevention of soil degradation. Wetlands are the main point of dispute; namely, which wetlands are included under WOTUS and are therefore subject to Section 404 permitting. The following paragraphs briefly describe the history of the interpretation of WOTUS, leading to the recent promulgation of a new regulation defining the term. Implementation of this new regulation is presently stayed as litigation determines the validity of the new provision.

The United States federalist system of governance limits federal authority to those powers that can be found in the United States Constitution. The United States Congress bases authority to regulate "waters of the United States" upon the Commerce Clause of the United States Constitution: "The Congress shall have Power… To regulate Commerce with foreign Nations, and among the several States, and with the Indian Tribes…"[73] Hence, Section 404 limits jurisdiction to "navigable waters". The definition of "waters of the United States" is therefore limited by the reach of the Commerce Clause.[74]

The prior regulation under this provision interpreted "the waters of the United States" to include not only traditional navigable waters, 33 CFR § 328.3(a)(1), but also other defined waters, § 328.3(a)(2), (3); "[t]ributaries" of such waters, § 328.3 (a)(5); and wetlands "adjacent" to such waters and tributaries, § 328.3(a)(7). "[A] djacent" wetlands include those "bordering, contiguous [to], or neighboring" waters

---

[71]33 U.S.C. §§ 1344.

[72]Altvater (2013), p. 59.

[73]United States Constitution, Article I, Section 8.

[74]For a review of the history of Section 404 and the use of the Commerce Clause to reach conservation purposes, see 34.

of the United States even when they are "separated from [such] waters ... by man-made dikes ... and the like." § 328.3(c).

Three United States Supreme Court decisions provide key interpretation of the outlines of the definition and the legality of the regulation. First, in *United States v. Riverside Bayview Homes, Inc.*, the Court found that the USACE holds jurisdiction over wetlands adjacent to navigable or interstate waters, even if the adjacent waters do not inundate or frequently flood the wetland.[75] In this case, groundwater caused the wetland conditions, so the Court held that USACE had acted reasonably in asserting jurisdiction and requiring a permit for discharge.[76]

Next, in *Solid Waste Agency of Northern Cook County v. the United States Army Corps of Engineers*, ("*SWANCC*"), the United States Supreme Court addressed the "isolated waters" issue.[77] Although a wetland was not involved in this case, the analysis applies to wetlands not "adjacent" to or somehow connected to a navigable water body. Invoking the "Migratory Bird Rule", the USACE had asserted jurisdiction over an abandoned sand and gravel pit that provided habitat for migratory birds. The Court found that the Section 404 permitting provisions (33 U.S.C. § 1344 (a)) do not apply to this "water" in the absence of a connection to navigable waters. Therefore, the USACE lacked jurisdiction. Thus, the Court emphasized the focus on the definitional connection to navigable waters rather than another way in which the water could have an interstate connection. So finding, the Court did not address whether the United States Congress has authority under the Commerce Clause, U.S. Const., Art. I, § 8, cl. 3 to enact provisions that would grant the USACE jurisdiction over the gravel pit.

The reasoning in *SWANCC* examined the language of the Clean Water Act in an attempt to glean the intent of Congress. The Court found that Congress clearly expressed an intent not to include isolated waterways within the jurisdiction of the Clean Water Act.[78]

The *SWANCC* decision essentially put the onus on the United States Congress to amend the Clean Water Act to clearly display its intentions for the scope of "waters of the United States". However, a gridlocked Congress introduced several bills, but none could pass. In addition, the USACE failed to amend the regulations, despite providing notice of a proposed rulemaking in light of *SWANCC*, 68 Fed. Reg.1991 (2003).[79] However, the USACE continued to assert jurisdiction over waters "'neighboring'" traditional navigable waters and their tributaries. 68 Fed. Reg.1997 (quoting 33 CFR § 328.3(c) (2002)).[80]

---

[75]United States v. Riverside Bayview Homes, Inc., 106 S.Ct. 455 (1985).

[76]Riverside Bayview, p. 131.

[77]Solid Waste Agency of Northern Cook County v. the United States Army Corps of Engineers, 530 U.S. 159 (2001).

[78]SWANNC, invoking the Chevron deference test, *Chevron U.S.A. Inc. v. Natural Resources Defense Council, Inc.*, 467 U.S. 837, 104 S.Ct. 2778, 81 L.Ed.2d 694 (1984).

[79]Rapanos v. United States, 126 S.Ct. 2208, 2217 (2006).

[80]*Ibid.*

> Even after *SWANCC*, the lower courts have continued to uphold the Corps' sweeping assertions of jurisdiction over ephemeral channels and drains as "tributaries." For example, courts have held that jurisdictional "tributaries" include the "intermittent flow of surface water through approximately 2.4 miles of natural streams and manmade ditches (paralleling and crossing under I–64)," *Treacy v. Newdunn Assoc.*, 344 F.3d 407, 410 (C.A.4 2003); a "roadside ditch" whose water took "a winding, thirty-two-mile path to the Chesapeake Bay," *United States v. Deaton*, 332 F.3d 698, 702 (C.A.4 2003); irrigation ditches and drains that intermittently connect to covered waters, *Community Assn. for Restoration of Environment v. Henry Bosma Dairy*, 305 F.3d 943, 954–955 (C.A.9 2002); *Headwaters, Inc. v. Talent Irrigation Dist.*, 243 F.3d 526, 534 (C.A.9 2001); and (most implausibly of all) the "washes and arroyos" of an "arid development site," located in the middle of the desert, through which "water courses . . . during periods of heavy rain," *Save Our Sonoran, Inc. v. Flowers*, 408 F.3d 1113, 1118 (C.A.9 2005).[81]

The latest United States Supreme Court pronouncement on the issue, *Rapanos v. United States*, involved one of these "tributary" cases with the closest navigable waterway "11-20 miles away".[82] The Court failed to produce a majority opinion, splitting 4-1-4. Justice Scalia wrote the plurality opinion for 4 justices vacating the decision below and remanding the case, and Justice Kennedy concurred but in a separate opinion laying out a different test. The plurality concluded that the USACE's interpretation was not reasonable. The plurality's test defines "the waters of the United States" as including "only those relatively permanent, standing, or continuously flowing bodies of water 'forming geographic features' that are described in ordinary parlance as 'streams,' 'oceans, rivers, [and] lakes. . .'"[83] The term "does not include channels through which water flows intermittently or ephemerally, or channels that periodically provide drainage for rainfall".[84]

Justice Kennedy's test for jurisdiction "depends upon the existence of a significant nexus between the wetlands in question and navigable waters in the traditional sense".[85] "The required nexus is met 'if the wetlands, either alone or in combination with similarly situated lands in the region, significantly affect the chemical, physical, and biological integrity of other covered waters more readily understood as 'navigable.' When, in contrast, wetlands' effects on water quality are speculative or insubstantial, they fall outside the zone fairly encompassed by the statutory term 'navigable waters.'"[86] The four dissenting justices, in a decision written by Justice Stevens, maintained that the court should defer to the judgment of the USACE as a reasonable interpretation of the Clean Water Act.

The divided court in *Rapanos* left the jurisdiction of the USACE confused and uncertain. Courts have differed in terms of which opinion is applicable, but they have generally upheld jurisdiction where either the plurality's test or Justice

[81]Rapanos, p. 2218.

[82]*Ibid.*, p. 2214.

[83]*Ibid.*, p. 2225.

[84]*Ibid.*

[85]*Ibid.*, p. 2248.

[86]*Ibid.*

Kennedy's test was met.[87] Precisely to avoid conflicting decisions, however, opinions from the *Rapanos* decision recommended that the regulations be revised if Congress does not take action.[88]

In April of 2014, the USACE and EPA introduced a revised rule that, depending upon the reader, "clarified" or "expanded" the definition of "waters of the United States". The rule attempted to codify Justice Kennedy's "significant nexus" test, which remained far from clear when applied by the lower courts to cases decided between the *Rapanos* decision and the rulemaking. Despite much controversy and a plethora of public comments that alternatively praised or condemned the new rule, the USACE finalized the rule in May 2015. The final rule includes some specific, distance-based limitations not included in the draft rule.[89] The National Association of Counties (NACO), which opposed the proposed rule, compiled a table comparing the old rule, the proposed rule, and the new rule. The table is available at: http://www.naco.org/sites/default/files/June2%20Chart%20Master%20WOTUS_FINAL%20rule%20chart.pdf.

Several lawsuits were initiated immediately after approval of the final rule. Plaintiffs included states, environmental groups, landowners and others. Therefore, a variety of groups that generally are not aligned find themselves all opposing the present rule, but for different reasons.

On February 22, 2016, the United States Court of Appeals for the Sixth Circuit resolved the dispute over which court holds jurisdiction over this matter (at least for now). The court ruled that the United States Court of Appeals holds jurisdiction rather than the United States District Court, supporting the position of the United States Department of the Interior and the EPA.[90] Based in large part upon the United States Supreme Court ruling in *E.I du Pont de Nemours Co. v. Train*, the court found that a "functional approach" to the statutory grants of authority gives the Courts of Appeals jurisdiction under 33 U.S.C. §1369(b)(1)(F).[91] Subsection (F) of that statute gives the Courts of Appeals jurisdiction over issues which "deny any permit under section 1342 [of the Clean Water Act]".

However, the court split on the issue 2-1, and the concurring opinion, which provided the deciding vote, contains some qualifications that make an en banc ruling by all judges on the Sixth Circuit likely, and an appeal may follow. Justice Griffin concurred in the judgment only because he felt "required" to do so based upon *National Cotton Council of America v. U.S. E.P.A.*[92] "Were it not for *National Cotton*, I would grant the motions to dismiss".[93]

[87] Meltz and Copeland (2015), pp. 7–8.

[88] See opinions of Justice Kennedy, Justice Breyer, and Chief Justice Roberts.

[89] In Re. EPA (2015), p. 807.

[90] In re: U.S. Department of Defense & U.S. Envtl. Protection Agency (2016).

[91] In re: U.S. Department of Defense & U.S. Envtl. Protection Agency (2016), pp. 17–18.

[92] National Cotton Council of America v. U.S. Environmental Protection Agency, 553 F.3d 927 (6th Cir. 2009).

[93] Id, Justice Griffin concurring, p. 19.

In *National Cotton*, the Sixth Circuit held that 33 U.S.C. §1369(b)(1)(F) authorizes Circuit Court jurisdiction in actions involving "regulations governing the issuance of permits".[94] This jurisdiction means that the case skips the United States District Court, which may streamline the process. The concurring opinion notes that a panel of the Sixth Circuit lacks authority to overrule its own published opinion in *National Cotton,* which found jurisdiction in the Circuit Court "absent an en banc review or an intervening and binding change in the state of the law".[95] Thus, an en banc review could potentially find that the Circuit Court does not have jurisdiction without the case first being heard in District Court and send it down to the District Court as the court of first instance.

In summary, the definition of "waters of the United States" remains unclear. A resolution likely will not come for years as the present litigation makes its way through the court system. Consequently, protection of wetlands, and soils, remains uncertain.

5 Des Moines Water Works Lawsuit

The Board of Water Works Trustees of the City of Des Moines, Iowa ("Des Moines Water Works" or "DMWW") filed a lawsuit in United States District Court in March 2015 against 13 drainage districts in the State of Iowa. The lawsuit includes claims under state law and the Clean Water Act. The crux of the complaint centers upon allegations that the drainage districts constitute "point sources" of pollution, and are required to have permits under the CWA's National Pollution Discharge Elimination System (NPDES) permit system and the State of Iowa's NPDES program. Specifically, DMWW alleges that the activities of the drainage districts contaminated the source for the public water supply with nitrates.

Des Moines Water Works provides drinking water to about half a million citizens in central Iowa, including the City of Des Moines. The Safe Drinking Water Act requires that drinking water contain no more than established levels of certain contaminates, including nitrates. Des Moines Water Works alleges that the activities of the drainage districts, in collecting and conveying water containing high levels of nitrates from agricultural drainage, through an artificial drainage system into the Raccoon River, the water source for Des Moines, causes damage to DMWW.[96] The lawsuit also alleges nuisance, trespass, negligence and an unconstitutional taking of property.[97]

---

[94]In re: U.S. Department of Defense & U.S. Envtl. Protection Agency (2016), p. 12, citing National Cotton, p. 933.

[95]*Ibid.*, Justice Griffin concurring, p. 30.

[96]Des Moines Water Works Complaint, paragraph 10 (2015).

[97]Des Moines Water Works Complaint, paragraph 14 (2015).

The New York Times referred to the case as a "conflict over soil and water quality".[98]

> The nitrate issue is, in many ways, an unfortunate side effect of one of Iowa's great assets: the nutrient-rich dirt that makes for some of the world's most productive cropland. Though that nitrogen-filled soil helps Mr. Johnson and others grow prodigious amounts of corn and soybeans, a significant rainstorm can wash many of those nutrients, along with nitrates applied as fertilizer, into tributaries of the Raccoon River. The Raccoon is one of two rivers that provide drinking water for Des Moines, the state's capital and urban center.[99]

On January 11, 2016, the United States District Court certified four questions of Iowa law to the Iowa Supreme Court that must be answered before the District Court can proceed. In the United States legal system, the state supreme court is the final arbiter of state law issues. Federal courts may send, or "certify", questions of state law to the state supreme court for resolution prior to deliberation by the federal court. The questions certified in this case focus on whether state law grants immunity from the lawsuit to the drainage districts and on state constitutional issues.[100]

The federal court explained that its decision will rely upon the Iowa Supreme Court's answers to the certified questions.

The Des Moines Water Works lawsuit raises more questions than it provides answers. However, the lawsuit likely will have a significant impact on the management of nitrates in Iowa as the resolution will impact the relationship between soil management on a broad scale and water quality. Additionally, if this case sets a precedent for administrative areas (e.g., the drainage districts) comprised of multiple non-point sources (i.e., individual farms) being collectively considered a point source, that holding could have far-reaching effects throughout the United States. The lawsuit focuses on the nature of the discharges from the drainage district. The drainage districts maintain that the discharges constitute "agricultural stormwater" exempt from regulation under 33 U.S.C. § 1362(14).[101] However, the lawsuit alleges that the discharges mainly consist of groundwater, not exempt from regulation.[102] If successful in this recharacterization, the lawsuit would make individual farms "point sources", requiring each farm to acquire a CWA NPDES permit. The permit would require, in essence, more intensive soil management by the farms to prevent runoff of nitrates.

Farm groups prefer a voluntary approach to reducing nitrate runoff.[103] Producer groups have formed associations to address water quality. The executive director of

[98]Smith (2015).

[99]Smith (2015).

[100]Order certifying questions (2016), page 3.

[101]33 U.S.C. § 1362(14) defines "point source". See, e.g., Alt v. Environmental Protection Agency, and the cases cited herein, for the contentious debate on the distinction between "point source" and "agricultural stormwater".

[102]Des Moines Water Works Complaint, paragraphs 10, 28, 77, 113, 130, 131, 135, 139, 151, 153, 154, 155, 222, and 223.

[103]Smith (2015).

one of those groups, Sean McMahon of the Iowa Agriculture Water Alliance, "said that many farmers were eager to employ conservation practices, but that education and time were needed to see more results. Money, he said, would be better spent on outreach and cost-sharing programs than on lawyers for the lawsuit".[104] Environmental groups allege that agricultural pollution heavily contributes to impaired water quality and that these drainage districts should be covered under the CWA rather than continuously granted immunity and given a "license to pollute".[105]

The future of nitrate management remains unclear until the lawsuit concludes. Doubts exist as to the extent of authority held by drainage districts to require landowners to take action to minimize nitrate runoff.[106] Can drainage districts require vegetative buffers or water treatment? Can drainage districts require landowners to create wetlands to treat drainage, or purchase land to be used as wetlands? The law is unclear. In addition, the Soil and Water Conservation Districts have, to this point, taken little or no initiative in using their authority to affect management of soils to reduce negative impacts to water quality.[107]

Some also now doubt the efficacy of Iowa's Nitrate Reduction Strategy (NRS) as a viable contribution to solving this issue.[108] The strategy lacks standards, timelines and measurements, and is entirely voluntary.[109] Perhaps more importantly, no funding exists for the strategy.[110] Some legislative solutions have been introduced, including a proposed increase in the sales tax, earmarked to nitrate reduction, to fund cost-sharing efforts.[111]

6 Conclusions

Although policy in the United States has not focused directly on soil degradation in some time, three recent developments may cause significant changes in soil policies in the future. The 2014 Farm Bill eliminates direct payments, the counter-cyclical price (CCP) program and the Average Crop Revenue Election (ACRE) program. As conservation compliance for wetlands and highly-erodible soils had been linked to direct payments, its continuation depended upon linking it to the newly altered crop insurance provisions. The Conservation Title II in the 2014 Farm Bill consolidated the number of programs available to farmers for voluntary incentives to conserve their soil, but it also reduced mandatory funding by almost 4 billion USD over the

[104]Smith (2015).
[105]9.
[106]29.
[107]29.
[108]29.
[109]29.
[110]29.
[111]29.

10-year baseline. However, a new program was created called the Agricultural Conservation Easement Program (ACEP) and it provides federal funding for the purchase of conservation easements to permanently prevent conversion or specific uses of covered soils. This could potentially offer longer term protection over private US farmland and wetland resources, but since allocation of funding may only be made to lands which have applied for selection, it does not guarantee that the highest quality wetlands or agricultural lands are being permanently preserved.

Federal agencies recently released a new regulation to define "waters of the United States" under the CWA following subsequent Supreme Court decisions on the issue. This definition provides a key component of the determination of federal jurisdiction over wetlands and permitting for dredging and filling activities on them, which impacts soil degradation. A number of lawsuits were immediately filed challenging the new regulation, and the regulation is presently stayed as the courts determine which court holds jurisdiction to hear the cases. The present boundaries of federal jurisdiction over private land use in relation to wetlands remain unclear, so clarification is needed. However, the present status of the rule, along with a gridlocked federal legislature, suggests that clarification will not come soon.

Finally, a lawsuit by the Des Moines Water Works against several drainage districts holds the potential to drastically change the policies on soil management in the United States. The suit pits water quality versus soil quality as the rich soils of Iowa drain nitrates into the drinking water for the City of Des Moines. The Des Moines Water Works treatment facility seeks to hold the drainage districts and, indirectly, farmers, responsible for managing the runoff through a novel mechanism—claiming that the collective runoff from farms is a "point source" of pollution. Regardless of who prevails in the lawsuit, the result will set a new path forward for soil management in the United States.

References

7 Code of Federal Regulations (C.F.R.) § 1410.7

40 C.F.R. part 450

16 United States Code (U.S.C.) §§ 21–40c; 41–79-1; 460mm–460mm-4; 460y-4; 3811 et seq.; 3838d

33 U.S.C. §§ 1342; 1344; 1362; 1369

42 U.S.C. §§ 6901-6992 k

Agricultural Act of 2014, Public Law 113-79

Alt v. United States Environmental Protection Agency, 979 F.Supp.2d 701 (U.S.D.C. N.D. W.Va. 2013)

Altvater S (2013) Gutachten "Kosten und Nutzen einer Europäischen Bodenschutzrichtlinie für Deutschland"– FKZ: 3712 14 230 (Study for the Federal Environmental Agency), not published

Benschoter v. Hakes, 232 Iowa 1354, 1361, 8 N.W.2d 481 (1943)

Borchers A (2015) Major land uses. USDA Economic Research Service. http://www.ers.usda.gov/topics/farm-economy/land-use,-land-value-tenure/major-land-uses.aspx. Accessed 29 Feb 2016

Burke IC, Yonker CM, Parton WJ, Cole CV, Flach K, Schimel DS (1989) Texture, climate, and cultivation effects on soil organic matter content in U.S. grassland soils. Soil Sci Soc Am J 53:800–805

California Government Code, Sections 65000-66037

Chevron U.S.A. Inc. v. Natural Resources Defense Council, Inc., 467 U.S. 837, 104 S.Ct. 2778, 81 L.Ed.2d 694 (1984)

Emili LA, Greene RP (2014) New cropland on former rangeland and lost cropland from urban development: the "replacement land" debate. Land 3:658–674

Environmental Protection Agency (2016a) Superfund: national priorities list (NPL). https://www.epa.gov/superfund/superfund-national-priorities-list-npl. Accessed 4 Apr 2016

Environmental Protection Agency (2016b) Construction and development effluent guidelines. https://www.epa.gov/eg/construction-and-development-effluent-guidelines. Accessed 6 Apr 2016).

Federal Register (2015) Conservation reserve program: overview of changes to CRP regulations. https://www.federalregister.gov/articles/2015/07/16/2015-17317/conservation-reserve-program#h-15. Accessed 23 Feb 2016

In re E.P.A., 803 F.3d 804 (2015)

In re: U.S. Dep't of Defense & U.S. Envtl. Protection Agency Final Rule: Clean Water Rule, Nos. 15-3751, et al. (U.S. Ct. App. 6th Cir., February 22, 2016)

Iowa Code Chapter 161A; §§ 455B.307; 562.7

Lucas v. South Carolina Coastal Council, 505 U.S. 1003 (1992)

McGranahan DA et al (2013) A historical primer on the US Farm Bill: supply management and conservation policy. J Soil Water Conserv 68:67A–73A

Meltz R, Copeland C (2015) The Wetlands Coverage of the Clean Water Act (CWA): *Rapanos* and Beyond. Congressional Research Service report RL33263. Available at http://nationalaglawcenter.org/wp-content/uploads/assets/crs/RL33263.pdf. Accessed 24 Feb 2016.

Minnesota Statutes 2015 § 103 F.48

National Cotton Council of America v. U.S. E.P.A., 553 F.3d 927 (6th Cir. 2009)

Natural Resources Conservation Service (NRCS) (2007) Natural resources inventory: soil erosion on cropland. United States Department of Agriculture (USDA) report. Available at http://www.nrcs.usda.gov/wps/portal/nrcs/detail/national/technical/nra/nri/?cid=stelprdb1041887. Accessed 18 Feb 2016

Nature Conservancy (2016) The conservation easements: all about conservation easements. http://www.nature.org/about-us/private-lands-conservation/conservation-easements/all-about-conservation-easements.xml. Accessed 21 Feb 2016

Nickerson C, Ebel R, Borchers A, Carriazo F (2011) Major uses of land in the United States, 2007. Economic Research Service, USDA, Economic Information Bulletin No. 89. Available at http://www.ers.usda.gov/media/188404/eib89_2_.pdf. Accessed 18 Feb 2016

Nickerson C, Borchers A, Harper M, Henrie C, Mayberry R, Shimmin S, Smith B, Smith J (2015) Land use and land cover estimates for the United States. USDA Economic Research Service, http://www.ers.usda.gov/about-ers/strengthening-statistics-through-the-interagency-council-on-agricultural-rural-statistics/land-use-and-land-cover-estimates-for-the-united-states.aspx. Accessed 29 Feb 2016

Oregon Department of Land Conservation and Development (2000) Oregon land use laws update – urban growth boundaries and future growth. Inside the boundaries. https://www.oregon.gov/LCD/docs/publications/insideboundaries.pdf. Accessed 6 Apr 2016

Pryor WH Jr (2001) Madison's double security: in defense of federalism, the separation of powers, and the Rehnquist court. Alabama Law Rev 53:1167–1182

Rapanos v. United States, 126 S.Ct. 2208, 2217 (2006)

Rees WE (2016) North American soils and world food. In: Ginzky H, Heuser IL, Qin T, Ruppel OC, Wegerdt P (eds) International yearbook of soil law and policy 2016. Springer, Heidelberg (in this volume). doi:10.1007/978-3-319-42508-5

Smith M (2015) Conflict over soil and water quality puts Iowa nice to the test, New York Times (April 18, 2015). Available at http://www.nytimes.com/2015/04/19/us/conflict-over-soil-and-water-quality-puts-iowa-nice-to-a-test.html?_r=0

Solid Waste Agency of Northern Cook County v. United States Army Corps of Engineers, 530 U.S.159 (2001)

United States v. Riverside Bayview Homes, Inc., 106 S.Ct. 455 (1985)

U.S. Census Bureau (2015) National totals: vintage 2015. http://www.census.gov/popest/data/national/totals/2015/index.html. Accessed 29 Feb 2016

U.S. Department of Agriculture (USDA) (2012) 2012 Census of agriculture, Table 11. http://www.agcensus.usda.gov/Publications/2012/Full_Report/Volume_1,_Chapter_1_US/usv1.pdf. Accessed 5 Apr 2016

USDA (2013) Summary report: 2010 National Resources Inventory. Natural Resources Conservation Service, Washington, DC, and Center for Survey Statistics and Methodology, Iowa State University, Ames, Iowa. Available at http://www.nrcs.usda.gov/Internet/FSE_DOCUMENTS/stelprdb1167354.pdf. Accessed 18 Feb 2016

Virginia Code § 10.1-1011

Woodbury County Soil Conservation Dist. v. Ortner, 279 N.W.2d 276 (1979)

Worster D (1982) Dust Bowl: the southern plains in the 1930s. Oxford University Press, Oxford

MIX
Papier aus verantwortungsvollen Quellen
Paper from responsible sources
FSC® C105338
www.fsc.org

If you have any concerns about our products,
you can contact us on
ProductSafety@springernature.com

In case Publisher is established outside the EU,
the EU authorized representative is:
Springer Nature Customer Service Center GmbH
Europaplatz 3, 69115 Heidelberg, Germany

Printed by Libri Plureos GmbH
in Hamburg, Germany